Lecture Notes in Computer Science 12773

Founding Editors

Gerhard Goos
Karlsruhe Institute of Technology, Karlsruhe, Germany
Juris Hartmanis
Cornell University, Ithaca, NY, USA

Editorial Board Members

Elisa Bertino
Purdue University, West Lafayette, IN, USA
Wen Gao
Peking University, Beijing, China
Bernhard Steffen
TU Dortmund University, Dortmund, Germany
Gerhard Woeginger
RWTH Aachen, Aachen, Germany
Moti Yung
Columbia University, New York, NY, USA

T0214093

More information about this subseries at http://www.springer.com/series/7409

Pei-Luen Patrick Rau (Ed.)

Cross-Cultural Design

Applications in Cultural Heritage, Tourism, Autonomous Vehicles, and Intelligent Agents

13th International Conference, CCD 2021
Held as Part of the 23rd HCI International Conference, HCII 2021
Virtual Event, July 24–29, 2021
Proceedings, Part III

Springer

Editor
Pei-Luen Patrick Rau
Tsinghua University
Beijing, China

ISSN 0302-9743 ISSN 1611-3349 (electronic)
Lecture Notes in Computer Science
ISBN 978-3-030-77079-2 ISBN 978-3-030-77080-8 (eBook)
https://doi.org/10.1007/978-3-030-77080-8

LNCS Sublibrary: SL3 – Information Systems and Applications, incl. Internet/Web, and HCI

This Springer imprint is published by the registered company Springer Nature Switzerland AG
The registered company address is: Gewerbestrasse 11, 6330 Cham, Switzerland

Foreword

Human-Computer Interaction (HCI) is acquiring an ever-increasing scientific and industrial importance, and having more impact on people's everyday life, as an ever-growing number of human activities are progressively moving from the physical to the digital world. This process, which has been ongoing for some time now, has been dramatically accelerated by the COVID-19 pandemic. The HCI International (HCII) conference series, held yearly, aims to respond to the compelling need to advance the exchange of knowledge and research and development efforts on the human aspects of design and use of computing systems.

The 23rd International Conference on Human-Computer Interaction, HCI International 2021 (HCII 2021), was planned to be held at the Washington Hilton Hotel, Washington DC, USA, during July 24–29, 2021. Due to the COVID-19 pandemic and with everyone's health and safety in mind, HCII 2021 was organized and run as a virtual conference. It incorporated the 21 thematic areas and affiliated conferences listed on the following page.

A total of 5222 individuals from academia, research institutes, industry, and governmental agencies from 81 countries submitted contributions, and 1276 papers and 241 posters were included in the proceedings to appear just before the start of the conference. The contributions thoroughly cover the entire field of HCI, addressing major advances in knowledge and effective use of computers in a variety of application areas. These papers provide academics, researchers, engineers, scientists, practitioners, and students with state-of-the-art information on the most recent advances in HCI. The volumes constituting the set of proceedings to appear before the start of the conference are listed in the following pages.

The HCI International (HCII) conference also offers the option of 'Late Breaking Work' which applies both for papers and posters, and the corresponding volume(s) of the proceedings will appear after the conference. Full papers will be included in the 'HCII 2021 - Late Breaking Papers' volumes of the proceedings to be published in the Springer LNCS series, while 'Poster Extended Abstracts' will be included as short research papers in the 'HCII 2021 - Late Breaking Posters' volumes to be published in the Springer CCIS series.

The present volume contains papers submitted and presented in the context of the 13th International Conference on Cross-Cultural Design (CCD 2021) affiliated conference to HCII 2021. I would like to thank the Chair, Pei-Luen Patrick Rau, for his invaluable contribution in its organization and the preparation of the Proceedings, as well as the members of the program board for their contributions and support. This year, the CCD affiliated conference has focused on topics related to cross-cultural experience and product design, cultural differences and cross-cultural communication, as well as design case studies in domains such as learning and creativity, well-being, social change and social development, cultural heritage and tourism, autonomous vehicles, virtual agents, robots and intelligent assistants.

I would also like to thank the Program Board Chairs and the members of the Program Boards of all thematic areas and affiliated conferences for their contribution towards the highest scientific quality and overall success of the HCI International 2021 conference.

This conference would not have been possible without the continuous and unwavering support and advice of Gavriel Salvendy, founder, General Chair Emeritus, and Scientific Advisor. For his outstanding efforts, I would like to express my appreciation to Abbas Moallem, Communications Chair and Editor of HCI International News.

July 2021   Constantine Stephanidis

HCI International 2021 Thematic Areas and Affiliated Conferences

Thematic Areas

- HCI: Human-Computer Interaction
- HIMI: Human Interface and the Management of Information

Affiliated Conferences

- EPCE: 18th International Conference on Engineering Psychology and Cognitive Ergonomics
- UAHCI: 15th International Conference on Universal Access in Human-Computer Interaction
- VAMR: 13th International Conference on Virtual, Augmented and Mixed Reality
- CCD: 13th International Conference on Cross-Cultural Design
- SCSM: 13th International Conference on Social Computing and Social Media
- AC: 15th International Conference on Augmented Cognition
- DHM: 12th International Conference on Digital Human Modeling and Applications in Health, Safety, Ergonomics and Risk Management
- DUXU: 10th International Conference on Design, User Experience, and Usability
- DAPI: 9th International Conference on Distributed, Ambient and Pervasive Interactions
- HCIBGO: 8th International Conference on HCI in Business, Government and Organizations
- LCT: 8th International Conference on Learning and Collaboration Technologies
- ITAP: 7th International Conference on Human Aspects of IT for the Aged Population
- HCI-CPT: 3rd International Conference on HCI for Cybersecurity, Privacy and Trust
- HCI-Games: 3rd International Conference on HCI in Games
- MobiTAS: 3rd International Conference on HCI in Mobility, Transport and Automotive Systems
- AIS: 3rd International Conference on Adaptive Instructional Systems
- C&C: 9th International Conference on Culture and Computing
- MOBILE: 2nd International Conference on Design, Operation and Evaluation of Mobile Communications
- AI-HCI: 2nd International Conference on Artificial Intelligence in HCI

List of Conference Proceedings Volumes Appearing Before the Conference

1. LNCS 12762, Human-Computer Interaction: Theory, Methods and Tools (Part I), edited by Masaaki Kurosu
2. LNCS 12763, Human-Computer Interaction: Interaction Techniques and Novel Applications (Part II), edited by Masaaki Kurosu
3. LNCS 12764, Human-Computer Interaction: Design and User Experience Case Studies (Part III), edited by Masaaki Kurosu
4. LNCS 12765, Human Interface and the Management of Information: Information Presentation and Visualization (Part I), edited by Sakae Yamamoto and Hirohiko Mori
5. LNCS 12766, Human Interface and the Management of Information: Information-rich and Intelligent Environments (Part II), edited by Sakae Yamamoto and Hirohiko Mori
6. LNAI 12767, Engineering Psychology and Cognitive Ergonomics, edited by Don Harris and Wen-Chin Li
7. LNCS 12768, Universal Access in Human-Computer Interaction: Design Methods and User Experience (Part I), edited by Margherita Antona and Constantine Stephanidis
8. LNCS 12769, Universal Access in Human-Computer Interaction: Access to Media, Learning and Assistive Environments (Part II), edited by Margherita Antona and Constantine Stephanidis
9. LNCS 12770, Virtual, Augmented and Mixed Reality, edited by Jessie Y. C. Chen and Gino Fragomeni
10. LNCS 12771, Cross-Cultural Design: Experience and Product Design Across Cultures (Part I), edited by P. L. Patrick Rau
11. LNCS 12772, Cross-Cultural Design: Applications in Arts, Learning, Well-being, and Social Development (Part II), edited by P. L. Patrick Rau
12. LNCS 12773, Cross-Cultural Design: Applications in Cultural Heritage, Tourism, Autonomous Vehicles, and Intelligent Agents (Part III), edited by P. L. Patrick Rau
13. LNCS 12774, Social Computing and Social Media: Experience Design and Social Network Analysis (Part I), edited by Gabriele Meiselwitz
14. LNCS 12775, Social Computing and Social Media: Applications in Marketing, Learning, and Health (Part II), edited by Gabriele Meiselwitz
15. LNAI 12776, Augmented Cognition, edited by Dylan D. Schmorrow and Cali M. Fidopiastis
16. LNCS 12777, Digital Human Modeling and Applications in Health, Safety, Ergonomics and Risk Management: Human Body, Motion and Behavior (Part I), edited by Vincent G. Duffy
17. LNCS 12778, Digital Human Modeling and Applications in Health, Safety, Ergonomics and Risk Management: AI, Product and Service (Part II), edited by Vincent G. Duffy

18. LNCS 12779, Design, User Experience, and Usability: UX Research and Design (Part I), edited by Marcelo Soares, Elizabeth Rosenzweig, and Aaron Marcus
19. LNCS 12780, Design, User Experience, and Usability: Design for Diversity, Well-being, and Social Development (Part II), edited by Marcelo M. Soares, Elizabeth Rosenzweig, and Aaron Marcus
20. LNCS 12781, Design, User Experience, and Usability: Design for Contemporary Technological Environments (Part III), edited by Marcelo M. Soares, Elizabeth Rosenzweig, and Aaron Marcus
21. LNCS 12782, Distributed, Ambient and Pervasive Interactions, edited by Norbert Streitz and Shin'ichi Konomi
22. LNCS 12783, HCI in Business, Government and Organizations, edited by Fiona Fui-Hoon Nah and Keng Siau
23. LNCS 12784, Learning and Collaboration Technologies: New Challenges and Learning Experiences (Part I), edited by Panayiotis Zaphiris and Andri Ioannou
24. LNCS 12785, Learning and Collaboration Technologies: Games and Virtual Environments for Learning (Part II), edited by Panayiotis Zaphiris and Andri Ioannou
25. LNCS 12786, Human Aspects of IT for the Aged Population: Technology Design and Acceptance (Part I), edited by Qin Gao and Jia Zhou
26. LNCS 12787, Human Aspects of IT for the Aged Population: Supporting Everyday Life Activities (Part II), edited by Qin Gao and Jia Zhou
27. LNCS 12788, HCI for Cybersecurity, Privacy and Trust, edited by Abbas Moallem
28. LNCS 12789, HCI in Games: Experience Design and Game Mechanics (Part I), edited by Xiaowen Fang
29. LNCS 12790, HCI in Games: Serious and Immersive Games (Part II), edited by Xiaowen Fang
30. LNCS 12791, HCI in Mobility, Transport and Automotive Systems, edited by Heidi Krömker
31. LNCS 12792, Adaptive Instructional Systems: Design and Evaluation (Part I), edited by Robert A. Sottilare and Jessica Schwarz
32. LNCS 12793, Adaptive Instructional Systems: Adaptation Strategies and Methods (Part II), edited by Robert A. Sottilare and Jessica Schwarz
33. LNCS 12794, Culture and Computing: Interactive Cultural Heritage and Arts (Part I), edited by Matthias Rauterberg
34. LNCS 12795, Culture and Computing: Design Thinking and Cultural Computing (Part II), edited by Matthias Rauterberg
35. LNCS 12796, Design, Operation and Evaluation of Mobile Communications, edited by Gavriel Salvendy and June Wei
36. LNAI 12797, Artificial Intelligence in HCI, edited by Helmut Degen and Stavroula Ntoa
37. CCIS 1419, HCI International 2021 Posters - Part I, edited by Constantine Stephanidis, Margherita Antona, and Stavroula Ntoa

38. CCIS 1420, HCI International 2021 Posters - Part II, edited by Constantine Stephanidis, Margherita Antona, and Stavroula Ntoa
39. CCIS 1421, HCI International 2021 Posters - Part III, edited by Constantine Stephanidis, Margherita Antona, and Stavroula Ntoa

http://2021.hci.international/proceedings

13th International Conference on Cross-Cultural Design (CCD 2021)

Program Board Chair: **Pei-Luen Patrick Rau,** *Tsinghua University, China*

- Kuohsiang Chen, China
- Na Chen, China
- Wen-Ko Chiou, Taiwan
- Zhiyong Fu, China
- Toshikazu Kato, Japan
- Sheau-Farn Max Liang, Taiwan
- Rungtai Lin, Taiwan
- Wei Lin, Taiwan
- Dyi-Yih Michael Lin, Taiwan
- Robert T. P. Lu, China
- Xingda Qu, China
- Chun-Yi (Danny) Shen, Taiwan
- Hao Tan, China
- Pei-Lee Teh, Malaysia
- Lin Wang, Korea
- Hsiu-Ping Yueh, Taiwan
- Run-Ting Zhong, China

The full list with the Program Board Chairs and the members of the Program Boards of all thematic areas and affiliated conferences is available online at:

http://www.hci.international/board-members-2021.php

HCI International 2022

The 24th International Conference on Human-Computer Interaction, HCI International 2022, will be held jointly with the affiliated conferences at the Gothia Towers Hotel and Swedish Exhibition & Congress Centre, Gothenburg, Sweden, June 26 – July 1, 2022. It will cover a broad spectrum of themes related to Human-Computer Interaction, including theoretical issues, methods, tools, processes, and case studies in HCI design, as well as novel interaction techniques, interfaces, and applications. The proceedings will be published by Springer. More information will be available on the conference website: http://2022.hci.international/:

General Chair
Prof. Constantine Stephanidis
University of Crete and ICS-FORTH
Heraklion, Crete, Greece
Email: general_chair@hcii2022.org

http://2022.hci.international/

Contents – Part III

CCD in Cultural Heritage and Tourism

The Unorthodox Use of Bamboo in Fashion Styling Design   3
   Tuck-Fai Cheng, Hsiu-Wen Teng, and Po-Hsien Lin

Application of Four-Chain Integration Theory on Cultural Derivative
Design — A Case Study of Shanghai History Museum   14
   Wei Ding, Xinyao Huang, Qianyu Zhang, Xiaolin Li, and Dadi An

Inheritance Model and Innovative Design of Chinese Southern Ivory
Carving Culture and Craft   23
   Ya Juan Gao, Hao Wu Chen, Min Ling Huang, and Rungtai Lin

Objective Evaluations Based on Urban Soundscape in Waterfront
Recreation Spaces   33
   Wei Lin, Yi-Ming Wu, Hui-Zhong Zhang, and Hsuan Lin

Generating Travel Recommendations for Older Adults Based on Their
Social Media Activities   44
   Yuhong Lu, Yuta Taniguchi, and Shin'ichi Konomi

Research on the Inheritance and Innovation Path of Intangible Cultural
Heritage from the Perspective of Consumer Sociology — Take Changsha
Kiln as an Example.   56
   Wen Lu and Yulu Ouyang

Behavioral Mapping: A Patch of the User Research Method in the Cruise
Tourists Preference Research   68
   Jiangyan Lu, Xiaolei Guo, Lu Ding, Zhenyu (Cheryl) Qian,
   and Yingjie (Victor) Chen

A Study of Sightseeing Illustration Map Design   80
   Wai Kit Ng and Jing Cao

Brand Construction of Chinese Traditional Handicrafts in the We-Media
era—A Case Study of "Rushanming", a Ru Ware Brand   90
   Shuang Ou, Minghong Shi, Xin Wen, and Rungtai Lin

The Interweaving of Memory and Recollection: A Case Study of Memorial
House "Qiyun Residence"   103
   Yikang Sun and Jianping Huang

Disseminating Intangible Cultural Heritage Through Gamified Learning
Experiences and Service Design   116
   Yunpeng Xiang, Jingzhi Wang, Jing Fa, Naixiao Gu,
   and Cheng-Hung Lo

Exploring the Creation of Substandard Stones in Fuzhou Shoushan Stone . . .   129
   Xi Xu

Meet the Local Through Storytelling: A Design Framework for the
Authenticity of Local Tourist Experience   139
   Wenlin Zhang

The Living Inheritance and Protection of Intangible Cultural Heritage
Lingnan Tide Embroidery in the Context of New Media   155
   Shujun Zheng

CCD in Autonomous Vehicles and Driving

Driver's Perception of A-Pillar Blind Area: Comparison of Two Different
Auditory Feedback   171
   Chenxi Cao, Jialing Wei, Xiangyi Wang, and Hao Tan

Automated Driving: Acceptance and Chances for Young People   182
   Shiying Cheng, Huimin Dong, Yifei Yue, and Hao Tan

Analyze the Impact of Human Desire on the Development of Vehicle
Navigation Systems   195
   Feng Lan, Chunman Qiu, Weiheng Qin, Peifang Du, and Hao Tan

Acceptance Factors for Younger Passengers in Shared
Autonomous Vehicles   212
   Hao Li, Sisi Yu, Jiatai Zheng, Xue Zhao, Peifang Du, and Hao Tan

Where is the Best Autonomous Vehicle Interactive Display Place When
Meeting a Manual Driving Vehicle in Intersection?   225
   Junzhang Li, Haowen Guo, Shuyu Pan, and Hao Tan

User-Friendliness of Different Pitches of Auditory Cues in Autonomous
Vehicle Scenarios   240
   Xinrui Ren, Yimeng Luan, Xue Zhao, Peifang Du, and Hao Tan

Evaluation of Haptic Feedback Cues on Steering Wheel Based
on Blind Spot Obstacle Avoidance   253
   Jini Tao, Duannaiyu Wang, and Enyi Zhu

The Study of the User Preferences of the Request Channel on Taking Over
During Level-3 Automated Vehicles' Driving Process   267
   Qiao Yan, Yujing Wang, and Jiaru Chen

A Study for Evaluations of Automobile Digital Dashboard Layouts Based
on Cognition Electroencephalogram. 281
   Hao Yang, Jitao Zhang, and Ruoyu Jia

Explore Acceptable Sound Thresholds for Car Navigation
in Different Environments 296
   Yulu Yang, Boxian Qiu, and Xuan Liu

Effects of Multimodal Warning Types on Driver's Task Performance,
Physiological Data and User Experience 304
   Yiqiao Zhang and Hao Tan

CCD in Virtual Agents, Robots and Intelligent Assistants

Cross-Cultural Design and Evaluation of Robot Prototypes Based
on Kawaii (Cute) Attributes 319
   Dave Berque, Hiroko Chiba, Tipporn Laohakangvalvit,
   Michiko Ohkura, Peeraya Sripian, Midori Sugaya, Kevin Bautista,
   Jordyn Blakey, Feng Chen, Wenkang Huang, Shun Imura,
   Kento Murayama, Eric Spehlmann, and Cade Wright

Towards Effective Robot-Assisted Photo Reminiscence: Personalizing
Interactions Through Visual Understanding and Inferring 335
   Edwinn Gamborino, Alberto Herrera Ruiz, Jing-Fen Wang,
   Tsung-Yuan Tseng, Su-Ling Yeh, and Li-Chen Fu

Lost in Interpretation? The Role of Culture on Rating the Emotional
Nonverbal Behaviors of a Virtual Agent 350
   Adineh Hosseinpanah and Nicole C. Krämer

Manage Your Agents: An Automatic Tool for Classification of Voice
Intelligent Agents 369
   Xiang Ji, Jingyu Zhao, and Pei-Luen Patrick Rau

Identifying Design Feature Factors Critical to Acceptance of Smart
Voice Assistant. 384
   Na Liu, Ruoxuan Liu, and Wentao Li

NEXT! Toaster: Promoting Design Process with a Smart Assistant 396
   Qing Xia and Zhiyong Fu

Effects of Gender Matching on Performance in Human-Robot Teams
and Acceptance of Robots 410
   Yanan Zhai, Na Chen, and Jiajia Cao

Author Index 423

CCD in Cultural Heritage and Tourism

The Unorthodox Use of Bamboo in Fashion Styling Design

Tuck-Fai Cheng[1,2]([✉]), Hsiu-Wen Teng[2], and Po-Hsien Lin[1]

[1] National Taiwan University of Arts, New Taipei City 22058, Taiwan
t0131@ntua.edu.tw
[2] Hungkuang University, Taichung City 43302, Taiwan
kenneth1831@hk.edu.tw

Abstract. The avant-garde fashion styling is based on the spirit of fashion pioneer, rejecting the taste of "mainstream value", and based on exploring any possibility of "boundary value" in fashion styling design. Its experimental and innovative design concepts trigger forward-looking thinking and form new work forms, and create unique visual charm works with non-traditional design styles. This article focuses on the expression of "bamboo" with unorthodox material, and expands its artistic expression and understanding of the overall image design through its alternative applications. This article explores the position of "Heterogeneity" in the non-mainstream "boundary value" of works based on: avant-garde art theory, information dissemination theory and hierarchy of needs theory, fusion of the theoretical construction of aesthetic pleasure perception mode, and expands on bamboo material with unorthodoxy performance applied to fashion styling design. In addition, the "questionnaire survey" reflects the public's feelings, reactions and aesthetic value of the use of unorthodox bamboo works. According to the results of statistical analysis, the "fashion craftsmanship" (Q5) (fashion integration) in the works showed a satisfaction score of 8.43 and 8.95 with the "Bamboo Weaving Traditional Cultural Characteristics" level. Moreover, the degree of creativity in the work is 8.36, and among the "Material--Media Expression" (F1), "Craftsmanship--Formation" (F2), and "Style--Aesthetics" (F3) values. "Style--Aesthetics" is the reason that affects the main characteristics of the work. The application and performance of bamboo work in the creative world of unorthodox materials will also carry out a new experiment in fashion aesthetics and follow-up exploration of related extended topics in the form, concept, historical view, and aesthetics of the work.

Keywords: Avant-garde · Unorthodox materials · Bamboo fashion styling

1 Introduction

Nantou county in Taiwan where bamboo craftsmanship is well-known overseas in Zhushan town. By exploring the bamboo culture, it inspires any possibility for traditional bamboo culture craftsmanship and fashionable overall design, which makes the development of bamboo characteristics in design industry to make the traditional skills

© Springer Nature Switzerland AG 2021
P.-L. P. Rau (Ed.): HCII 2021, LNCS 12773, pp. 3–13, 2021.
https://doi.org/10.1007/978-3-030-77080-8_1

into a new style in the era. This research combines bamboo materials with the installation techniques of fashion styling design, and using various bamboo weaving techniques to show the new aesthetic style of contemporary creation of overall fashion styling.

With the development of society, people have surpassed the instinctive needs of animals, and the emergence of spiritual needs has produced a wealth of spiritual life and consciousness. Considering the stage of traditional bamboo handicrafts transforming from its practicality into aesthetics thinking, Breaking-through bamboo handicrafts should no longer be limited to making traditional practical living utensils. In order to combine bamboo craft weaving with fashion styling design, it is hoped that bamboo products can also be promoted in the industry and used creatively. Instead, it should be expressed in different art forms, breaking away from the tradition and creating a different new style.

A questionnaire survey was derived from teaching experience mainly implemented from the three directions(Main-factors) of "Materials", "Craftsmanship" and "Style", which focusing on 9 sub-factors as follows: "Alternative Materials" (Q1), "Technological Novelty" (Q2), "Visual Expression" (Q3), "Material Emotions" (Q4), "Fashion Application" (Q5), "Symbolic Representation" (Q6), "Model Shaping" (Q7), "Innovative Performance" (Q8), "Taste Cultivation" (Q9), to explore the relationship and value of "Material--Media expression" (F1), "Craftsmanship --Formation" (F2), "Style--Aesthetics" (F3) (see Fig. 1).

The results of the study were expected to reinforce theoretical support for bamboo as unorthodox used materials, especially in fashion styling design. Therefore, the research purpose can be briefly described as follows:

1. Discussing the cognition of bamboo as unorthodox materials in fashion styling design.
2. Explore the artistic expression of bamboo as a medium for avant-garde fashion styling.
3. Incorporating the essence of traditional bamboo culture into contemporary craftsmanship, technology and aesthetic creation.

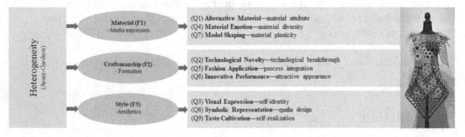

Fig. 1. The evaluation of creative "heterogeneity" in design works

2 Literature Review

2.1 Avant-Gardism

Using bamboo and traditional weaving craftsmanship, this article uses avant-gardism to deconstruct the artistic ideology, to expose and change the new style of bamboo works, and become a creation with the status of traditional works and a new discourse mode. Under the thinking of avant-gardism [2], the profound thinking of using bamboo as the fashion styling design-behind the action, the artistic rules of revealing and covering are a purpose with profound significance [7]. The self-discipline of creative aesthetics under avant-gardism can bring about the core principles of works, and the goal is to sublate the "cynicism" of art and life. Avant-gardism questions the visible reality and reveals that reality does not exist as a "total" as characteristic. It reveals reality as a kind of construction, and the existence of reality is basically irregular, fractured and contradictory without the characteristics of "integrity". It serves as a preliminary task for the exploration of postmodernist aesthetics, continuing to involve themselves in contradictions in artistic norms, and use deconstruction as the definition, norms and conditions of art popular to create "indescribable themes" [5]. This fashion styling design work created with bamboo materials will later respond to the value of its work with statistical analysis. *The World of Wearable Art Awards Competition of New Zealand* (WOW) artworks are the fusion of creativity and avant-garde creation presented in the global avant-garde Wearable art award competition [8].

2.2 Aesthetic Experience

Hekkert based on the original aesthetic experience map of Leder et al. (2004), proposed the perception model of aesthetic pleasure as shown in Fig. 2, showing the emotional state of aesthetic experience, from the self-operational work to the aesthetic judgment of the perceiver (Aesthetic judgment) or aesthetic experience pleasure. The process of which starts with perceptual analysis, implicit information integration, explicit classification, and cognitive mastering. After the evaluation, an aesthetic judgment or aesthetic pleasure experience is produced. This model to be used in bamboo styling works, is to clarify the perception of aesthetic pleasure and is deeply influenced by personal past experience [1]. The schematic relationship between fashion styling design [6] and theoretical framework of aesthetic experience can be also shown in Fig. 2.

3 Research Methodology

3.1 Research Process

This research was based on the graduation project of graduate students' works of Applied Cosmetology Department of Hungkuang University, Taiwan. An online questionnaire was developed to examine the standard of the "Heterogeneity" avant-garde concept of using bamboo as the physical materials to being transformed by the designer into appearance materials or appearance design works. A total of 91 questionnaires had been returned. After obtaining the data required for style analysis, a statistical technique of multiple regression analysis was employed to explore and evaluate the significance of these relations in the design works (see Fig. 3).

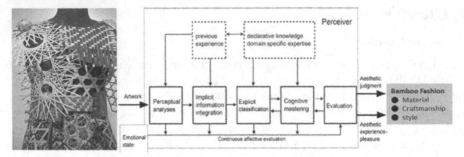

Fig. 2. Converted schematic diagram of the relationship between bamboo fashion styling design and theoretical framework of aesthetic experience from Leder et al.

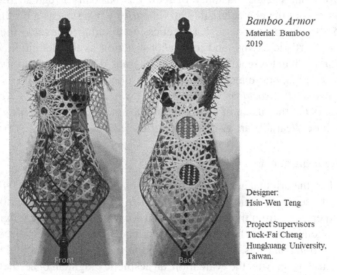

Bamboo Armor
Material: Bamboo
2019

Designer:
Hsiu-Wen Teng

Project Supervisors
Tuck-Fai Cheng
Hungkuang University,
Taiwan.

Fig. 3. Bamboo armor, bamboo works on fashion styling design

3.2 Research Framework

This research framework mainly illustrates how the case study could be used to develop the process to reach the conclusion from artist to audience. The figure illustrates the 9 items (questions) in the questionnaire content (from Q1 to Q9). The purpose was to decode the coding of the designer or creator's work to audiences. The decoding process was to understand the conversion from the technical level of the work to the semantic level, to reach the effectiveness to get the meaning. In addition, it also involved from the emotional material, through the aesthetic process, to create the emotional image and finally aesthetical experience would be obtained. Meanwhile, 9 items (questions) in the questionnaire were used to have further analysis (see Fig. 4).

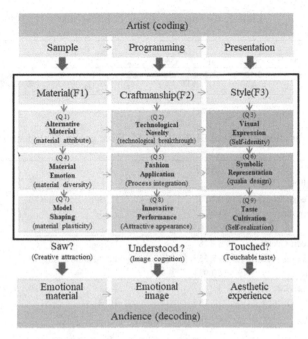

Fig. 4. Research framework for case study

3.3 Research Stimuli

These 9 sub-factors (Q1–Q9) as a test to discuss the topic, were being consultation with experts and scholars. The participants of questionnaire survey were composed of graduate and undergraduate students. A total of 91 participants completed the questionnaire.

4 Results and Discussions

4.1 Analysis the Degree of Preference

This study analyzed 91 participants' impression towards design works. Evaluation was used on the degrees of 9 sub-factors (Q1–Q9) demonstrated in the design works. A preference and mean scores were for evaluating overall impression and the outcome was shown in Table 2. The first rank was "fashion Application" for "process integration" (Q5), and the last would be "Taste Cultivation" for "Self-realization" (Q9) (Table 1).

4.2 Variation Analysis for Three Groups of Students

F1, F2, F3 showed no significant differences for the three groups of questionnaire survey test (Table 2).

Table 1. The ranking degree of preference

Rank	1	2	3	4	5	6	7	8	9
Question	Q5	Q3	Q1	Q6	Q8	Q7	Q4	Q2	Q9
Mean scores	8.43	8.38	8.37	8.36	8.32	8.31	8.23	8.19	8.18
SD	1.3796	1.4675	1.2717	1.5281	1.3345	1.4946	1.4196	1.4211	1.3792

Table 2. Variation analysis of F1, F2, F3,M for three groups of students (anova)

	Source of variation	SS	Df	MS	F	Scheffe method
F1	between Groups	3.578	2	1.789	1.340	
	Within Groups	117.527	88	1.336		
	Total	121.105	90			
F2	between Groups	3.378	2	1.689	1.264	
	Within Groups	117.553	88	1.336		
	Total	120.931	90			
F3	between Groups	3.246	2	1.623	1.219	
	Within Groups	117.183	88	1.332		
	Total	120.429	90			

4.3 T Test of the Effect Gender

Gender differences slightly different for F1, F2, F3. In other words, the average score of male was generally higher than the average score of female but not up to par (Table 3).

Table 3. T test of the effect gender on F1, F2, and F3 in the study sample (N = 91)

Variable	item	N	M	SD	t
Gender	Female	83			
	Male	8			
	F1	83	8.29	.429	-1.040
		8	8.73	.590	-.756
	F2	83	8.27	.428	-1.200
		8	8.79	.558	-.920
	F3	83	8.26	.426	-1.441
		8	8.88	.520	-1.181

4.4 Construct Validity for F1, F2, F3

Construct Validity for F1, F2, F3, showed that communalities were higher than .5. The elgenvalue was higher than 1.0, factor loading was higher than .7, % of variance was higher than 50%. This conclusion showed that the way to constitute F1, F2, F3 from Q1 to Q9 were acceptable (Table 4).

Table 4. Construct validity for F1, F2, F3

Sub-scale	item	Factor loading	communalities	Elgenvalue	% of variance
F1	Q1. Q4. Q7	.718, .703, .745	.716, .731, .686	2.435	71.109
F2	Q2. Q5. Q8.	.707, .720, .864	.816, .806, .806	2.256	75.187
F3	Q3. Q6. Q9.	.748, .889, .843	.894, .742, .795	2.431	81.031

4.5 Reliability Analysis on F1, F2, F3

From this reliability analysis, the whole scale of F1, F2, and F3 values were .936.and α deletion of 9 items (Q1–Q9) in lower than .936, which showed the reliability). For the overall reliability analysis, the Cronbach α and the average value of F1, F2, and F3 were all together greater than 0.7. Therefore, this reliability analysis was reliability predictive (Table 5).

Table 5. Summary for reliability analysis on F1, F2, F3

Subscale	item	α if item deleted	α
F1	Q1. Q4. Q7	.718, .703, .745	.797
F2	Q2. Q5, Q8.	.707, .720, .864	.833
F3	Q3. Q6, Q9.	.748, .889, .843	.882
Whole scale			.936

4.6 Multiple Regression Analysis on Creativity Level

The F1,F2 and F3 in the multiple regression analysis model to "Creativity Level" showed that F value of the overall regression model reached 59.642 (p < .05), which showed that a significant correlation between the independent and dependent variables. Produced R^2 = .673, F = .820, suggested a statistically significant association between independent variables and the dependent variable (p < .05). It could be seen in Table 7, F3 (Visual Expression) scales had significant positive regression weight, indicated the design works with higher scores on the F3 was expected to have the strongest significant to "Creativity Level" (Table 6).

4.7 Multiple Regression Analysis on Preference Level

The F1, F2 and F3 in the multiple regression analysis model to "Preference Level" showed that F value of the overall regression model reached 71.106 (p < .05), which showed that a significant correlation between the independent and dependent variables. Produced R^2 = .710, F = .843, suggested a statistically significant association between independent variables and the dependent variable (p < .05). It could be seen in Table 8, F3 (Visual Expression) scales had significant positive regression weight, indicated the design works with higher scores on the F3 was expected to have the strongest significant to "Preference Level" (Table 7).

Table 6. Multiple regression analyses with fundamental relationsas the dependent variable (creativity level)

Dependent Variable	Independent Variable	B	SE	β	t
Creativity	F1	.790	.464	.690	1.701
Level	F2	-.892	.727	-.777	-1.226
	F3	1.049	.468	.913	2.241*
R=.820			R2=.673		F=59.642***

Table 7. Multiple regression analyses with fundamental relations as the dependent variable (preference level)

Dependent Variable	Independent Variable	B	SE	β	t
Preference	F1	.839	.478	.669	1.755
Level	F2	-.939	.748	-.749	-1.254
	F3	1.165	.481	.927	2.419*
R=.843			R2=.710		F=71.106***

4.8 Multiple Regression Analysis on Tradition Bamboo Culture Level

The F1, F2 and F3 in the multiple regression analysis model to "Tradition Bamboo Culture Level" showed that F value of the overall regression model reached 18.020 ($p < .05$), which showed that a significant correlation between the independent and dependent variables. Produced $R^2 = .383$, F = .619, suggested a closed statistically significant association between independent variables and the dependent variable ($p < .05$). It could be seen in Table 8, F3 (Visual Expression) scales had significant positive regression weight, indicated the design works with higher scores on the F3 was expected to have the strongest significant to "Tradition Bamboo Culture Level" (Table 8).

Table 8. Multiple regression analyses with fundamental relations as the dependent variable (tradition bamboo culture level)

Dependent Variable	Independent Variable	B	SE	β	t
Tradition	F1	.253	.565	.250	.449
bamboo	F2	-1.133	.884	-1.116	-1.281
culture	F3	1.491	.569	1.465	2.620*
R=.619			R2=.383		F=18.020***

4.9 Multiple Regression Analysis on Color Contemporary Application Level

The F1, F2 and F3 in the multiple regression analysis model to "Color Contemporary Application Level" showed that F value of the overall regression model reached 33.500 (p < .05), which showed that a significant correlation between the independent and dependent variables. Produced $R^2 = .536$, F = .732, suggested a statistically significant association between independent variables and the dependent variable (p < .05). As could be seen in Table 8, F2 (Q2, Q5, Q8) scales had significant positive regression weight, indicated the design works with higher scores on the F2 (Craftmanship) was expected to have the strongest significant to "Color Contemporary Application Level" (Table 9).

Table 9. Multiple regression analyses with fundamental relations as the dependent variable (color contemporary application level)

Dependent Variable	Independent Variable	B	SE	β	t
Color	F1	-1.067	.660	-.780	-1.615
Contemporary	F2	2.515	1.034	1.836	2.431*
Application	F3	-.478	.665	-.348	-.718
	R=.732		R2=.536		F=33.500***

5 Results and Discussions

The degree of "creative level" in the works of Q1–Q9 is prominent in F3 (Q3, Q6, Q9). This research selected 9 items (sub-factors) from "Alternative Material", "Material Emotion", "Model Shaping", "Technological novelty", "Fashion Application", "Innovative Performance", "Visual Expression", "Symbolic Representation" and "Taste Cultivation", to evaluate the degree of the of bamboo used in design work. The results of related research through quantitative analysis after the questionnaire surveys showed that 9 sub-factor of "Fashion Application" (Q5) (8.43) is the highest scores, and "Taste Cultivation" is the lowest from 9 items.

Generally, the scores of 9 items all exceed 8 points, indicating that the scores of the 9 items of the works are not too far apart. Besides, the "Fashion Application" (Q5) (8.43), "Visual Expression" (Q3) (8.38) and "Alternative Material" (Q1) (8.37) are the first three important ranking to the design works from 9 sub-factors. Mean of F1, F2, F3 (or Q1–Q9) was 8.30. Simultaneously, 9 items are distributed into 3 factors of "Material", "Craftmanship", and "Style", showing that the sub-factors that affect the level of "creative", "preference", "traditional bamboo weaving cultural characteristics", in the work are from "Visual Expression" (Q3), "Symbolic Representation" (Q6) [9], and "Taste Cultivation" (Q9). Another sub-factors affected for "Contemporary application of colors" are "Technological Novelty" (Q2), "Fashion Application" (Q5), "Innovative Performance" (Q8).

The result can be used as a reference for subsequent performance research in bamboo. F1 for "Material-Media expression" (Q1, Q4, Q7), F2 for "Craftsmanship-Formation"

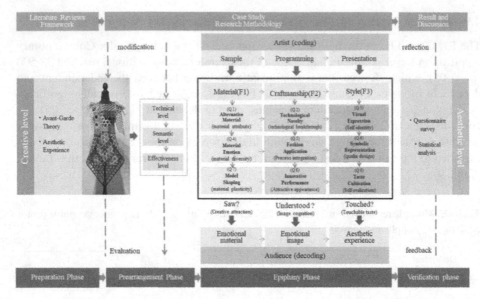

Fig. 5. The schematic diagram of bamboo fashion design and research model

(Q2, Q5, Q8), F3 for "Style-Aesthetics" (Q3, Q6, Q9) showed significant differences for the three groups of questionnaire test. Through variation analysis for T Test of the effect gender, the average score of male was generally higher than the average score of female, but not up to par. Construct Validity for F1, F2, F3 tested could be acceptable. Reliability Analysis was reliability predictive. Multiple Regression Analyses on level of "Creative", "Preference", "traditional bamboo weaving cultural characteristics" and "Color Contemporary Application" had showed the strongest significant to the research. Through the development of the theme research, a complete research framework diagram was established as follow: (Fig. 5).

Among the many unorthodox materials to engage in image styling creation, bamboo was one of the few that used on overall image design products. Most of the reasons focusing on bamboo was due to its toughness physical function to aesthetic function [3]. Through the questionnaire survey we concluded that bamboo materials can be used as the material or design in fashion styling industry. It provided an aesthetic discussion of the application of unconventional materials as a media with pleasure images [4] in human image appearance design.

Acknowledgements. Special thanks to the experts and scholars of the Creative Industry Design, Gradual school of National Taiwan University of Arts for their assistance and suggestions in this article.

References

1. Blijlevens, J., Thurgood, C., Hekkert, P., Chen, L., Leder, H., Whitfield, T.W.: the aesthetic pleasure in design scale: the development of a scale to measure aesthetic pleasure for designed artifacts. Psychol. Aesthetics Creativity Arts **11**(1), 86–98 (2017). https://doi.org/10.1037/aca0000098
2. Calinescu, M.: Five Faces of Modernity: Modernism, Avant-garde, Decadence, Kitsch, Postmodernism. Duke University Press, Durham (1987)
3. Hsieh, T.S.: Introduction to the Art, p. 11. Wer Far Books, Taipei (2004)
4. Hsiao, K.A., Chen, Y.P.: Cognition and shape features of pleasure images. Des. J. **15**(2), 1–17 (2010). https://doi.org/10.6381/JD.201006.0001
5. Murphy, R.: Theorizing the Avant-Garde: Modernism, Expressionism, and the Problem of Postmodernity, 1st edn. Cambridge University Press, Cambridge (2010)
6. Ocvirk, O.G., Stinson, R.E., Wigg, P.R., Bone, R.O., Cayton, D.L.: Art Fundamentals: Theory and Pratice, 8th edn. McGraw-Hill education, New York (1997)
7. Gongini, B.I.: https://barbaraigongini.com/, Accessed 12 Feb 2016
8. WOW. https://www.worldofwearableart.com/, Accessed 12 Feb 2021
9. Yen, H.Y., Lin, P.H., Lin, R.: Qualia characteristics of cultural and creative products. J. Kansei **2**(1), 34–61 (2014)

Application of Four-Chain Integration Theory on Cultural Derivative Design — A Case Study of Shanghai History Museum

Wei Ding, Xinyao Huang, Qianyu Zhang, Xiaolin Li, and Dadi An[✉]

School of Art Design and Media, East China University of Science and Technology, 130 Meilong Rd., Xuhui District, Shanghai, China

Abstract. The development of society, the increase of people's income and other factors have driven the upgrading of tourism consumption.In recent years, with the rapid development of social new media, there are more and more ways to spread culture, and the innovative development of cultural derivatives has become one of the important links. Excellent cultural derivatives can greatly improve the status of culture in social groups and establish regional cultural brands. Due to the impact of modern new technologies and new models, the design of cultural derivatives is also facing new opportunities and challenges in integrating historical features, creating scale experience and providing good services. As to the problems and insufficiency in derivatives innovation and development, this study takes Shanghai History Museum (SHM) culture derivatives design as an example, by adopting evaluation principle and Semantic Differential Method, to evaluate viewers feedback in products, interactive, service, brand four dimensions on the basis of investigation and questionnaire research. With the integration of the above principles, this study defines cultural derivative design with a new logic, clarifies the development paradigm of cultural heritage inheritance and innovation, to improve the realistic value of cultural heritage.

Keywords: Four-Chain Integration Theory (FCIT) · Cultural and Creative Product Design · Semantic Differential Method (SD) · Design evaluation

1 Background

Shanghai is an international first-tier city with rich cultural heritages. Since opening its port in 1843, Shanghai has rapidly developed from an offshore fishing village into the most prosperous port and economic and financial center in the Far East at that time, known as the "Ten-mile Foreign Market" [1]. Cities which known as global cities, without exception, have integrated historical traditions and cultural heritages into urban development planning and spiritual shaping. This preserves and protects a large number of intangible cultural heritages, such as knowledge, ideas, objects, skills, ceremonies and festivals, thus highlighting the uniqueness of the city.

The city history museum is responsible for protecting the regional heritage, building a window for cultural exchange and promoting the development of local culture.

© Springer Nature Switzerland AG 2021
P.-L. P. Rau (Ed.): HCII 2021, LNCS 12773, pp. 14–22, 2021.
https://doi.org/10.1007/978-3-030-77080-8_2

However, due to the new landmark semantics brought by museum construction and the impact of new technologies and new models brought by modernization, museum cultural creation is often faced with challenges and opportunities in integrating historical features, creating humanized scale experience and providing good services. Therefore, as a medium to maintain museum and external space, cultural and creative derivatives of museums urgently need to build design paradigm according to people's needs and objective environment, and promote the orderly and dynamic development of cultural heritage.

The domestic some museum or institution on the living condition of the traditional culture conducted a series of exploration and attempt, most are of the type of tour visit, and the same souvenir sales, unable to form a complete theoretical framework. This paper taking SHM as an example, aim at cultural and creative derivatives of the Museum as the research object, and sets twelve indicators for the practicality and representativeness of the products, the humanization of the use experience, the completeness of the service system and the degree of brand communication, which are oriented to every tourist.

2 Research Design

2.1 Four-Chain Integration Theory

Product chain emphasizes the physical logic of design, through the resonance of new materials, new processes, new technologies, new models, to bring visitors unique products. Faced China's rich cultural heritage, a series of product designs for relevant cultures can be carried out to help cultural heritage better transform into economic value.

Interaction chain emphasizes the behavior logic, achieving the optimized experience through the link of information, and bring users experience value. With the development of science and technology and the Internet, interaction design has been integrated into life in various forms, including product design, user experience, service design, new media design, installation art, space interaction and other directions.

Service chain emphasizes the system logic of the whole process. By redefining the service process design, it connects products, experiences and service systems. Explore the needs of each stakeholder in the system, construct an overall service framework, and design all kinds of contacts in the service framework. It uses services to create better experience and value for users and other stakeholders in the system.

Brand chain emphasizes image logic, through visual transmission of brand connotation, so as to establish the communication between the brand and users. Brand chain puts more emphasis on the IP building of relevant brands and emphasizes the re-innovation of IP elements. At the same time, it helps enterprises to build their image, form brand value, deepen the memory of consumers, and enhance the sense of belonging and identity of consumers by combining the accumulated innovation experience for a long time.

2.2 SD Research Method

SD method is a psychological measurement method proposed by American psychologist C.E.Os good in 1957 [2]. By constructing semantic psychological test, the audience's perception of the product and experience is collected quantitatively, and using the language scale to measure psychological feeling. The measured results were presented by Likert scale and analyzed by certain statistical methods. The SD method usually selects multiple groups of adjectives with opposite meanings related to the research object. The gradient of each pair of phrases is generally 5–11. By drawing the collected data into a chart, the semantic difference can be obtained and the psychological perception characteristics of the environment can be intuitively obtained, thus to provide a basis for further research.

2.3 Research Goal

Tourists score concepts and things according to their own subjective consciousness and feelings through the SD scale, and then extract emotional semantic vocabulary according to the emotional needs of target users. The evaluation data and statistical analysis are integrated to establish a database of tourists' emotions, and the results are converted into a reference source of modeling language used for designing products, interactions, services and brands [3].

Based on the SD curve, actual investigation and interview, this paper makes a comparative analysis of the cultural derivatives of SHM from four dimensions: product chain, interaction chain, service chain and brand chain. Then explore the advantages and disadvantages of cultural and creative derivatives of historical museums in these four dimensions, demonstrate the feasibility and development value of the FCIT in the cultural derivatives industry, and provide data analysis support for the follow-up research.

3   Results

3.1   Evaluation Factor Selection

The selection process of emotional words mainly includes four steps [4]. In the first step, we collected 365 emotion-related words from more than 30 articles of previous scholars on cultural creation derivatives. The second step is to discuss with industry experts and select and retain 120. Step 3: Eliminate repeated words and combine similar words to keep 90. The fourth step is to select the antonyms and synonyms of each word. Finally, through discussion, a total of 16 pairs (32 pairs) are derived as comprehensive evaluation factors (Table 1). The process of selecting evaluation factors is shown in Fig. 1.

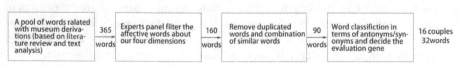

Fig. 1. Emotional words collection

Table 1. Evaluation factor

Value dimension	Evaluation gene	Adjective couple	
Product value	Product innovation	Rigescent	Vivid
	Product uniqueness	Ordinary	Personalized
	Product friendliness	Vulgar	Elegant
	Product line diversity	Simplex	Multiple
Interactive value	Thinking mode	Contemplative	Enlightening
	Guide way	Dedicated	Emanative
	Focus degree	Central	Dispersive
	Science sense	Traditional	Technological
Service value	Service comfort level	Ice-cold	Genial
	Service opening level	Private	Open
	Facility level	Rough	Complete
	Atmosphere creating	Hidebound	Relaxed
Brand value	Economy efficiency	Low-cost performance	High-cost performance
	Artistry	Conservative	Original
	Cordial degree	Approachable	Stately
	Connotation degree	Unitary and boring	Sincere

3.2 Questionnaire

The SD questionnaire consists of three parts: basic information, evaluation form and recommendation. Combined with domestic and foreign experience, four groups of 16 pairs of evaluation factors are finally determined, reflecting the four dimensions of product design, interaction design, service design and brand building. In order to improve the fineness of the questionnaire, an appropriate five-layer gradient was adopted in the evaluation table. The two poles were distinguished by "very", "comparison" and "general", with values of $-2, -1, 0, 1$ and 2 [5].

The questionnaire is divided into two groups: the professional group, which is dominated by art majors, and the non-professional group, which is dominated by tourists and residents. In the pre-investigation, it was found that the professional group had a high integrity and clarity of cognition on the cultural derivatives system. However, for non-professional groups, it is difficult to complete the questionnaire, and there will be a certain scrap rate [6]. During the whole experiment, a total of 120 questionnaires were sent out, including 40 from the professional group and 80 from the non-professional group. The recovery rate of 40 from the professional group was 100%, and 74 from the non-professional group was 93%.

3.3   Experiment Result

The score values of each factor of 114 valid questionnaires were grouped into groups to get the average score. Then the score values were used as the coordinate axis to draw the visual radar chart. It can intuitively show the respondents' comprehensive evaluation of the product (Fig. 2).

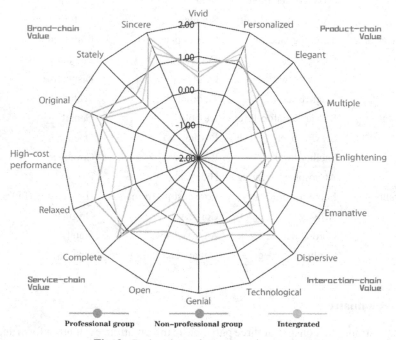

Fig. 2.   Radar chart of questionnaire results

As can be seen from the radar map, professional for each factor scored an average of four dimensions above the number 0, it scored more than 1 point in six evaluation factors, including personalization, dispersion, completeness and ease, sincerity and novelty, the other is 0–1 points. On the perception of cultural derivatives for professional evaluation is satisfied. Non-professional group for the average scores of most factors above the number 0. It scored negative in three evaluation factors, such as diversity, divergence and service openness, and more than 1 in personalization, completeness, novelty and sincerity, other factor score between 0–1. It indicated that the non-professional group's perception evaluation of cultural derivatives was between the mean and satisfaction, and did not reach a relatively satisfactory evaluation result.

4   Discuss

4.1   Product Chain

Product chain is market-oriented, use cultural heritage as the material to design and manage products. This whole set of processes will have a greater impact on a cultural heritage. It is not only for the inheritance skills of cultural heritage, but also for the innovation, inheritance and related development methods of culture. Excellent serialized product design can even have a greater radiation effect on the development of the entire cultural industry and related industries.

The cultural derivatives of SHM can be divided into travel, practical, entertainment and aesthetics categories according to their categories. The four categories are basically evenly distributed, among which memorial and jewelry account for a slightly higher proportion, while other categories of daily necessities, school supplies, postcards, electronic toys and handicrafts are evenly matched (Fig. 3).

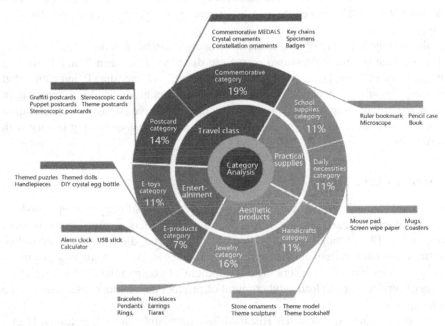

Fig. 3. Cultural derivative products of Shanghai History Museum

From the SD average score of product dimension, it can be seen that users are satisfied with the "personalization" and "vivability" of the product, and think that this series of cultural derivatives can better highlight the characteristics of the museum. However, the product diversity score is relatively low, even is significantly lower than 0, indicating that there is still room for improvement in product line richness. In conclusion, the perceptual evaluation of product value dimension is generally moderate.

4.2 Interaction Chain

After product development, interactive chain focuses on taking a cultural heritage as an integral cultural industry to graft to other industries, and developed as an important resource. With the market as the final guide, the overall layout of cultural heritage is carried out, so that cultural heritage and related products from the inside out to the market.

Interaction design has penetrated into every aspect of our life, and even fields unrelated to design are experiencing a drastic industrial revolution of interaction, digitalization and informatization [7]. The interactive chain needs to accurately determine the corresponding category based on offline research and online big data analysis, and find the interactive contact points between users and products, so as to continuously optimize the relevant interactive experience and meet the positioning requirements of different channels.

The derivative series of SHM mainly extracts the cultural elements of Shanghai to make the prototype selection, so as to determine the style intention and design concept. To find the contact point between Shanghai history and modern popular culture and truly break down the "wall" between the masses, especially the young people and Shanghai history.

In the average SD score of interaction dimension, it can be seen that only two factors score more than 0, and the rest two scores are distributed between 0 and 1. Among them, the score of "divergence" and "technology sense" is lower than 0, indicating that more energy should be spent on the application of new technology and new function. In addition, the enlightening score is lower than the comprehensive average score, indicating that there is still a great room for improvement in the process of linking history with modern trends.

4.3 Service Chain

Service chain is mainly aimed at the systematic design of the strategy, concept, product, management, process, service, use, recovery and other aspects involved in the product service system [8]. Through the integration of offline space, display layout and online platform, establish a system conducive to unified business. At the same time, focus on "drainage + experience" to ensure the development of commercial ecology. Relevant big data of service sales to lead another round of product development and redefine the product appeal.

The service chain of Shanghai History Museum mainly links four levels: brand, design team, supplier and platform, so as to complete the whole process of product authorization, design, manufacturing and marketing. SHM provides copyright and content, and is responsible for sample review; The design team completed product screening and content mining, and completed the final design through element extraction, process definition, color definition and so on; Supplier is responsible for product proofing and mass production; The final platform is responsible for traffic import, online sales and online promotion. To sum up, complete the whole process management of the service chain (Fig. 4).

In the SD average score of the service dimension, the scores of three factors are more than 0, the scores of "completeness" and "ease" are both higher than the overall average score, indicating that the service chain basically meets the needs of buying and experiencing pleasure. However, the score of the friendliness is low, and the score of the openness is negative, indicating that the building of the atmosphere of the purchased space still needs to be improved.

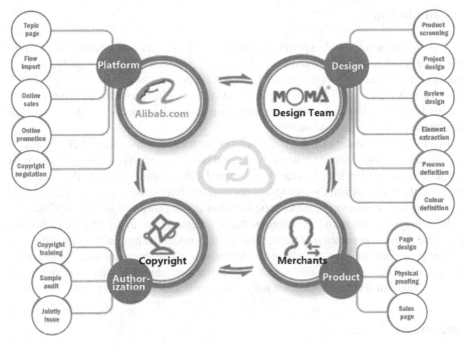

Fig. 4. Service chain of Shanghai history museum

4.4 Brand Chain

Brand chain is centered on brand cultural resources. By systematically sorting out vision, mission and values, the core of brand value is extracted to make regional cultural brands more consumer oriented and enhance brand affinity and popularity [9]. Through all kinds of promotion means, to achieve the perfect communication between the brand and consumers.

Among the elements and features of the cultural derivative design of SHM, it not only takes the folk and interesting stories of old Shanghai as the theme, but also shapes the IP elements of the life of old Shanghai. Also render young cartoon IP elements with fantasy myths about Shanghai as the theme. Overall derivatives are presented at high, medium and low levels. According to different consumer groups, IP design is completed in three directions: classic IP elements, expanded IP elements, and IP elements combined with current affairs.

In the average SD score of brand dimension, the four factors are all higher than 0, "novelty" and "sincerity" scores are all higher than 1, significantly higher than the overall average, indicating that the general public is very satisfied with the brand image. However, it has a low score in terms of cost performance, which also provides a direction for the later design direction and supply chain management.

5   Conclusion

Through questionnaire and experimental analysis, the results emphasize that FCIT in design period of cultural derivatives reveals a strong practical value. This study explored related patterns trials about the product chain, interaction chain, service chain, brand chain four dimensions. With the empirical case study of Shanghai History Museum, we indicated that these four aspects play an important role in the process of cultural heritage protection and inheritance innovation.

In addition, this study clarified the development paradigm of cultural heritage inheritance and innovation, driving cultural heritage develop in orderly and dynamic way, as well as to improve the social value of traditional culture and to realize the economic value. Furthermore, this study attempts to serve as a reference for other cultural heritage inheritance and innovation in the context of cultural soft power improving of China.

With respect to research methodology, the SD method used in this study can directly obtain the tourists' perception of the cultural derivatives of the museum, so as to directly feel the propagation effect of culture. However, due to different interviewees' understanding of adjectives, the results cannot fully reflect the objective situation. Moreover, the lack of questionnaire respondents led to another limitation in this study.

References

1. Jie, F.: Re-mention of '"Shanghai Culture"': external communication of Shanghai's city image. South. Media Res. **05**, 105–111 (2020)
2. Li, M., Lu, Z., Huang, L.: Research on vehicle modeling image cognition based on eye movement experiment and semantic difference. Art Educ. **2016**(04), 212–214 (2016)
3. Yan, J.: Study of the evaluation of external space of museums in the historical district based on SD method: A case study of the art museum of cantonese opera. Xi'an University of Architecture and Technology. A New idea for starting point of the silk road: urban and rural design for human. In: Proceedings of the International Conference on Environment-Behavior Studies (EBRA 2020). Chinese Society of Environmental Behavior, Xi'an University of Architecture and Technology, Huazhong University of Science and Technology Press Co., Ltd., vol. 7 (2020)
4. Shin, G., Park, S., Kim, Y., Lee, Y., Yun, M.: Comparing semantic differential methods in affective engineering processes: a case study on vehicle instrument panels. Appl. Sci. **10**(14), 4751 (2020)
5. Klettner, S.: Affective Communication of map symbols: a semantic differential analysis. ISPRS Int. J. Geo-Inf. **9**(5), 289 (2020)
6. Min, L., Ke, Z., Jiahua, Z., Zhiwei, H.: Study on post-evaluation of underground city link space based on SD method: a case study of Shanghai Wujiaochang. New Build. **6**, 15–20 (2019)
7. Xu, Z.: Research on digital display media of intangible cultural heritage. Pack. Eng. **36**(10), 20–23+48 (2015)
8. Li, H., Wei, D.: Research on shared product system design based on service design concept. Design **20**, 100–101 (2017)
9. Wang, J.: Research on the design of creative intangible cultural heritage products based on transboundary. Pack. Eng. **40**(22), 253–259 (2019)

Inheritance Model and Innovative Design of Chinese Southern Ivory Carving Culture and Craft

Ya Juan Gao[1], Hao Wu Chen[1(✉)], Min Ling Huang[1], and Rungtai Lin[2]

[1] School of Fine Arts and Design, Guangzhou University, Guangzhou City,
Guangdong Province, China
[2] Graduate School of Creative Industry Design, National Taiwan University of Arts, Ban Ciao,
New Taipei City, Taiwan
rtlin@mail.ntua.edu.tw

Abstract. The Southern ivory carving is a folk carving art and craft represented by the Guangzhou area of China, which began in the Qin and Han dynasties and flourished during the Ming and Qing dynasties. The Southern ivory carving skill is exquisite and famous for its hollowed and pierced carving skills. With the banning of ivory carving materials, it has gradually declined. It was included in China's National Intangible Cultural Heritage List in 2006, emphasizing the preservation of exquisite skills and the live transmission. The study analyzes the three levels of the Southern ivory carving Culture, Outer level, Middle level, and Inner level, form a design model that corresponds to the design of cultural products with a three levels cultural transformation of appearance, function, and meaning, combining modern digital technology and design to verify the design model with case design, and analyze the inheritance model of the Southern ivory carving culture and skills in the process of modern life. Make the communication and design of intangible cultural heritage full of vitality.

Keywords: Chinese Southern ivory carving · Inheritance model · Innovative design

1 Introduction

From ancient times to modern times, ivory carving has a long history and is one of the representatives of traditional handicrafts. It is a local traditional handicraft that uses ivory as raw material for carving, and has high cultural and artistic value. It has been the subject of attention and research all over the world (Johnston 1983; KE 1994; Jeanmichel 1999; Gansell et al. 2014; Lin et al. 2017). Its development history can be traced back to the tomb of the Nanyue King, 2100 years ago. It contains ivory carved seals and other items. According to the "Book of Jin", (晋书)during the Wei and Jin Dynasties, artisans in Guangzhou had already made ivory seats. In the Tang Dynasty, Guangzhou tooth sculptures were mostly stationery, combs, vases and other utensils, and the industrialization of tooth sculptures was obvious. An ivory carving workshop is formed at the intersection

© Springer Nature Switzerland AG 2021
P.-L. P. Rau (Ed.): HCII 2021, LNCS 12773, pp. 23–32, 2021.
https://doi.org/10.1007/978-3-030-77080-8_3

of Beijing Road and Huifu Road in Guangzhou. The art of hollowing out during the Song, Yuan and Ming Dynasties developed rapidly, from two-layer hollowed-out ivory ball carving to multiple layers, and a large number of ivory chopsticks and ornaments entered the daily lives of citizens. During the Qing Dynasty, Guangzhou's dental sculptures developed to their peak. After Guangzhou opened to the outside world for trade, the production of dental sculptures ranked first in the country. There are hundreds of ivory carving shops in Daxin Street, Guangzhou, employing as many as 2,780 people. As shown in Fig. 1, the Guangzhou ivory carving craftsman photographed by Scottish photographer John Thomson. Dental carvers are world-renowned for their unique skills such as carving and inlay.

Fig. 1. Guangzhou ivory carving craftsman during the Qing Dynasty (photographed by Scottish photographer John Thomson).

Southern ivory carving gradually formed a complete set of exquisite carving craftsmanship in long-term craft practice, and became one of the outstanding representatives of traditional Chinese handicrafts, forming a unique carving style and characteristics. Southern ivory carving works focus on carving craftsmanship, carving, inlay and weaving techniques, and absorb a large number of Persian styles and European "full carving" craftsmanship, and pay attention to bleaching the ivory, making the works exquisitely polished. It is famous for its hollow and deep carving techniques. It has a variety of creative themes and varieties, and is closely related to the utensils used in daily life at home and abroad, such as pierced flower folding fans, necklaces, chess, cigarette holders, powder boxes and tooth combs.

Therefore, it can be seen that during the development process of Guangzhou dental sculpture, not only the craftsmanship developed earlier, the historical and cultural heritage is deep; but also the rapid development, the craftsmanship industrialization model is obvious, the variety is integrated into daily life, and the development of maritime trade will form a better circulation spread. Most domestic and foreign research scholars pay attention to the history of the rise and fall of Guangzhou dental sculptures, the carving crafts, the appreciation of fine products, and the identification of materials. It is a very traditional inheritance study and is no longer suitable for the spread and development of today's dental sculpture culture. Because ivory materials are special and precious engraving materials, and the requirements of environmental protection, ivory materials are restricted. Not only is the output reduced, but the number of people who inherit and learn is decreasing, and related research is also gradually decreasing. The protection and inheritance of dental carving skills is very difficult. How to inherit dental carving craftsmanship and innovative development is a problem worth exploring and researching.

2 The Dilemma of Inheritance and Communication of Southern Ivory Carving Culture and Craft

Nowadays, Southern ivory carving is in three dilemmas: there is no tooth to carve, Lack successors, and the skill is about to be lost. Under the global awareness of environmental protection, ivory and sculpture works are gradually banned, and raw materials are no longer available for carving. There are fewer and fewer people learning to sculpture, and the transaction is gradually decreasing. Modern heirs are all over 60 years old, and can no longer sculpt new works after their eyesight has deteriorated. However, their children are unwilling to continue learning dental carving techniques. The techniques of ivory weaving and silk splitting in the southern tooth carving craft have been lost, and the two handicrafts of micro carving and open carving tooth balls are about to be lost.

After limiting the ivory materials, the inheritors of Guangzhou ivory carving have been exploring and trying new materials to replace African ivory. It is hoped that through the transformation of materials, the dental carving craftsmanship can be preserved, so that the aesthetics and subtle skills of dental carving works can be continuously inherited and developed. The master craftsman is constantly exploring new ways of communication and development. The master of craftsmanship Zhang Minhui replaced the buffalo bone with the innovation of ivory materials, and borrowed from the Ming and Qing wood mortise and tenon craftsmanship and traditional mosaic craftsmanship, so that the inlaid crafts have been reborn. During the period, he created countless new works, such as "Happiness as Immense as the Eastern Sea (福如东海)", "Litchi Bay Area of the Regional Customs", "Everything goes Well (吉祥如意)", "The New Image of Yuexiu Area (越秀新晖)", "Icing on the Cake (锦上添花)" and "Four Seasons", etc. As shown in Fig. 2. Bone as a material fusion dental carving craftsmanship, won the gold award in the provincial or national competitions (He et al. 2013).

Fig. 2. The bone carving work "The New Image of Yuexiu Area" (越秀新晖)based on the ivory carving technique innovation

However, the characteristics of buffalo bone materials are different from ivory materials. They still only solve part of the problems, and some crafts and works cannot be realized. Just like the carved ivory ball, the most prominent feature of Guangzhou dental carving, it cannot be carved with buffalo bones. Only ivory can be used to carve a multi-layer rotating ball. Because of the extremely skillful uncanny workmanship extremely finecraft, it is called "Superlative Craftsmanship Ball". When making him, it has to go through many complicated processes such as cutting, embossing, drilling, leveling, and carving. An ivory ball can be carved dozens of layers, each layer can be rotated, each layer has carved patterns. The 30-story ivory sculpture ball won the gold medal at the 1915 "Panama-Pacific International Exhibition" (Zheng and Lu 2008). It later developed to the 60th floor, and now it can no longer be passed on and developed. The wonderful craftsmanship of ivory carving balls is on the verge of being lost.

3 Inheritance Model and Innovative Communication of Southern Ivory Carving

When discussing the issue of cultural inheritance, using innovative design methods and establishing good communication channels will be conducive to the inheritance of ivory carving culture. How to innovate and inherit the culture starts with analyzing the cultural level. Culture can be classified into three layers: (1) Physical or material culture, including food, garments, and transportation related objects, (2) Social or behavioral culture, including human relationships and social organization, and (3) Spiritual or ideal culture, including art and religion, as shown in the bottom of Fig. 3 (Lin 2007). These three culture layers can be fitted into Leong's three culture levels given above (Leong and Clark 2003). Since cultural objects can be incorporated into cultural design, three design features can be identified as follows: (1) the inner-level containing special content such as stories, emotion, and cultural features, (2) the mid- level containing function, operational concerns, usability, and safety, and (3) the outer-level dealing with colors, texture, form, decoration, surface pattern, line quality, and details (Leong and Clark 2003; Lin 2007; Lin et al. 2017; Gao et al. 2018).

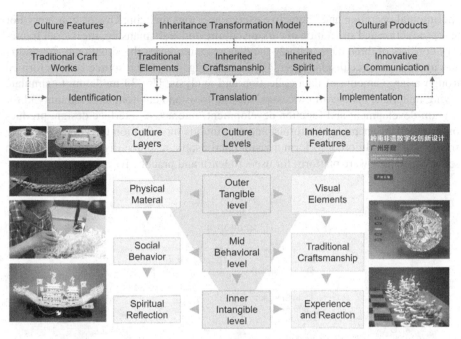

Fig. 3. Cultural product and inheritance model

As shown in Fig. 3,based on the three levels of culture, the outer, middle, and inner layers, the characteristics of traditional culture are also analyzed, and the three aspects of traditional elements, inherited craftsmanship and inherited spirit are transformed (Lin and Lin 2010). In the process of inheritance and transformation, it is expressed as visual elements, traditional craftsmanship, experience and reaction. Further realize digital innovative design and innovative communication methods. Take the Southern style ivory carving culture as an example. The culture will be inherited from the three levels of visually visible ivory carving works, special handicrafts, story and spirit. Combining modern technology with traditional culture, truly adapt to the cognition and learning methods of modern life, and meet the design needs of contemporary life. It is an effective way to inherit the traditional ivory carving culture.

4   Cases of Using Virtual Technology for Innovative Communication and Design

Combining modern science and technology with traditional crafts and culture, using virtual reality technology to integrate southern ivory carving. Virtual simulation technology is not limited by ivory materials, and it is simple and fast to develop experimental experience and innovative design, increasing interactive experience and learning fun. Through the joint realization of 3D modeling, environmental simulation and calculation processing, the experience can watch dental sculpture works from multiple angles as a general audience, visit the craft scene, and can also be used as a virtual dental craftsman for experimental operation and interactive experience, thus realizing Guangzhou Ivory For the reproduction and simulation of carving culture and craftsmanship, the overall framework consists of three parts, namely the data resource layer, the situational simulation layer and the resource display layer, as shown in Fig. 4, a data resource layer is formed through a large amount of data collection and analysis. Use 3d modeling to simulate carving materials, handicraft processes, carving tools and environment, use unity3d for program development and function development, interactive interface design, production of practical operating platforms, and continuous modification to form a real ivory carving situation. In order to form a visible and usable resource display layer. We can use the virtual software platform for users to learn and practice ivory carving.

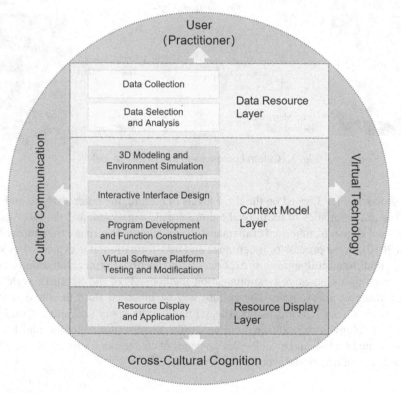

Fig. 4. The framework and implementation process of applying virtual reality technology to Guangzhou ivory carving

The digital recording and protection of Guangzhou dental carving craft requires real data collection and data analysis. Through in-depth interviews with the inheritors, the national representative inheritors Zhang Minghui, Weng Yaoxiang and his son Weng Zhanxuan were selected as objects. Through many in-depth interviews, the operation process of traditional and modern ivory carving techniques was recorded, and a large number of digital shootings were completed, and technical details were broken down.

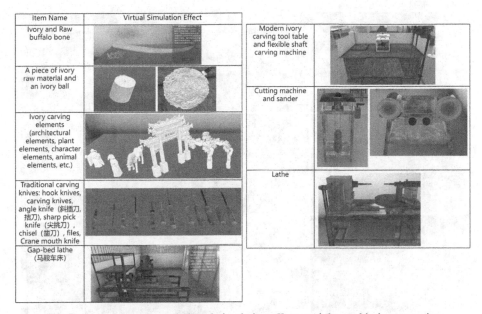

Fig. 5. Material names and virtual simulation effects mainly used in ivory carving

Choose the most representative ivory ball of southern ivory carving, and use its production process to show the production process. Organize and edit traditional and modern carving materials and tools, as shown in Fig. 5, and list the main names and virtual simulation effects. Such as traditional ivory and buffalo bones, carved ivory balls, ivory carving elements, traditional carving knives, gap-bed lathe, Modern ivory carving tool table and flexible shaft carving machine, etc.

The virtual simulation platform interface shown in Fig. 6. First, learn the history and culture of southern ivory carving, and then appreciate the representative works. The multi-layered ivory ball can be rotated freely to appreciate the beauty of the carving in all directions. Next is to enter the ivory carving workshop to learn about materials and tools step by step. Then enter the experimental link, turn an ivory segment into a spherical shape, and evenly open 14 holes of the same size on the sphere, the closer the center of the sphere, the smaller the hole, which is a cone-shaped hole. Next is the most difficult layering. Divide an 18 cm sphere into dozens of layers (the thickness of each layer can be determined according to calculations). It is also when the ivory ball is the easiest to break, you must choose the right tool and be very careful. The next step is to carve each layer inside the ivory ball, make it hollow and transparent, and complete an exquisite carving of ivory. It not only shows the special carving skills of the ivory ball, but also shows the new ivory carving work completed by the inlay process and the buffalo bone material. Finally, the innovative combination method can be freely combined according to the theme that the creator will express to complete new works, which is conducive to stimulating the theme creativity and design innovation of traditional works.

The constructed virtual simulation platform promotes the cultivation of inheritors of Guangzhou ivory carving and the spread of culture. Incorporated into the courses of

Fig. 6. Virtual simulation platform interface and example of flow chart for making ivory ball

Guangzhou University, such as "Lingnan Heritage Digital Innovative Design" and "Lingnan Traditional Carving Art" courses, adding sculpture practice links. Two semesters of teaching have been carried out, and 120 people have participated in the practice, experiencing the beauty of ivory carving works and craftsmanship through interaction, and experiencing the spirit of craftsmanship and traditional culture. At the same time, carry out inheritor training courses to provide the society with resource sharing, promote talent training and inheritance, protect the endangered traditional crafts and processes, and promote and disseminate the importance of Lingnan regional culture and innovative design.

5   Conclusions

The protection of intangible cultural heritage based on digital virtual technology has become a new trend. For those endangered and non-copyable folk art and culture, the use of text and multimedia means to record them in a complete and vivid manner is conducive to their protection and inheritance. This research constructs a cross-cultural communication and application model, quickly analyzes the value characteristics of cultural objects from the three main levels of outer, middle and inner, and guides design innovation and realization (Gao et al. 2018; Lin et al. 2016). Take China Southern School Ivory Sculpture (Guangzhou Ivory Sculpture) as a case, integrate virtual reality technology to achieve the following goals:

1. The existing text data, image data, oral data and digital publications have been sorted out to protect the hollow carving skills that are on the verge of loss. It is more intuitive,

visual and logical to clearly express the process, forming a cultural communication and experience mode that can be emotionally interactive.

2. Cross-cultural, cross-temporal, cross-regional, multi-perspective exhibition of dental sculpture aesthetics, interactive and dissemination, in-depth understanding of Chinese culture and folk crafts.

3. Build a digital virtual simulation experiment platform for dental sculpture technology, expand the audience, integrate into college education, and promote the transformation of the training model of inherited talents.

4. Inspire creative design with inheritance mode and cultural product mode, promote the transformation of dental carving craft products, and meet the needs of cross-cultural and international users.

The ivory sculpture cultural protection project that integrates virtual reality technology will be further promoted to other schools and social groups, using the Internet to increase the speed and scope of dissemination, enhance the influence of intangible culture, protect the handicrafts that are about to die, inherit the spirit of craftsmanship, and carry forward the regional characteristics Culture, increase the influence of cultural exchanges. This research only takes ivory carving as the object, and subsequent research can also be extended to other cultural heritage.

References

Gansell, A.R., Meent, J.W.V.D., Zairis, S., Wiggins, C.H.: Stylistic clusters and the syrian/south syrian tradition of first-millennium bce levantine ivory carving: a machine learning approach. J. Archaeol. Sci. **44**, 194–205 (2014)

Gao, Y.J., Chang, W.C., Fang, W.T., Lin, R.: Acculturation in human culture interaction - a case study of culture meaning in cultural product design. Ergon. Int. J. **2**(1), 1–10 (2018)

Gao, Y.J., Fang, W., Gao, Y., Lin, R.: Conceptual framework and case study of China's womanese scripts used in culture product design. J. Arts Hum. **7**(3), 57–67 (2018)

He, X.Q., Wang, Z.F., Liu, W.: Master of Chinese Art and Crafts Zhang Minhui: GuanZhou ivory carving. Jiangsu Fine Arts Publishing House, 27–33 (2013)

Lin, P.-H., Gao, Y.-J., Lan, T., Wang, X.: Integration and innovation: learning by exchanging views - a report of the cross-cultural design workshop for stone craving. In: Rau, Pei-Luen Patrick (ed.) CCD 2017. LNCS, vol. 10281, pp. 181–191. Springer, Cham (2017). https://doi.org/10. 1007/978-3-319-57931-3_15

Leong, D., Clark, H.: Culture -based knowledge towards new design thinking and practice - a dialogue. Des. Issues **19**(3), 48–58 (2003)

Lin, R.: Transforming taiwan aboriginal cultural features into modern product design: a case study of a cross-cultural product design model. Int. J. Des. **1**, 45–53 (2007)

Lin, R. Lin, C.L.: From digital archive to e-business: a case study of turning "art" to "e-business". In: Proceedings of 2010 International Conference on E- Business (2010)

Lin, C.L., Chen, S.J., Hsiao, W.H., Lin, R.: Cultural ergonomics in interactional and experiential design: conceptual framework and case study of the Taiwanese twin cup. Appl. Ergon. **52**, 242–252 (2016)

Johnston, W.R.: Nineteenth-century french ivories at the walters art gallery. Magazine Antiques (1983)

Jeanmichel, S.: A. cutler, late antique and byzantine ivory carving, ashgate, variorum, 1998, 312 p. Bull. Monumental **157**, 399–400 (1999)

KE. Carved tusk and finial. Metropolitan Museum of Art Bulletin (1994)

Zheng, Y.F., Lu, S.G.: Perfect craftsmanship: GuanZhou ivory carving. Guangdong Education Publishing House, 41–44 (2008)

Objective Evaluations Based on Urban Soundscape in Waterfront Recreation Spaces

Wei Lin[1]([✉]), Yi-Ming Wu[2], Hui-Zhong Zhang[2], and Hsuan Lin[3]

[1] School of Architecture, Feng Chia University, Taichung, Taiwan
wlin@fcu.edu.tw
[2] Program for Civil Engineering, Water Resources Engineering, and Infrastructure Planning, Feng Chia University, Taichung, Taiwan
hz@mail.fcu.edu.tw
[3] Department of Product Design, Tainan University of Technology, Tainan, Taiwan
te0038@mail.tut.edu.tw

Abstract. Space and form play an important role in the current geometry of recreational space and cognition of sound field. Therefore, there is different objective physical quantity information of sound energy. These concerned sounds are related to urban development and historical and cultural characteristics and will be highly recognizable sounds. It can be seen from the above that in urban recreational streets, through the objective physical characteristics of the street sound field sorted out, the positive evaluation can be predictably given, and the correspondence between the subjective feelings of the people will be explored. If the physical quantity of sound energy discusses the homogeneity of normalization of the soundscape and comprehensive physical quantity information, it is afraid to ignore the characteristics of spatial qualitative and internal variability. The currently research results propose the urban soundscape acoustic environment factors of waterfront recreation spaces, confirm and draw distinctive sound energy distribution maps and sound field perception. Understand the public's sensitivity and preference for sound, and explore the correlation and consistency of measurement and computer simulation correspondence.

Keywords: Soundscape perception · Objective field measurements

1 Introduction

From the second half of the 20th century [1], among the environmental indicators of urban street development, the awareness of the impact of sound energy on the lives of residents has gradually increased, especially in recreational areas. Residents are increasingly eager for a pleasant and accessible soundscape environment. Control and suppression measures are processed to pursue a high-quality acoustic environment. Nowadays, the metropolis is narrow and densely populated. In addition to the acoustic energy generated by transportation, the exposure of life sounds, and recreational activities, the auditory perception of walking in the urban recreational spaces is the direction indicator of current soundscape research. Residents also yearn for a quiet living environment and a dynamic soundscape

© Springer Nature Switzerland AG 2021
P.-L. P. Rau (Ed.): HCII 2021, LNCS 12773, pp. 33–43, 2021.
https://doi.org/10.1007/978-3-030-77080-8_4

field. However, urban acoustic energy is dominated by road-mobile vehicles, which is an unavoidable potential influence factor in the urban development acoustic environment. In recent years, in relevant economic activity data or social science research, most international environmental policies focus on noise control and have considerable research norms in the energy time domain and spectrum domain. Although reducing noise levels does not necessarily improve the quality of life of citizens [2–5]. The World Health Organization and its national indicators try to use the high-decibel sound energy produced by transportation and industrial production to improve compliance with regional regulations (World Health Organization, 1999 [6]; European Parliament and Council, 2002 [7]; World Health Organization, 2011 [8]), especially in the urbanization development indicators, actively pursue the creation of a high-quality living environment with low sound energy, as an important reference for a livable city. Noise source processing or passive noise suppression structure hides the multi-dimensional sound energy characteristics in the sound field of the recreational space. If ignored, the identity of the urban acoustic environment will not be presented. The pattern and rhythm of residents' life [9], the distribution of urban development, change the intensity and distribution of sound energy, and alleviate the annoying noise energy in the city by creating urban soundscapes, thereby shaping the acoustic environment and urban identity. Government departments encourage urban residents to walk and take transportation, which is widely recognized for reducing urban carbon emissions and enhancing sustainable maintenance, while also slowing down the annoying acoustic energy generated by transportation [10]. Such as road traffic, natural and human sounds, which have meaning and characteristics in the acoustic environment and reflect relevant information, and their dynamics must be considered. Therefore, "soundscape" is different from "acoustic environment"; the former is mainly composed of human perception, the latter refers to physical phenomena, and most importantly, the soundscape is presented through human perception of the sound environment. Furthermore, the correlation between the soundscape and the environment of the street recreational space tends to be complicated. In addition to the record of objective physical basic data, the relationship between the pleasant sound generated by the environment and the physical quantity of the sound field during the walking space; Pleasantness of the sound environment, which induces the sound field quality index to define the high-quality urban sound scene field [11, 12]. Based on the sound diversity record between the cultures and folk customs of various countries, focusing on the actual feelings and hearing experience of the senses, urban sound energy is regarded as a usable "resource" rather than a "pollution". The soundscape features are on the verge of being submerged in the homogenized urban sound field. While re-examining the urban development context and street recreational space planning, the specific soundscape can indicate the characteristics of a regional community separately, enhancing its identity, and becoming more important. A series of related studies led by Professor Jian Kang from the Department of Architecture of the University of London in the United Kingdom have sprung up as relevant research topics in EU countries [13]. Leading the European Soundscape Research Alliance (Soundscape of European Cities and Landscapes) and the UK Noise Futures Network (UK Noise futures Network) and other research institutions, in 2009 held a seminar on the future development of an urban acoustic environment in the city, focusing on discussions Solve the problems of regional soundscape characteristics and future soundscape trends. Under the impetus of this conference, the center of Sheffield was used as a research pilot, based on the quantitative and qualitative analysis of the soundscape characteristics of public spaces, based on the source, space, people, and the environment is defined by four factors [14], among which space factor and sound source factor are the most widely discussed. The spatial factors in the sound field include spatial patterns, boundary materials, street objects and furniture, and landscape elements; the sound source factors are waterscapes, bird calls, and church bells that are most popular among the people. For residents, speech sounds are the main source of relatively annoying sounds. Related research has provided the British urban soundscape features as a direction for protection. Human factors mainly clearly describe social and demographic trends, including gender, age, place of residence (such as local residents or from other urban residents), and cultural and educational background. Among them, the acoustic experience is more important, especially the acoustic environment at home and workplace. Finally, the environmental factors must be described in terms of the actual conditions of the physical environment, especially the auditory preferences are easily affected by visual effects, and the interaction of environmental atmosphere and spatial context. From the perspective of sound source factors, the sound generated in the open space of urban recreational streets and the pleasant sense of hearing can be regarded as a soundscape, although the sound produced by traffic equipment can be disturbing sounds. Therefore, the sound energy level is measured using A-weighted sound energy dB(A), which is very similar to the frequency response of the human hearing, except that the equivalent average sound volume LAeq (dB) and the maximum equivalent sound volume LAFmax (dB) are used to explain the total sound In addition to the basic characteristic physical quantities of energy strength, in order to find the description of dynamic sound energy and time sequence, the cumulative time of a certain sound pressure level value is interpreted as a percentage (n-Percent Exceeded Level), the symbol is Ln, and the percentage value is LA_10 (dB), LA_50 (dB), and LA_90 (dB). Therefore, there are different objective physical quantity information of sound energy. These concerned sounds are related to urban development and historical and cultural characteristics, and they are all highly recognizable sounds. It can be seen from the above research that in urban recreational streets, through the objective physical characteristics of the street sound field, the positive evaluation can be predictably given, and the corresponding relationship between subjective evaluation and objective measurement can be conducted. What if only the physical quantity of sound energy is used to explore the sound normalization of homogeneity, comprehensive physical quantity information, it may ignore the characteristics of spatial qualitative and internal variability. In 2014, the International Organization for Standardization (ISO) released the soundscape international standard ISO 12913 Part 1 [15], which defines "soundscape" as "a kind of sound environment that is perceived, experienced and understood by people". Regarding the definition of open space soundscape from the perspective of human hearing, and further discussing the types and interfaces that affect hearing in street recreational spaces, in addition to effectively controlling noise sources, the research team of Professor Kang Jian and Professor Francesco Aletta [16], through To understand the impact of the subjective soundscape perception of the recreational space by the sound produced by walking on different materials. The research methods are based on the different perceptual influences of the sounds produced by self-walking and listening to the sounds of other people's walking, and the subjective questionnaire on preference was implemented in Weston Park and Valley Gardens Park in the United Kingdom. Four materials considered to be recreational walking spaces may be used for discussion; Grass, wood, stones, and gravel. The research results show that the nuisance and sound characteristics of walking on the surface of the material have a significant impact. Grass is the most accepted material, and gravel is poorly evaluated. Both are highly correlated with subjective perception. Corresponding to the annoying effect of the subjective perception of a single material, people who produce sounds by themselves by walking have poor tolerance when just listening to the sounds of walking. This part does not evaluate the significant effects of the overall evaluation factors. This is the subjective perception of soundscapes affected by urban recreational space types, pedestrian street materials, and interfaces, and is an important direction for discussion. In addition, the noise is blocked to enhance the features in the soundscape of the neighborhood, such as combining water feature facilities and noise barriers, embedded in the open recreational space of the city, and using sound barriers to block noise sources and waterscapes acoustic masks disturbing background noises to improve the perception of the characteristics of the sound environment. Different soundscape features provide spectral features and ranges, combined with street furniture and display billboards, blending local history and culture, and creating urban self-evidence. For another example, the fountain water and metal noise barriers in Sheffield city in the United Kingdom. The fountain represents the river and the metal represents the steel industry. It symbolizes the two important historical contexts of Sheffield's industrial development and urban planning and enhances residents' and tourists' rest and enjoyment, which are rich in cultural and educational significance. In 2020, the research team of Professor Kang Jian and Professor Francesco Aletta [17] continued to conduct physical measurements and observations on the sound field of 11 urban open spaces in London in spring during emergencies; under the influence of the Covid-19 incident, The difference in sound energy of open space before (2019) and after Lockdown (2020) in London. Based on the urban soundscape theory, this research defines the current sound source types of urban soundscapes and explores the acoustic environment planning corresponding to the spatial geometry of urban recreational pedestrian streets. The research object is the waterfront trails along with the recreational space. The sound source collection of the soundscape is mainly based on the human voice of the retail street, the surrounding environmental noise, the mobile sound source, and the fixed sound source are mainly discussed. The field is drawn from the connection of walk path along with waterfront which distributed from dots, lines to planes. A characteristic sound energy distribution map, with a view to reinterpreting the self-explanatory nature of the green belt urban style of the potential commercial district and its transportation nodes. Preliminary research has been conducted through domestic and foreign research on urban waterscape-related papers and materials. At present, on-site sound field measurement has been carried out to confirm and draw distinctive sound energy distribution maps, to understand the sensitivity and preference of the public to sound, and to explore objective correspondence and consistency with subjective evaluation.

2 Description of Urban Waterfront Recreation Walkway

In this study, the Liuchuan Waterfront walkway in the central district of Taichung City was used as the acoustic environment field of the urban pedestrian space, as shown in Fig. 1. Liuchuan is one of the four rivers in the center of Taichung City. After its renovation, it became the first landscape river bank developed by sustainable ecological construction methods in the urban area. History changed to the Japanese occupation period, so the chessboard street between Liuchuan and Taichung Park was built and developed on the scale of a small Kyoto, the water ditches were renovated, and willow trees were planted on both sides of the bank, so it was called in Chinese pronunciation as Liuchuan.

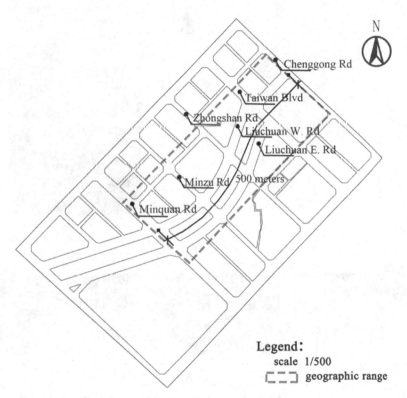

Fig. 1. Figure of study geographic range for Liuchuan urban waterfront recreation walkway

Following large number of immigrants cut down the willow trees on both sides of the Liuchuan river and built temporary houses, destroying the original landscape. Recreational walking space is attractive, the soundscape characteristics of the waterfront walking space must be clarified. Through on-site measurement, it can provide complete and different sound field distribution data from commercial recreational walking. In the current measurement process, the measurement of sound energy on the waterfront walk will provide a basis for the complete sound field distribution. The Liuchuan West Road and East Road section at 500 m in length and 10 m in width. The height ratio between the building and the street is 1:0.5 to 1:2, and the height ratio of the river edge and road surface is 1:2, and the entire range of the riverfront walkway is trumpet-shaped. The measurement time is carried out on weekdays and weekends, with a total of 36 sets of measurement points, and each measurement point is evaluated and recorded in LAeq (dB). The measurement time is 1 min to obtain an objective evaluation of the regional soundscape. This information will become important basic data for the next stage of soundscape perception discussion. A schematic diagram of the on-site measurement distribution and photos of each measurement point is shown in Fig. 2.

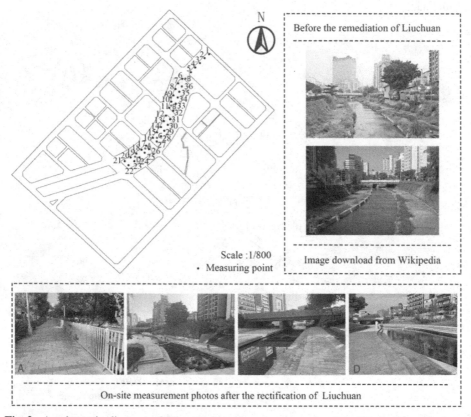

Fig. 2. A schematic diagram of 36 sets of on-site measurement points distribution and photos from the measurement perspective, Photo A and B refer to Wikipedia [17]

3   Results of on-Site Measurement

The physical characteristics of the spatial sound field of urban recreational streets. For reflection and diffusion boundaries, the reverberation is usually much shorter, and the sound attenuation of the distance between the sound source and the receiving source is greater than the sound energy attenuation of the geometric reflection boundary. In the free sound field of the long walkway profile, when the length and width are doubled, the sound pressure level SPL will be attenuated by 6–9 dB, and will attenuate as the distance squared increases; if the boundary sound absorption coefficient increases, the sound absorption coefficient increases from 0.1 At 0.9, the sound energy attenuation is usually 12 dB. LAeq (dB) measurement results of the sound field characteristics were analyzed which is the critical point in Liuchuan river, especially the spatial pattern belongs to the acoustic canyon type, which needs to be confirmed in the future study, and realizing the specificity of the soundscape environment integrated with the waterscape of sound field characteristics are explained. General conditions of the on-site measurement are list in Table 1.

Table 1. List of general conditions of the on-site measurement

Site	Liuchuan W. St (Water front trails)
Date	18th and 22th Oct. 2019
Time	Morning (10:00 AM), Afternoon (3:00 PM), Evening (9:00 PM)
Weather condition	Temperature: 28 C, Humidity: 55–68%
Measurement apparatus	Rion
Quantities of measurement point	35 points
Sampling time	1 min for each time
Frequency band	20–20000 Hz
Method	Environmental sound energy measurement according to NIEA method

The measurement results with the equivalent average volume *LAeq* (dB) as the evaluation index and the A-weighted noise intermittently exposed in a certain period of time in the selected position in the sound field are averaged by the energy. The parameter index formula is shown in Eq. 1.

$$LAeq = 10log\frac{1}{T}\int_{t}^{t+T}\left(\frac{Pt}{P0}\right)^{2}dt \tag{1}$$

LAeq: A-weighted average energy level dB (A) in period time;
T: measurement time in seconds;
Pt: measure sound pressure in Pa;
P0: reference sound pressure, based on 20 μPa

For the main pedestrian street and waterfront walkway of Liuchuan West Road, the sound field distribution was discussed for different time nodes. 21 test points (1–21) were selected for the pedestrian space along the river bank, and 15 test points (22–36) Measure at different times and fixed positions. The measurement time is divided into 10:00 (morning), 15:00 and 9:00 in the evening, and the main section of Liuchuan West

Road (7–16), the measurement result shows (6–21) the whole day Leq near the river embankment, the average LAeq (dB) is 61 dB(A), and (22–36) the LAeq (dB) of the whole day on the water shore trails which the energy distribution is formed as U-shaped, the LAeq (dB) of the middle section is lower, and the LAeq (dB) of the two endpoints is relatively high, which may be affected by the traffic on Taiwan Avenue and Minquan Road. The intersection of Taiwan Avenue and Minquan Road is the main traffic arterial road in Taichung City, and its LAeq (dB) value is slightly higher in the afternoon than in the evening, which indicates that the traffic is more concentrated in the afternoon due to rush hour. The measurement results of 36 sets of measurement points are shown in Fig. 3.

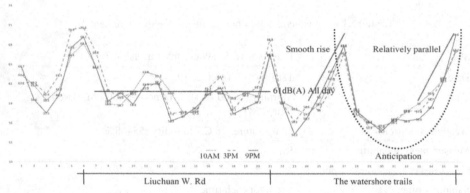

Fig. 3. 36 sets of measurement results of LAeq (dB) distribution at Liuchuan W. Rd. and water shore trails

4 Visualization of Sound Distribution

The relative software was be presented, it can reduce the time needed for the creation of noise contour maps considerably. The development of a fit for purpose software for mapping contours based on measurements. Scanned maps and CAD drawings can be used as backgrounds for future evaluation. The actual measurement is used to draw the sound energy distribution map, and the NoiseAtWork software is used to analyze and draw the distribution curve. The results can be used to establish the optimization plan for the overall distribution of the sound field. Individual environmental factors can be used to verify and hypothesize to construct the model, through the implementation of measurement correspondence, further draws sound pressure level images and verifies sound field models, predicts future scenes and benchmarks for subjective evaluation, and then takes measures. Using the computer simulation of the sound field environment results, the calculation method to perform grid calculations, draw the sound energy distribution curve diagram, which has an intuitive trend and visual representation, and can evaluate the sound field in which it is located under different conditions surroundings. The simulation results provide a lot of information, including the location and strength of sound energy. In some cases, this method has the function of trend prediction, especially under stable sound sources, such as fixed noise sources (fixed mechanical sound), and multiple sound source are conducted for the measurements. For the simulation of the direct influence on the plane on both sides of the Liuchuan River and the simulation under the enclosure of the building, it can be clearly seen that the sound energy is focused to 68 dB(A) under the enclosure of the building, while also ensuring the integrity of the waterfront landscape. The waterfront walkway is less affected by the surrounding of the river embankment, which shields vehicle noise, and the concave terrain maintains the sound field environment of the waterfront walkway. The simulation results of the sound field at Liuchuan East Street and waterfront walkway are shown in Fig. 4.

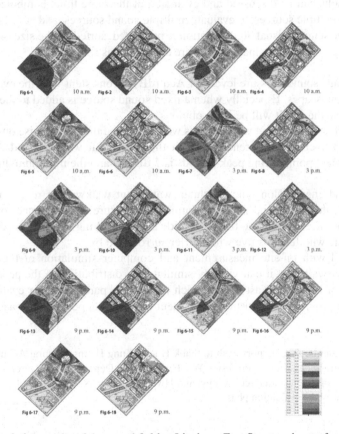

Fig. 4. Simulation results of the sound field at Liuchuan East Street and waterfront walkway

5 Discussion

This study uses the actual measured sound energy of the recreational trail along the waterfront, and continues to analyze and plot the LAeq (dB) sound energy distribution with NoiseAtWork computer software, and the distribution of the waterfront and surrounding trail areas and the overall result, in order to be used as urban soundscape planning basis. Construct the actual sound field model based on spatial factors, and further draw the sound pressure level distribution through the corresponding relationship between measurement and simulation, and the basic data for future soundscape and subjective and objective evaluation. The sound field environment results generated by computer simulation, and the calculation method is grid calculation, which has an intuitive trend and a visual presentation. Under different conditions, evaluate the sound field environment you are in, including the position of sound energy and the strength of sound energy, which has the function of trend prediction. Under stable sound sources, such as fixed noise sources (fixed mechanical sound), multiple sound sources The measurement results can be discussed and evaluated at the same time, combined with the influence of multiple sources, to evaluate multiple sound sources, and at the same time, the immersive state of visual domestication is presented, sorted by the size of the sound energy results. Some preliminary results are abstracted as followed:

1. The average sound pressure level of LAeq (dB) is consistent with the evaluation of urban soundscapes. Especially when a fixed sound source is added to the computer simulation, the result will be more obvious.
2. The Liuchuan waterfront walkway has well alleviated the traffic noise on both sides of it. Sound energy difference between the waterfront walkway and the ground measurement point during peak periods is 7 dB (A) and the minimum difference is 3 dB (A).
3. The sound energy along the Liuchuan waterfront walkway of reveal a U-shape, indicating that the sound field environment and nature are integrated to provide a good acoustic environment. It may provide a good walking space for local residents and people who feel positive with barrier surrounded by.
4. Combined with on-site measurement and computer simulation of the Liuchuan waterfront walkway, it can visually simulate the distribution of the acoustic environment of Liuchuan East Road, which has on-site parameters for evaluating and further improving the acoustic environment of the urban pedestrian space.

Acknowledgements. The authors wish to thank Huang Kung Huang, Natural Acoustic Co. for the helps of software assistance, Professor Wei-Hwa Chiang, Dep. of Architecture, National Taiwan University of Science and Technology, and Hwa Hsia University of Technology the kindly assistances during the evaluation phase.

References

1. Rueb, E.S.: Many Pleas for Quiet, but City Still Thunders, New York edition (2013)
2. Alves, S., Est_evez-Mauriz, L., Aletta, F., Echevarria-Sanchez, G.M., Puyana Romero, V.: Towards the integration of urban sound planning in urban development processes: the study of four test sites within the SONORUS project. Noise Mapp. 2(1), 57e85 (2015)
3. Andringa, T.C., Weber, M., Payne, S.R., Krijnders, J.D., Dixon, M.N., Linden, R., et al.: Positioning soundscape research and management. J. Acoust. Soc. Am. 134(4), 2739e2747 (2013)

4. Asdrubali, F.: New frontiers in environmental noise research. Noise Mapp. 1, 1e2 (2014)
5. van Kempen, E., Devilee, J., Swart, W., van Kamp, I.: Characterizing urban areas with good sound quality: development of a research protocol. Noise Health, 16(73), 380e387 (2014)
6. World Health Organization, Guidelines for community noise, Geneve: World Health Organization (1999)
7. European Parliament, & Council, Directive 2002/49/EC relating to the assessment and management of environmental noise, Brussels: Publications Office of the European Union (2002)
8. World Health Organization, Burden of disease from environmental noise. Copenhagen: WHO Regional Office for Europe, (2011)
9. Bauman, Z.: Liquid modernity, p 240 (2000). ISBN-13:978-0745624105
10. King, E.A., Murphy, E., McNabola, A.: Reducing pedestrian exposure to environmental pollutants: a combined noise exposure and air quality analysis approach. Transp. Res. Transp. Environ. 14, 309–316 (2009)
11. Lavandier, C., Defréville, B.: The contribution of sound source characteristics in the assessment of urban soundscapes. Acta Acust. U. Acust. 92, 912–921 (2006)
12. Ricciardi, P., Delaitre, P., Lavandier, C., Torchia, F., Aumond, P.: Sound quality indicators for urban places in Paris cross-validated by Milan data. J. Acoust. Soc. Am. 138, 2337–2348 (2015)
13. Kang, J, et al.: Ten questions on the soundscapes of the built environment. Build. Environ. 108, 284e294 (2016)
14. Kang, J.: Noise management: soundscape approach, reference module in earth systems and environmental sciences, encyclopedia of environmental health, 174–184 (2011)
15. International Organization for Standardization. ISO12913-1:2014acoustics—soundscape—part 1: definition and conceptual framework. Geneva: ISO(2014)
16. Aletta, F., Kang, J., Astolfi, A., Fuda, S.: Differences in soundscape appreciation of walking sounds from different footpath materials in urban parks. Sustain. Cities Soc. Vol. 27(November), 367–376 (2016)
17. Aletta, F., Oberman, T., Mitchell, A., Tong, H., Kang, J.: Assessing the changing urban sound environment during the COVID-19 lockdown period using short-term acoustic measurements. Noise Mapp. 7, 123–134 (2020)
18. https://zh.m.wikipedia.org/wiki/%E6%9F%B3%E5%B7%9D

Generating Travel Recommendations for Older Adults Based on Their Social Media Activities

Yuhong Lu[1]($\boxtimes$), Yuta Taniguchi[2], and Shin'ichi Konomi[3]

[1] Graduate School of Information Science and Electrical Engineering, Kyushu University, 744, Motooka, Nishi-ku, Fukuoka 819-0395, Japan
lu.yuhong.678@s.kyushu-u.ac.jp
[2] Ressearch Institute for Information Technology, Kyushu University, 744, Motooka, Nishi-ku, Fukuoka 819-0395, Japan
[3] Faculty of Arts and Science, Kyushu University, 744, Motooka, Nishi-ku, Fukuoka 819-0395, Japan

Abstract. The declining birthrate and the increasing aging population can exacerbate various societal issues such as social isolation, which can have a serious impact on the mental and physical health of older adults. Increased frequency of going out can reduce the possibility of future social isolation and facilitate recovery from social isolation. In this paper, we propose a novel method for generating travel recommendations for older adults to increase their frequency of going out. The proposed method builds a travel-recommendation model based on social media posts by older adults. The modelling process exploits the semi-supervised Latent Dirichlet Allocation (ssLDA) and object detection techniques to extract the interests of older adults by analyzing latent topics in textual and visual messages. Travel recommendations can be generated by matching the latent topics and the online information about travel destinations. Our feasibility study demonstrates a higher recall in predicting relevant topics for older adults compared to a baseline method that relies on the conventional Latent Dirichlet Allocation (LDA) model.

Keywords: Older adults · Travel recommendation · Social media · SNS · LDA · Image processing

1 Introduction

The declining birthrate and the increasing aging population are exacerbating various societal issues in different countries. In Japan, the ratio of older adults who are sixty-five years old or older exceeded twenty-one percent in 2007, and the country is described as "super-aged" since then. With the continued aging of the population, there is the increasing number of older adults living alone without sufficient opportunities to communicate with others nor build and maintain social relationships. One of the most critical issues in such "super-aged" societies is the negative effect of social isolation as poor social relationships can affect mental and physical health of older adults. There is the increasing awareness about this critical issue, and researchers have conducted studies to analyze relevant factors. Ito, Goto, and Yamamura (2019) have analyzed the association

© Springer Nature Switzerland AG 2021
P.-L. P. Rau (Ed.): HCII 2021, LNCS 12773, pp. 44–55, 2021.
https://doi.org/10.1007/978-3-030-77080-8_5

between loneliness and everyday activities of older adults in Nerima Ward of Tokyo, and shown that everyday leisure outings may alleviate the loneliness of older adults [1]. Takahashi, et al. (2020) have conducted a two-year cohort study in Ota Ward of Tokyo to analyze factors relating to social isolation of older adults, and shown that a lower frequency of going out is one of the factors that predicted future social isolation, and that a higher frequency of going out is one of the factors that predicted recovery from social isolation [2]. We build on these studies and propose computational techniques for promoting leisure outings to alleviate the problems of social isolation in "super-aged" societies.

With the spread of smartphones and other interactive computing devices in recent years, many people share information about various topics by using social media services. Given the sheer amount of information that is shared through social media services, some of which closely reflects users' interests, one could analyze and model interests of the target population according to social media posts. Approximately half of the Japanese older adults use social networking services (SNS) and the percentage is increasing at a high speed [3]. Thus, we could use the social media posts by older adults to analyze and model their interests. Providing travel recommendations based on such analysis and modeling can promote everyday leisure outings, thereby reducing social isolation as well as the feeling of loneliness.

In this paper, we propose a novel method for generating travel recommendations for older adults to increase their frequency of going out. As shown in Fig. 1, the proposed method builds a travel-recommendation model based on a large amount of textual and visual social media posts by older adults. Our method first extracts the topics of their interests, and then recommends places to visit according to the features extracted from the textual descriptions of relevant travel destinations. In particular, the topic-extraction process exploits the semi-supervised Latent Dirichlet Allocation (ssLDA) and object detection techniques to extract the interests of older adults by analyzing latent topics in textual and visual messages. Travel recommendations can be generated by matching the latent topics and the online information about travel destinations.

One of the advantages of using social media data in generating travel recommendations is the ability to reflect the dynamic and real-time information in the modeling process. For example, the model can be constructed to capture trending, common topics of interest among older adults. We also believe that the use of social media data could have some benefits in creating conditions that facilitates the creation of meaningful social ties among older adults.

Our feasibility study demonstrates a higher recall in predicting the topics of older adults' interests compared to a baseline method that relies on the conventional Latent Dirichlet Allocation (LDA) model.

2   Related Works

In this section, we discuss different techniques for modelling the topics of users' interests based on SNS data.

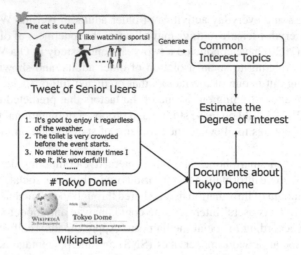

Fig. 1. Overview of the proposed method

2.1 Extracting the Topic Categories of Interest

Kume, et al. model users' interests based on topic categories to recommend relevant Twitter accounts [4]. Their technique detects a category for each keyword by using an existing public API called Hatena Keyword Automatic Link API [5], and use the categorical information to infer the degrees of users' interests in different categories including sports and music. Users whose tweets often include keywords of a certain topic category, such as sports, are considered to have high degrees of interest in the category. However, this approach makes a simplified assumption that each keyword belongs to one category although we may want to consider multiple categories of a keyword to model users' interests accurately.

2.2 Weighted LDA Models

We can model users' interests by using topic modeling that models the characteristics of sentences as latent topics. Latent Dirichlet allocation (LDA) is a commonly-used topic model, which allows a keyword to belong to multiple topics. Wang, et al. use topic modeling to predict users' interests based on Twitter information [6]. Their method first extracts topics of users' interests using topic modeling, and then prioritize extracted topics based on users' 'likes', profiles, followers, and tweet contents. In particular, they consider that tweets with many 'likes' closely reflect users' interests, and assign them large weights in determining topic priorities.

2.3 Implicit and Explicit Interests of Users

Han, et al. model users' interests by considering both implicit and explicit interests [7]. Their method first exploit category information from news media sites and extract implicit and explicit features of news articles in different categories such as sports and politics. Implicit features are derived based on topic modeling, and explicit features are generated by finding keywords with high *term frequency-inverse document frequency* (TF-IDF) values. Characteristic words representing users are generated based on a topic model and TF-IDF statistic with their sentences. Finally, topic models and TF-IDF values are compared between users' sentences and news articles from different categories to calculate the degrees of users' interests.

2.4 Expanding Recommendation Lists Based on Relevant Text and Images

Oohigasi, et al. use SNS data to recommend points of interests (POIs) in geographical spaces. They propose to combine visual and textual information to remedy the problem of limited coverage with the POI recommendation lists when the amount of textual information about POIs is small. Their method exploits so-called Bags of Visual Words (BoVW), which is based on scale-invariant feature transform (SIFT) features, as the key features of POIs. Visual and textual features can be combined by linking BoVW and text data from relevant POIs. This approach relies on location-visit histories of users, and can be affected by the cold start problem, i.e., it cannot capture users' interests sufficiently when there is only a small amount of location-visit histories.

3   Proposed Approach

Twitter users may follow user accounts that are relevant to their interests. We thus propose a topic modeling-based approach to derive latent topics from older adults' tweets, predict their interests, and recommend tourist spots that match shared interests of groups. Our approach improves the accuracy of recommendation by augmenting and guiding the modeling process. This approach can derive relevant topics based on limited tweet contents and match derived topics with users' topics of interest. The approach can also improve the understandability of recommended topics. In particular, we augment the information used in the modeling process by incorporating images included in tweets as well as their textual contents, and guide the modeling processes based on semi-supervised Latent Dirichlet allocation. The derived topics are matched with the features of tourist POIs to recommend tourist spots to older adults effectively.

Figure 2 shows the processing flow of the proposed method. The details of each step is as follows:

1. Extract text and image data from the tweets posted by older adults.
2. Extract visually represented information by using object detection techniques, and convert it to text. This allows for integrated processing of visual and textual information, and we obtain Bag of Words (BoW) for each older-adult user, including their frequency counts.
3. Collect tweets with hash tags to obtain representative words or seed words for each interest. Calculate TF-IDF values of the words, and select the words with high values as representative words. The selected words will be used to characterize interests in each topic category.

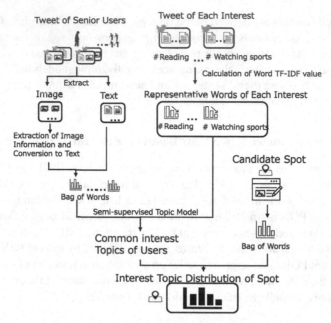

Fig. 2. Processing flow of the proposed method.

4. Apply semi-supervised topic modeling algorithm to the set of BoW from multiple older-adult users. Guide the modeling process by assigning higher weights to the representative words, and predict topics of interest to many older adults.
5. Extract BoW from existing documents about tourist spots, and calculate the topic distributions of different tourist spots based on topic-wise distribution of words and word-wise distribution of topics.
6. Compare the topic distributions of tourist spots and older adult's tweets, and predict the degree of older adults' interest in different tourist spots.

We can then recommend the tourist spots that are highly probable to be relevant to the interests of older adults.

3.1 Preprocessing Tweets

In natural language processing, preprocessing is an extremely important part that greatly affects experimental results. Tweets have the characteristics that the number of characters is limited and that users can send them relatively easily. Therefore, tweets are overwhelmingly short compared to common documents such as newspaper articles and academic papers, and they tend to include may 'noise' words. If the topic model is applied directly without preprocessing, it will not function properly.

We next explain the process for applying the topic model to tweets. Firstly, we remove retweets from the data as the contents of retweets are normally created by some other users and may include many 'noise' words. Because tweets are very short, we regard all recent tweets of a user as a document, and do not consider user accounts without more than 100 tweets.

Since words are not separated by spaces in Japanese, morphological analysis of tweet sentences is performed to extract pairs of morphemes and parts of speech. In addition, we consider pictograms, symbols, line-feed characters, URLs, stop words, etc. as 'noise' and delete them from the data sets. Of the remaining words, we consider that the nouns in the tweets are more likely to represent topics related to the user's hobbies and tastes. We only retain general nouns, proper nouns, and nouns that can be connected with "suru" to form verbs, and construct vectors of posted contents.

3.2 Processing Images

In this research, we extract latent topics of interests based on the tweets from older adults. In doing so, we exploit image as well as text data that are embedded in tweets. In order to apply the topic model to the information about potential interest expressed as an image, we propose a text-image topic model that detects the information contained in images, converts it into text, and combines it with general text information. In this manner, we can augment the data that are fed to the topic modeling algorithm and enable multi-faceted prediction of interests.

We use the object detection model YOLOv3 [9] to detect the type of object from the input image. YOLOv3 is trained with a data set called MS-COCO, and 80 kinds of categories and their positions can be detected. YOLOv3 consists of 75 convolution layers, and extracts the features of the object. Since it does not use a fully connected layer, it can process input images of arbitrary size, and thus can handle tweet-embedded images of different sizes. In YOLOv3, the accuracy of object detection is greatly improved by simultaneously learning and predicting the box coordinates and their reliability scores.

Fig. 3. Example of information extraction from images.

Figure 3 shows the results of detecting objects in images posted by older adults using a pretrained object detection model. The predicted object names can be used to describe the image contents. These object names can complement tweets that do not include the object names, thereby facilitating detection of the topics that are of interest to older adults. It is necessary to specify a threshold value with detection reliability. Only when the score of the object is higher than the specified threshold value, we use the label of the object included in the target image.

3.3 Extracting Interests Through Text-Image Topic Modeling

Since SNS contents are diverse, we need to extract the topics of interest to older adults. We thus focus on tweets that are relevant to the daily lives of older adults, and model common interests for groups of older adults. In this section, we first describe the process to derive representative keywords from tweets to grasp the characteristics of the interest topics for each category. Next, we discuss a semi-supervised topic model that guides the topic generation process. Finally, we introduce a method for discovering common interest topics from older adults' tweets by using the representative keywords and the semi-supervised topic modeling technique.

Extraction of Representative Keywords. We extract the characteristics of each interest topic according to the features such as the frequency of terms and the structure of the representative words.

Tweet hashtags are used to classify tweets, and they allow users to easily follow the topic they are interested in. Therefore, in this research, we use tweets with hashtags to find content related to different interests. We use TF-IDF values to find the keywords representing each interest. For example, the word "Hanshin" frequently appears in tweets with the hashtag "baseball". Therefore, the word "Hanshin" helps to explain the interests related to "baseball." Characteristics of each category can be represented by the keywords that appear most frequently in the category. We calculate the TF-IDF values as follows:

$$tf(t, i) = \frac{n_{t,i}}{\sum_k n_{k,i}} \tag{1}$$

$$idf(t) = log \frac{|Interest|}{|\{interest : i \in interest\}|} \tag{2}$$

$n_{t,i}$ represents the number of appearances of term t in interest $interest_i$, $\Sigma_k n_{k,i}$ represents the sum of the number of appearances of all words in interest $interest_i$, $|Interest|$ represents the total number of interests, and $|\{interest: t \in interest\}|$ the number of interests that include word t. The higher the TF-IDF value of the term, the more important it is to explain the corresponding category.

Semi-supervised Topic Model. We next describe the semi-supervised topic modeling [10] technique that we use to guide the modeling process and extract the topics of interest to older adults. LDA is a probabilistic model that is often used in the natural language processing field, and is based on the idea that words appear in documents based on latent topics and not independently. Words that belong to the same topic may appear more frequently in the same document. LDA thus generates topic-wise appearance probability of words and document-wise probability of topics based on a given number of topics. As LDA is an unsupervised generative model and we cannot intervene the topic generation process, it may sometimes generate topics that could not be perceived as meaningful nor interesting by humans. Figure 4 shows the process of generating the relationship between topics and words.

```
For each document d in document set D:
  For each word w in document d:
    For each topic z:
      P(w|z) = count(across all document w that belongs to top-
               ic z
      P(z|d) = count(in d all words that belongs to topic z)
      tokes in z = total count of words in z
      P(z|w,d) = P(w|z)*P(z|d) / tokens in z
```

Fig. 4. Generating topic-word relationships.

In this study, we focus on topics related to the interests of older adults, and use a semi-supervised topic model [10] that guides the generation process of the relationship between topics and words. We establish the seed topics and the seed words belonging to the topics in advance. When generating the probability $P(w|z)$ of occurrence of word w in topic z, a higher weight W_s is given to the seed word to guide the process so that it likely appears in the seed topic. The frequency of the occurrence of the seed word of the seed topic is as follows:

$$P(w|z) = W_s * P(w|z) \tag{3}$$

The probability of assigning word w to topic z for document d is as follows:

$$P(z|w, d) = \frac{W_s * P(w|z) * P(z|d)}{total\,count\,of\,words\,in\,z} \propto P(w|z) \tag{4}$$

When deciding the topic of the seed word, the probability of it belonging to the seed topic is more likely the largest.

Extracting Topics by Also Combining Text and Image Information. Again, the representative words in each category are based on TF-IDF values, and we use them as seed words to guide the generation of topics. We also combine the information extracted from images and text. We then use a semi-supervised topic model to extract interest topics more accurately from the SNS posts by older adults.

3.4 Recommendation of Tourist Spots

Based on the topic distribution of common interest topics of older adults, the topic distribution of relevant tourist spots *doc_topic_distribution* is predicted based on the appearance frequency of each word in the documents about tourist spots. The more interest topics contained in tourist-spot documents, the higher the degree of interest by older adults. This is reflected in the higher recommendation score that is calculated based on the following equation:

$$Score = Max[doc\_topic\_distribution] \tag{5}$$

4 Experiments

In order to find topics of interest to older adults, we collected 4,770 Twitter accounts of Japanese older adults by using a crowdsourcing service called Lancers [11], and collected tweets using the Twitter API. We ignored users with 100 or less tweets, and from the remaining 4,156 accounts, we treat all tweets from a user as a document and apply the topic model.

4.1 Comprehensiveness of Generated Interest Topics

In order to validate the effectiveness of the proposed method, we first evaluate the proposed method by conducting a survey with 30 older adults about their recent and general interests. The survey questions asked them to choose multiple topics of their interest. Table 1 shows the ranking of interest topics based on the survey.

The ranking result also shows that the proposed method generated more interest topics in the ranking list than the straightforward method based on conventional LDA. We can calculate the recall of the proposed method and the conventional LDA-based method by manually finding the interest topics that matches the algorithmically generated topics:

$$recall = \frac{|\{generated\,topics\} \cap \{interest\,topics\}|}{|\{interested\,topics\}|} \tag{6}$$

Based on the results in Table 1, the recall of the proposed method is 0.71, and the recall of the conventional LDA-based method is 0.38. The conventional LDA-based method also generated many topics that are irrelevant to the interest topics. The proposed method is superior in generating topics that are related to the interests of the older adults.

4.2 Identifying the Characteristics of Tourist Spots

Capturing the characteristics of tourist spots is critical for recommending tourist spots successfully. This section discusses if the proposed method can capture the characteristics of tourist spots effectively based on the score and topics they generate for well-known tourist spots.

We first collect hash-tagged tweets and Wikipedia pages about well-known tourist spots in Japan, including "Tokyo Dome," "Mt. Takao," "Tsukiji Market," and "Meiji Jingu." We then treat them as documents about each tourist spot, and calculate the scores according to the method described in Sect. 3.4. Tables 2 and 3 show the results

Table 1. Ranking of topics that are of interest to the 30 older adults.

Interest topics	Proposed method	LDA
Sports (73.3%)	○	○
Watching movies (60.0%)	○	○
Reading (46.7%)	○	
Music/Musical Instruments (43.3%)	○	○
Travel (40.0%)	○	○
Walk (40.0%)	○	○
Pets (20.0%)	○	
Muscle training (20.0%)	○	○
Do-it-yourself (20.0%)		
Food Gourmet (16.7%)	○	○
Hot Spring Tour (16.7%)		
Fishing (13.3%)		
Horticulture (13.3%)	○	
Manga/Anime/Game (6.7%)	○	
Café (6.7%)		
Visiting shrines (3.3%)	○	
Idol (3.3%)	○	
Painting (3.3%)	○	
Haiku (3.3%)	○	○
Swimming (3.3%)		
Camp (3.3%)		

Table 2. Score and topics generated by the proposed method.

Tourist spots	Score	Topics
Tokyo Dome	0.81	# 45: Players, Matches, News, Winners, Cheers, Teams, Hiroshima, Directors, Expectations, Carp
Mt. Takao	0.46	# 40: Photos, Bicycles, Trains, Places, Trucks, Motorcycle, Bus, Airplane, Shooting, Mt. Fuji
Tsukiji Market	0.24	# 38: Donburi, Cup, Bottle, Chair, Meal, Lunch, Sales, Ramen, Spoon, Waiting
Meiji Jingu	0.22	# 1: Place, Shrine, Japan, Fortune-telling, History, Onmyoji, teacher, worship, era, Showa

Table 3. Score and topics generated by the conventional LDA-based method

Tourist spots	Score	Topics
Tokyo Dome	0.59	# 17: News, Sankei News, Players, Mainichi Shimbun, Hiroshima, Curve, Giant, Director, Match, Jiji
Mt. Takao	0.24	# 40: Train, Tokyo, Shooting, Mt. Fuji, Park, Home, Cultivation, Photograph, Camera, Nearby
Tsukiji Market	0.18	# 38: Representative, Opposition, Komei Party, Osaka, Restoration, Candidate, Communist Party, Citizen, Question, Participation
Meiji Jingu	0.23	# 20: End, Tokyo, Walking, Support,, Hour-minute, use, app, schedule, long time no see, Yokohama

for the proposed method and the conventional LDA-based method. The results suggest that the conventional LDA-based method can be problematic in that their topics are mixed and include apparently unrelated ones. Comparing the both results, one can argue that combining image and text information and guiding the topic modeling process can help extract better interest topics, thereby capturing the characteristics of different tourist spots effectively and enabling successful recommendations of tourist spots.

5 Conclusion and Future Work

Acknowledging the importance of the problems related to the loneliness and isolation of older adults, we have proposed a method to model the interests of older adults and recommend them relevant tourist spots effectively. We hope this method could play some role remedying the problems. The proposed method exploits the SNS contents posted by older adults and combine text and image information to augment the information for modeling. Furthermore, we proposed to use a semi-supervised topic model so that relevant topics can be generated. Compared to the conventional LDA-based method, the proposed method can predict interest topics and characteristics of tourist spots more effectively.

Our future work includes improvement of the accuracy of prediction and recommendation by extracting more information from images. Developing real-time mechanisms will enable just-in-time recommendation of tourist spots based on a large-scale data set.

Acknowledgements. This work was supported by JSPS KAKENHI Grant Number JP17KT0154.

References

1. Ito, H., Goto, H., Yamamura, S.: Association between Loneliness and Living Activities among Living-Alone Senior – A Case Study of Mutsumidai Apartment Complex in Nerima Ward, Tokyo. J. City Plann. Inst. Japan **54**(3), 1200–1207 (2019). (in Japanese)

2. Takahashi, T., Nonaka, K., Matsunaga, H., Hasebe, M., Murayama, H. Koike, T., Murayama, Y., Kobayashi, E., Fujiwara, Y.: Factors relating to social isolation in Urban Japanese Older People: A 2-year prospective cohort study. Archives Gerontol. Geriatrics **86**, 103936 (2020)

3. Utilization ratio of SNS by Older Adults. https://www.seniorlife-soken.com/archives/15336. Accessed 02 Oct 2021. (in Japanese)

4. Kume, Y., Uchiya, T., Takumi, I.: Recommendation of Twitter Account based on Category of Interest. In: Proceedings of the Multimedia, Distributed, Cooperative, and Mobile Symposium (DICOMO 2014), 9–11 July, 2014. pp. 1594–1598. (in Japanese)

5. Hatena Developer Center. https://developer.hatena.ne.jp/ja/documents/keyword. Accessed 14 Feb 2021

6. Wang, Y., Maeda, R.: A proposal of a method for estimating users' preferences using topic models based on Twitter-relevant information. In: Proceedings of the Form on Data Engineering and Information Management (DEIM2019), 4–6 March, 2019. C8–03. (in Japanese)

7. Han, J., Lee, H.: Characterizing the interests of social media users: refinement of a topic model for incorporating heterogeneous media. Inf. Sci. **358–359**, 112–128 (2016)

8. Oohigashi, Y., Ariyama, S., Nobuhara, H.: An extension of location information in social service based on relation between images and text and their application to coverage improvement of recommendation. IPSJ J. **58**(12), 2006–2014 (2017). (in Japanese)

9. Redmon, J., Farhadi, A.: YOLOv3: an incremental improvement. arXiv:1804.02767 [cs.CV] (2018)

10. GuidedLDA: A Python Package using Semi-Supervised Topic Modelling by Incorporating Lexical Priors. https://hasgeek.com/fifthelephant/2019/proposals/guidedlda-a-python-package-using-semi-supervised-t 9UU8Y8SKp7qk4bH8wHLQXV. Accessed 8 Feb 2021

11. Lancers. https://www.lancers.jp/. Accessed 14 Feb 2020

Research on the Inheritance and Innovation Path of Intangible Cultural Heritage from the Perspective of Consumer Sociology
— Take Changsha Kiln as an Example

Wen Lu[✉] and Yulu Ouyang

School of Fine Arts, Hunan Normal University, 36, Lushan Street, Changsha 410081, Hunan, People's Republic of China
luwen@hunnu.edu.Cn

Abstract. Based on the theory of consumption sociology, this article analyzes from the perspective of public taste and cultural consumption, the symbolization, product design process, and life cycle of ceramic crafts, and then inherently discusses the ceramic artistic inheritance and innovation of Changsha kiln under the market economy. The author believes that the life cycle of a product is closely related to consumers' hobbies, temperaments, and values, which are affected by their own economic conditions and social status. Therefore, studying consumers' lifestyles is of great value to product innovation and marketing.

Keywords: Cultural consumption · Public taste · Commodity symbolization · Product design · Changsha kiln

1 Introduction

The ceramic skills of Changsha kiln originated from the Tang Dynasty, inherited and innovated continuously in the long history of dynasties [1]. In 2009, it was included in the Hunan Provincial Intangible Cultural Heritage List; in 2011, it became one of the third batch of national intangible cultural heritage protection projects [2]. Before the reform and opening up, the production of Tongguan ceramics was mainly to meet the needs of daily life, and most of its product forms were cylinders, altars, and bowls. In the 1990s, Changsha kilns began to revive. Art creation workshops and pottery studios were established one after another, and artistic aesthetics became the dominant element of creative consideration. Tongguan's life ceramics not only have practical significance but also reflect cultural differences and the aesthetic concepts of potters.

In the past two years, the author has conducted field surveys in the Tongguan area of Changsha to study the inheritance, protection and development of Tongguan ceramic skills. This article starts from the perspective of public taste and cultural consumption and discusses how traditional ceramic craftsmanship can be inherited and developed under the market economy.

P.-L. P. Rau (Ed.): HCII 2021, LNCS 12773, pp. 56–67, 2021.
https://doi.org/10.1007/978-3-030-77080-8_6

2   Public Taste and Consumption Habits

The term "public taste" is inspired by the French sociologist Bourdieu's research in Distinction. This book believes that society divides different classes through cultural tastes [3]. Different classes use different lifestyles and symbols to show the field of mutual competition. In the field of production and taste, different classes use different strategies to maintain or change their positions and constantly compete. Behind this reflects the divisions and differences in society, and the various cultural consumption represented by cultural tastes reproduce these divisions and differences.

In fact, all consumption habits that imply public tastes are the internal manifestations of an individual's social status, class/class conditions, and life experience. Consumption habits reflect the individual's life trajectory and life experience. And the individual's life experience is always closely connected with the natural environment, social history, cultural psychology and so on. Even though there are great differences in individual consumption habits, by analyzing the consumption habits of different individuals, we can still see the traces of social structural factors. Therefore, consumption habits are the result of personal socialization. Once it is formed, they will become a relatively independent factor influencing consumption choices. As Simmel's study of fashion psychology argues, the top of the public's taste is what the fashion hero represents, and between the social instinct and the individual impulse, the fashion hero represents a true primitive balance of power [4]. It can be seen that the formation of public taste is actually a product of socialization.

The public taste of culture is an example of this. As Bourdieu mentioned: "Cultural needs are the product of education" [3]. Public taste is firstly related to the education level of social groups. Only when people have a certain level of education can they be able to appreciate and consume artworks. And this process of socialization is obviously closely related to the family's social origin. These needs and abilities are, to some extent, a reflection of the social hierarchy and class.

3   Cultural Consciousness and Material Cultural Consumption

Groups of different levels or strata in society will always have a continuous historical process of understanding their own culture. Their cognition of the culture largely affects their position in society. Just as Fei Xiaotong said, with "cultural awareness", people in it can autonomously master the ability of cultural transformation and have the position of independent choice of culture in adapting to the new environment and the new era [5]. However, it is also an arduous and long process from self-awareness, self-reflection to self-innovation, and the identity of the subject also changes in this process, from single to complex gradually. Take Changsha Kiln as an example. After the establishment of Tongguan Ceramics Company, it was divided into eight factory areas, each of which factory area had different products during production. The first and second factories mostly produce export stoneware and some tea utensils, the third and fourth factories produce daily water tanks, the sixth factory mainly produces glazed tiles, and the eighth factory mainly produces fine art ceramics. The ceramic workers are assigned to different positions to make ceramics, and they are paid according to their work, and the subject identity is single. In the 1990s, when the ceramic company ceased production and closed down, workers were forced to be laid off. Some began to make individual creations. They made antique pottery by themselves and took orders for sales. However, they were suppressed by the local government and faced the danger of fines and penalties. Under this pressure, the potters can only find another way out and rely on their own skills to support their families. Most of them go to Guangzhou, Shenzhen and other places to work in foreign ceramic companies and become migrant workers. At this stage, the ceramic workers understand the consumer needs of foreign markets and integrate foreign cultures into their original ceramic skills, gradually forming a new creative style, and also mastering some new production techniques. At the beginning of the 21st century, the cultural heritage phenomenon spread globally. Since the Chinese government became a signatory of the Convention for the Protection of the World Cultural and Natural Heritage, it has begun to do a lot of work for the protection of the folk cultural and artistic heritage of various ethnic groups in China. At the same time, the cultural administration department has shifted from focusing only on the culture of the upper echelon of the elite to focusing on the culture of the grassroots. The copper officials set up a leading group for the protective development of "intangible heritage" ceramic firing techniques, and set up special funds to support the development of the local ceramic industry. In this situation, ceramic craftsmen scattered around returned to their hometown, trying to rebuild and revive the local ceramic industry [6]. These potters have been transformed from migrant workers to craftsmen, "shou" artists who guard and pass on their skills. Their identities are complex. They have started corresponding action strategies in different contexts of policy, art and market.

In the context of cultural revival and rural revitalization, winning the title of non-genetic inheritor with a unique ceramic technique symbol is not only a personal honor but also provides an effective strategy for the development of local economy and culture. For example, when the non-genetic inheritor Liu Zhiguang made antique pottery, he used the theme of "Maritime Silk Road" and wrote poems on replica pottery, giving the poems unique interpretations, in line with the current "Belt and Road" development mainstream, and realized ceramics. The symbolic function makes consumption gain cultural meaning [7]. (As shown in Fig. 1)

The consumption of ceramics belongs to the category of material consumption culture. In the process of consumption, it is the symbolic attribute of consumption crafts. The ceramics produced by Changsha kilns not only exist as physical entities but also become part of the ceramic culture world as a symbol carrier governed by artistic creation rules and expressing meaning. This kind of material and culture are intertwined, which means that Changsha kilns are not only produced according to the laws of nature but also shaped according to the logic of culture. In the 1970s and 1980s, the production goal of Tongguan Ceramic Factory was to meet the needs of daily life. With the changes and development of daily life, cultural pursuits and artistic aesthetics have become increasingly diversified and refined, and the symbolization process of ceramics has become increasingly rich and complex. In the context of art, Tongguan Ceramics has gradually departed from the horizon of life. Craftsmen construct their own identities through art exhibitions, art auctions and other activities, interact closely with artists and investors, and express their artistic language through communication. For example, the potter

Fig. 1. Imitation Changsha Kiln Holding Pot made by Liu Zhiguang

Zeng Deguo used the traditional green glaze of Tongguan and the symbolic characteristics of the classic pot-holding device to integrate the Anhua moraine rock when making teapots, so that it satisfies functions and has aesthetic attributes. (As shown in Fig. 2) Liu Zhaoming, a ceramic artist sent by the academy, applied the principles of modern design into ceramic design, conveying strong artistic symbols in the visual, and exploring the "beauty of kiln change, texture and incompleteness" in ceramics. (As shown in Fig. 3) The potter Zhou Shihong integrated the symbolic features of Changsha kilns into antique and innovation. In the field of antiques, celadon and other symbol elements were used to completely replicate; and in the field of innovation, Mr. Zhou used the Chinese zodiac paper-cut patterns and Changsha symbolic elements such as grass, wood, fish and insects on the kiln pottery bowls of modern pottery works, using hand-kneaded and plated clay sticks to achieve natural effects. (As shown in Fig. 4)This artistic language is not only the materialized expression of ceramics but also the creator's mood and emotional communication across time and space. (As shown in Fig. 5). Through these works, we can see how creators poured their thoughts, intentions, and hopes into materials. Ceramics are integrated into the realm of meaning, become signs and symbols of meaning, and become carriers and tools of culture in the form of symbols. They are also a record of the inheritance and innovation process of intangible cultural heritage.

In the current market context, the development of the market economy has transformed Tongguan ceramics from the category of folk culture into important cultural resources. When the ceramic firing techniques of Changsha kilns became intangible cultural heritage, the cultural value of ceramics has greatly increased. Craftsmen from market awareness through the new market consumption chain and product production,

Fig. 2. Moraine rock tea set made by Zeng Deguo

Fig. 3. "New" Changsha Kiln works made by Liu Zhaoming

thereby constructing the identity of the production subject. In the local area, some artisans combine cultural and creative design with intangible heritage protection to form exclusive brands and have close contact with the market and consumers to realize the conversion of commodity symbols.

The potter Peng Wangqiu runs the "Fuxing Kiln" brand, whose name is derived from the one-stop kiln of Tongguan in the Qing Dynasty. On the display racks in the store, we witnessed Mr. Peng's design: From a big tank that satisfies daily life to a "cylinder" cup for tea art culture, the craftsman has not only changed the size of ceramics in material form during the process of innovation., but also has given unique connotation of Huxiang tea set in the cultural world. (As shown in Fig. 6) "Huxiang is the land of barbarians,

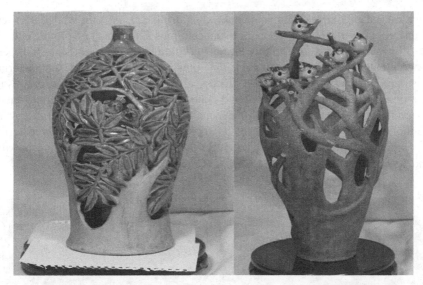

Fig. 4. Modern pottery work "Noisy Branch" made by Zhou Shihong

Fig. 5. Ceramic craftsmen in the creation

especially Meishan Chi. It has become a popular language in Hunan. Since the rise of Changsha kilns in the Tang Dynasty, the creative ideas of sticking to the use of tools have continued. For thousands of years, the big cylinder has been once one of the indispensable large items in the homes of the people in the Dongting Lake area. With the improvement of people's living standards, they have talked about their vision. But it is a testimony to the glory of the Changsha kiln. Complex, after making tea, the tea soup is hot and hot. The design idea of stabilizing the wall thickness of the large cylinder vessel is the design idea. The vessel does not lack its strength due to the size, as much as the person does not lose its strength due to age. The characteristics of loyalty, honesty and calmness can reproduce the feelings of Hunan people while drinking". In Peng Wangqiu's view, "the transfer of the meaning of copper official ceramics is to transfer from the cultural world". Tongguan Town is connected to the Shizhu culture of the Tang Dynasty and opens the modern cultural and creative market. The transformation from clay raw materials to consumer handicrafts is not only a process of material transformation but also a process of ceramics entering its social and cultural life track. The commercialization and marketization of ceramics reveal the social relationships behind the crafts, while the symbolization reveals the cultural significance of the crafts. From clay raw materials without cultural life to ceramic handicrafts endowed with cultural life, Changsha kilns are undergoing transformation in the context of modern commodity economy.

Fig. 6. "Cylinder Cup"set from "Fuxing Kiln" tea set products

In Tongguan, Fuyao is also a distinctive tea shop. The founder Wu Qi put forward the brand concept of "Life Aesthetics". Liu Yuedi of the Institute of Philosophy of the Chinese Academy of Social Sciences mentioned in the article that [8] this aesthetic concept was born in accordance with the new historical context [9]. With the upgrading of consumption, nowadays more and more attention have been attached to aesthetics. Aesthetics in life has become a manifestation of economic power, allowing people to obtain the enjoyment and pleasure of "beauty". When Wu Qi made teaware, he fully examined and considered the meaning, goal, value, cognition, concept and habits of consumers, and designed a crack cup with strong artistic sense and beautiful appearance. The Hunan culture is linked, and products such as the teapots of the Xiaoxiang Eight Views series are launched (Fig. 7), giving the "new" Changsha kiln ceramics cultural significance. This makes the ceramics that were used as ordinary daily necessities transformed into handicrafts representing a certain kind of public taste, from the general typology and the objective meaning suitable for common people to the subjective meaning suitable for a certain group of people and special groups. As a result, the consumption of ceramics has shifted to individualization, becoming a real "use for oneself" and "fitting with oneself" handicrafts for consumers, and gradually give the individual emotional and subjective meaning of ceramic handicrafts during the consumption process.

Fig. 7. "Lotus pond moonlight" set from "FuKiln"tea set products

In the fiercely competitive world of pottery and crafts, Tongguan craftsmen have rendered the crafts unique meaning and cultural characteristics, attracting the attention and attention of consumers [10]. Through the participation of consumers, the ceramics that give individual emotional meanings are symbolized, and together with the technical-ization and commercialization of ceramics, they constitute three different but mutually influential aspects of ceramic production. Pottery craftsmanship solves how to convert soil into handicrafts most economically, ceramic sales solves how to use handicrafts as a medium of the transaction to obtain ideal profits, and the symbolization of ceramics solves how to give the symbolic meaning of handicrafts to win consumers Favor.

After these ceramics are produced and put on the market, their physical life will be consumed, damaged, and ultimately out of use. It is generally seen that the social life of ceramic crafts is often shorter than its physical life. This is because ceramic handicrafts are symbols of cultural significance, and with the changes of social environment, their symbolic meaning will continue to change. This involves a product life cycle issue [11]. In the process of product design, traditional crafts are used as symbols of Changsha kilns and are used by creators, which also endow culture with certain energy [12]. These representative local folk crafts can present the local knowledge of Tongguan and increase the cultural confidence of the locals.

4   Product Design, Process and Life Cycle

As a commodity, ceramic handicrafts have a limited consumption cycle, that is, they have a life cycle just like people. Compared with the small-scale pre-industrial society, in modern society, the type of life biographies of commodities is more dependent on social competition and cultural taste [13]. Generally speaking, the product life cycle is divided into the natural life cycle and market life cycle. The former refers to the physical life and use function of the product, and the latter refers to the market acceptance and time limit of the product. The vast majority of products cannot always be accepted by the market. There are two main reasons: First, new products with better performance, cheaper prices or more in line with social trends will continue to appear; second, consumer tastes and requirements will continue to change. The competitiveness of the market and the socio-cultural nature of the period and scope of product acceptance determine that the product's market life cycle is limited [14]. In addition, the factors that affect the product's market life cycle often include changes in economic conditions, changes in lifestyles, and changes in values. In modern market conditions, consumers can determine the fate of products. From the perspective of consumer choice, the market life cycle of a product can be further divided into an objective life cycle and a subjective life cycle. The end of the objective life cycle of life ceramics means that the whole society no longer accepts it, so its market life cycle also ends. Just like at the beginning, due to the limitation of technical level and the tendency of public taste, the first consideration of the products of Tongguan Ceramic Factory is function and durability. The subjective life cycle of craft ceramics shows the process of consumers from choosing to giving up. The end of its subjective life cycle means that consumers are no longer satisfied and unwilling to use or preserve the attitude, and proclaim the end of their life subjectively. This is a process in which consumers gradually lose interest, which is often closely related to the temperament tendency of their public tastes.

Therefore, there are three specific issues to consider in the design and sales of Changsha kiln ceramic crafts: First, the acceptance speed of crafts. That is, how long does it take for Changsha kilns to be accepted by the market? Second, the acceptance period for crafts. In other words, how long will it last for consumers of crafts to go from appreciation, taste to boredom? Third, the acceptance range of crafts. Which consumer groups will be interested in them? Whether in terms of product acceptance speed, product acceptance period or product acceptance range, the subjective life cycle of a product is closely related to consumers' hobbies, temperaments, and values. These subjective factors of consumers are affected by factors such as their economic conditions and social status. As mentioned in the article "Consumer Culture and Fashion Consumption-From Jean Baudrillard's "Consumer Society", fashion consumption is a symbolic system with the nature of "social coding" through which people belong In all social classes, people can use symbols to identify the status of others or themselves in society [15]. Therefore, studying the subjective life cycle of a product is essentially studying consumers and their choices of actions. The cultural consumption that consumers choose is essential to choose a certain lifestyle and style of life.

In the process of product design, the symbolic value of Changsha kiln ceramics still exists in the differences between different ceramic crafts [16]. These differences are reflected in various production styles, reflecting the characteristics of local folk crafts. Such as colorful underglaze, high-temperature copper red, blanking, kneading, printing, applique decoration, scratching decoration, etc., And different styles have their own inheritors. These inheritors inherit cultural knowledge and enrich ceramic connotations through traditional crafts. As the unique characteristic glaze of the craftsman Hu

Fig. 8. "Lotus pot" made by Hu Wuqiang

Wuqiang, chicken blood red has become a symbol of Changsha kiln in ceramic shape; (As shown in Fig. 8) Due to the market demand for large copper pottery ceramics, it is necessary to use the pinching process to shape the utensils. Xie Fuxiang became a master of copper officials with his superb pinching skills; craftsman Yong Jiangang used the method of embossing to make irregular-shaped ceramics. This traditional craft is rare in copper officials. Only craftsmanship can solve the problem. As a result, the utensils

Fig. 9. "Family style" made by Liu Kunting

have more folk characteristics. As the inheritor of the national representative intangible cultural heritage, Liu Kuntin, whose works mostly reflect the culture of childishness and copper official fishing nets, so he chooses to create in the production process to realize the inheritance of the family. (As shown in Fig. 9) These differences constitute the uniqueness and difference of each craftsman's ceramics, which makes it possible to distinguish similar works [17]. Field survey data show that the production volume of handmade pottery is limited, and everyone's style is different, that is, "the work speaks, the market speaks". Therefore, the production of handicrafts is not so much the production of material products as it is the production of styles and symbols [18].

5 Conclusion

Consumers always try to create their own subjective culture and try to express their personality, interests and tastes through various consumption styles. After the public tastes of different classes are formed, their consumption habits reflect the operational logic of cultural consumption.

In the period of "ascetic society", the state suppressed consumption and advocated thrifty pride [19]. The products of ceramic factories were mainly practical and durable, and could be very popular; now in the period of transition to "consumer society", culture and art such as "The swallows under the eaves of Wang dao and Xie an have now flown into the homes of ordinary people". Pottery craftsmen have to face a large and critical-sighted consumer group, whose subjectivity and cultural choices imply public tastes of class attributes. The construction of an ecosystem has become an important path at this time in the attitude and implementation of the inheritance and innovation of Changsha kilns. The construction of the path system for the inheritance and innovative development of Changsha kilns should be based on the excavation and integration of various element resources, and build an innovative development body system with "shou" artists as the core and the joint participation of enterprise, government, academic media. Guided by market demand and cultural needs, supported by creative design and innovative technology, and focused on innovation and expansion of the industrial chain, the ultimate realization of the coordination and unity of economic, social, ecological and cultural benefits. In this ecosystem, by creating local characteristic brands, integrating "intangible cultural heritage" culture into it, telling stories that belong to the Tongguan; integrating consumer elements, activating local industry vitality, and finally realizing cultural significance in consumption; Inheriting ceramic skills and assisting rapid development, craftsmen use traditional crafts to inherit ceramic skills, closely contact consumers, designers, producers and sellers, coordinate and unify, explore the cultural value of "intangible heritage", and realize the foundation of consumer society Cultural self-examination and cultural consciousness condense the "intangible heritage" inheritance of Changsha kiln culture and extend new branches of cultural industry development.

Acknowledgment. This work was supported by the Scientific Research Project of Hunan Provincial Department of Education. (No. 19A309).

References

1. Changsha Kiln Research Group. Changsha Kiln. Beijing: Forbidden City Press (1996)
2. Lingyao, Li: Research on the Tourism Utilization of Changsha Tongguan Kiln Ruins from the Perspective of Living Protection. Xiangtan University, Xiangtan (2019)
3. (French) Pierre Bourdieu. Distinction (2013)
4. (Germany) Simmel, G.: The Sociology of Georg Simmel. The Free Press, Free Press House (1964)
5. Xiaotong, Fei: On Anthropology and Cultural Consciousness. Huaxia Publishing House, Beijing (2004)
6. Chen, P.: Analysis of the ceramic industry business model of cultural and creative industries. Taipei: National Taiwan University, (2006)
7. Haitao, Qi: Analysis of intangible cultural heritage visual elements innovative derivative design under semiotics. Packag. Eng. 41(20), 195–199 (2020)
8. Yuedi, Liu: The aestheticization of daily life and the daily life of aesthetics——on how "life aesthetics" is possible. Philos. Stud. 01, 107–111 (2005)
9. Yuedi, Liu: The rise of "Life Aesthetics" and the dusk of Kant's aesthetics. Literary Controversy 05, 12–20 (2010)
10. Crafting values: economies, ethics and aesthetics of artistic valuation. J. Cult. Econ. 13(6), 663–671 (2020)
11. Li, W.: Research on Innovative Design of Ceramic Products for Daily Use (2019)
12. Duan, J.: Daily ceramic design based on the concept of cultural creation. Front. Art Res. 1(2) (2019)
13. Appadurai, A., ed.: The Social Life of Things, Cambridge University Press (1986)
14. Ning, W.: Consumer Sociology (Second Edition). China Publishing House, Beijing (2004)
15. Linlin, Li: Consumer culture and fashion consumption——from jean Baudrillard's "consumer society", Decoration 09, 16–17 (2008)
16. Li, W-T., Ho, M-C., Yang, C.: A design thinking-based study of the prospect of the sustainable development of traditional handicrafts. Sustainability, 11(18) (2019)
17. Tao, T., Zou, X.: Differentiation Research of Daily Ceramic Products (2018)
18. Peng, F.: The transmission of symbolic meaning in modern ceramic product design. In: Proceedings of the 2017 2nd International Conference on Financial Innovation and Economic Development (ICFIED 2017), (2017)
19. Ning, W.: From an Ascetic Society to a Consumer Society: China's Urban Consumption System, Labor Incentives and the Transformation of Main Structure. Social Sciences Archives Press, Beijing (2009)

Behavioral Mapping: A Patch of the User Research Method in the Cruise Tourists Preference Research

Jiangyan Lu[1]($\boxtimes$), Xiaolei Guo[2], Lu Ding[2], Zhenyu (Cheryl) Qian[3],
and Yingjie (Victor) Chen[2]

[1] School of Art and Design, Wuhan University of Technology, Wuhan, Hubei, China
lujiangyan@whut.edu.cn
[2] Department of Computer Graphics Technology, Purdue University, West Lafayette, IN, USA
{guo579,ding241,victorchen}@purdue.edu
[3] Department of Art and Design, Purdue University, West Lafayette, IN, USA
qianz@purdue.edu

Abstract. In recent years, cruise travel has become an important branch of tourism. However, the existing cruise function system is designed based on the Western lifestyle and is not entirely suitable for Chinese tourists. Therefore, it is necessary to carry out user research on the usage preferences of Chinese tourists. Considering the specicularity and complexity of cruise space, this study adopts a novel user research method: Behavior Mapping (BM) method, which can be used to understand the unique behavior path of Chinese tourists and analyze their demands. In addition, the BM method was also used combined with the observation method and in-depth interview method in this study. The results show that BM is a good supplement to other research methods and can make the research results more comprehensive and accurate.

Keywords: User study method · Behavioral mapping · User experience design · Cruise ship deign

1 Introduction

The relationship between space design and human behavior has been an evident and critical issue in design proceeding (Zifferblatt 1972). Thus, an exploration of the routing, habits, and human perception corresponding with the physical environment is apt to gain attention in research. Especially for spaces with comprehensive usages, there is a stronger requisition to understanding user behavior and perceptions, which have not been filled by existing user study methods.

Behavioral Mapping (BM) is a systematic way to observe behavior without attracting the attention of the target object (Hanington and Martin 2012). Compared with traditional observation methods, BM's unique feature presents geographical occupation and space usability, which makes the environmental characteristics corresponding to observable behaviors. BM method based on the behavior setting concept (Cosco et al. 2010), which

P.-L. P. Rau (Ed.): HCII 2021, LNCS 12773, pp. 68–79, 2021.
https://doi.org/10.1007/978-3-030-77080-8_7

is understood as a natural ecological setting that actions occur in a certain pattern and repeat at a stable frequency, which is related to the environment in which they appear (Scott 2005).

BM method has been adopted in various research domains, such as environmental science, landscape planning, and architecture design. Bayramzadeh et al. (2018) has used BM to explore the influence of operating room layout on a group of nurses' working patterns and behavior routes. In the context of generating intervention measures or improvement plans for the space environment, Ngesan and Zubir (2015) planned urban park space by exploring and understanding user activity patterns with behavioral maps. Medical rehabilitation space was improved by studying the patient behaviors (Bernhardt and Denehy 2015). Railway station facilities were redesigned by understanding passengers Behavior (Zubair et al. 2019).

Every coin has two sides. The BM method collects and integrates the data subject ID, positions, and routes with timestamps for design analysis. However, it has shortcomings in obtaining the attitudes, viewpoints of the users, and meaning of their behavior (Sommer and Sommer 2002). There also exist arguments about the reliability of data collected by Behavioral Map and the accuracy of data interpretation. To amend these shortcomings, the Behavioral Mapping method usually needs to be used with other user research methods to gain more comprehensive and in-depth insights into user behavior and improve the accuracy of research results. The BM method has not formed a systematic combination with other user research methods at its infancy stage. In this study, the BM method is conducted to explore the Behavior of Chinese tourists in a cruise space. The authors also explain the detailed process of using BM in user studies and how it complements the other two user research methods to provide insights for researchers.

2 Behavioral Mapping Method

Currently, there are two BM methods used to explore the relationship between people and the environment: person-centered map and place-centered map, which was proposed by Sommer in 2002. The place-centered map focuses on how people use a space (Zadeh et al. 2014) and how the space and facilities affect people's behavior (Choi and Bosch 2013; Healy et al. 2015). To be specific, the place-centered map presents a graphical representation of the space and divides the space into different areas. As a static observation method, place-centered maps are able to record the behavior and location of people in each area. Therefore, the place-centered map is more suitable for the study of the use of specific physical spaces.

In contrast, the person-centered map has a dynamic observation scenario, which able to record behavior, location, and time by tracing the target user. Typically, a person-centered map contains multiple locations, and the activity route of the target user needs to be presented in the form of passenger mapping (Sommer and Sommer 2002). Besides, a more sophisticated map can also reveal how often spaces are used, when used, what types of people, and how long they stay at each location. Therefore, the person-centered map is suitable for studying the behavior patterns of individuals or groups, exploring the social activities of target groups, spatial adjacency, and layout on a larger spatial scale, which are not available in traditional observation methods.

For example, a Behavioral map has used to explore patients' behavior in the medical environment and gain in-depth insights into factors affecting patients' daily activities and social interactions, which has been proven to be reliable and effective (Berney et al. 2015; Hokstad et al. 2015; Lay et al. 2015). The advantage of person-centered map is that the patient's daily activity level, reallocation, and social interaction can be recorded every once in a while (Kramer et al. 2013). The recorded data can be used to explore the impact of spatial planning and environmental design on patient behavior and then to evaluate the architectural and environmental design (Joseph and Hamilton, 2008)

3 Study Design

In this study, we adopted the BM method to collect data from several cruise trips, analyzed the activity routes, spatial usage preferences, and needs of Chinese tourists. In addition, two traditional user research methods are employed: participant observation method and in-depth interview method.

3.1 Study Procedure

We focus on conducting the BM method in the user study and understanding how it complements the other two user research methods. There are three phases in the implementation of the BM methods:

Map Generation Phase: researchers got familiar with the floor plans of the Costa Venezia cruise ship, sketch out the maps, and labels the characteristics of the public spaces in a cruise ship. At this stage, researchers created a set of maps for conducting the study

BM Execution Phase: after the subject recruitment, the maps were distributed to participants to record their activity path on the map every day. Besides, participants were asked to rate their satisfaction increasingly, on a scale of 0 to 5, according to the space experience.

Data Analysis Phase: the maps of each participant was collected and re-examined to ensure data quality. Then researchers used computer software to redraw the routes of all participants based on the original plans. The map data of several consecutive days were overlayed to analyze and identify the participants' use frequency of each public space and their daily patterns in the cruise ship.

3.2 Study Setting

Because of the characteristics of cruise travel, participants were recruited in groups. The researchers surveyed seven groups of participants, and most of them were family groups. The tests lasted five days, participants were given a map per day when they got on the cruise ship, and the maps were taken back when they got off the cruise ship. Besides, during the five days, the researchers observed the flow of passengers in each functional space during a specific period of time. According to the result from the observational method, researchers drew a line chart of the passenger flow in each of the six functional spaces at different periods of times of each day, as well as the total passenger flow in each observed space over five days. On the last day of the test, the researchers conducted in-depth interviews with the participants after they handed in their maps. The interview questions included understanding the daily activities and routes, used to verify the accuracy of the results from BM method. And questions like whether the behavior map is intuitive and easy to use, were used to evaluate the usability of the BM method. The public spaces have investigated on the cruise include:

The Dining Space. Chinese restaurant, Western restaurant, other exceptional restaurants, cafeteria, coffee shops, stores, bars, etc.

The Entertainment Space. Dance halls, theaters, karaoke, electronic game room, etc.

The Leisure Space. Shopping, SPA, child care centers, lounge, museums, galleries, etc.

The Office Space. Conference center, library, etc.

The Sports Space. Swimming pool, fitness center, multi-functional sports hall, rock climbing, surfing, parent-child activities, etc.

In order to facilitate the follow-up investigation and analysis, the public space of the cruise ship is divided into three main types: catering space, entertainment space, and sports space.

4   Behavioral Mapping

4.1   Map Generation

In order to successfully create a 3-dimensional map for the cruise, a site inventory study stage for information gathering is required from the researchers. In this study, faculty members with a design background were assembled onto the cruise. They have been handed a single sheet of paper with a master plan of each floor of the cruise. Each researcher was in charge of marking up the landmarks of each floor with a pen. By the end of this stage, the author collected all the maps created by the faculties and overlapped the information accumulated from the maps. The duplicated names, similar expressions, missed landmarks identified on each map have been correctly merged and renamed, respectively, utilizing the overlapping. The correct names and locations of those landmarks have been represented on the final map. According to the record, the author also categorized the names of the landmarks by their characteristics of the function.

Once the base map has been drafted for the BM method, researchers took into account the ability of the average tourist to read and understand the map. Taking the Costa Venezia as an example, the three-dimensional layered display form is adopted. Each floor's deck is arranged from bottom to top, and different colors are used to distinguish the bases (Fig. 1). Besides, the cruise ship can be regarded as a complex high-rise building in which the movement of tourists depends on the transport of elevators. The location of elevators is critical to make it easier for visitors to identify the location of themselves. Therefore, the vertical space of the cruise ship is divided by three sets of elevators in BM. This makes the geographical position of each functional space in the cruise more clear and easy to understand. In addition, different types of spaces are represented by different colors. For example, the red color represents the dining space, the yellow color represents shopping space, and the green color represents leisure and entertainment space, etc. It does not only make it easier for visitors to find where they want to go, but also helps researchers to sort out the collected data later.

4.2   BM Execution and Data Collection

In execution phase, all participants were given printed maps and asked to record their route on the cruise. Taking Costa Venezia as an example, participants need to tick the landmarks they have experienced on each floor. Lines will be used to connect each tick with the order of visiting. And the direction of the route will be determined by arrows on the line. After 6 copies have been filled in a week, participants have also been required to rate their satisfaction from low to high, on a scale of 0 to 5, according to the space experience (Fig. 2). Also, the dates of use, and the types of visitors were recorded in each BM.

There is another alternative way for recording, handwriting. Here are three samples from the participants, and their handwritten content was translated as follows:

Participant 1: "Get on board → rescue drill on the fourth floor → go to the cabin through the sixth floor → go out for dinner from the tenth floor → look at the deck from the 11th floor → return to the cabin through the stairway".

Participant 2: "①6/ F inner cabin → stair → 3/ F breakfast → inner cabin

②Inner cabin → 4/F card → Staircase → 11 Downstairs sun deck → 3/F lunch → Inner cabin ④ Inside Cabin → Stairs to 15th floor → 11th floor → 10th floor Dinner → Stairs to 2nd floor → Stairs to 6th floor → 10th floor Watch sunset ⑤ Makeup Ballroom → Art Gallery → Stair back to the cabin"

Participant 3: "Take the stairs from the 6th floor to the 3rd floor for breakfast and the 10th floor for deck play. At 8 o'clock, walk down the stairs to the 6th floor and get ready to start. Get off the boat at 8:50 and go back to the entrance. At 5:00 PM, return to the 6th floor, take a rest and have dinner on the 10th floor.

Details of the route: It was too noisy to go back to the room at 8 pm, so I went to the customer service (because I didn't sleep well for three nights), but failed to solve the problem. I got three pairs of earplugs after two times of contact, so I gave zero points. There was no pick up of the phone many times, which push me to the customer service center. The answer is you go back to the room immediately someone arrived processing; 20 min still did not come, so had to look for customer service again, took three pairs of earplugs (middle landline contact)".

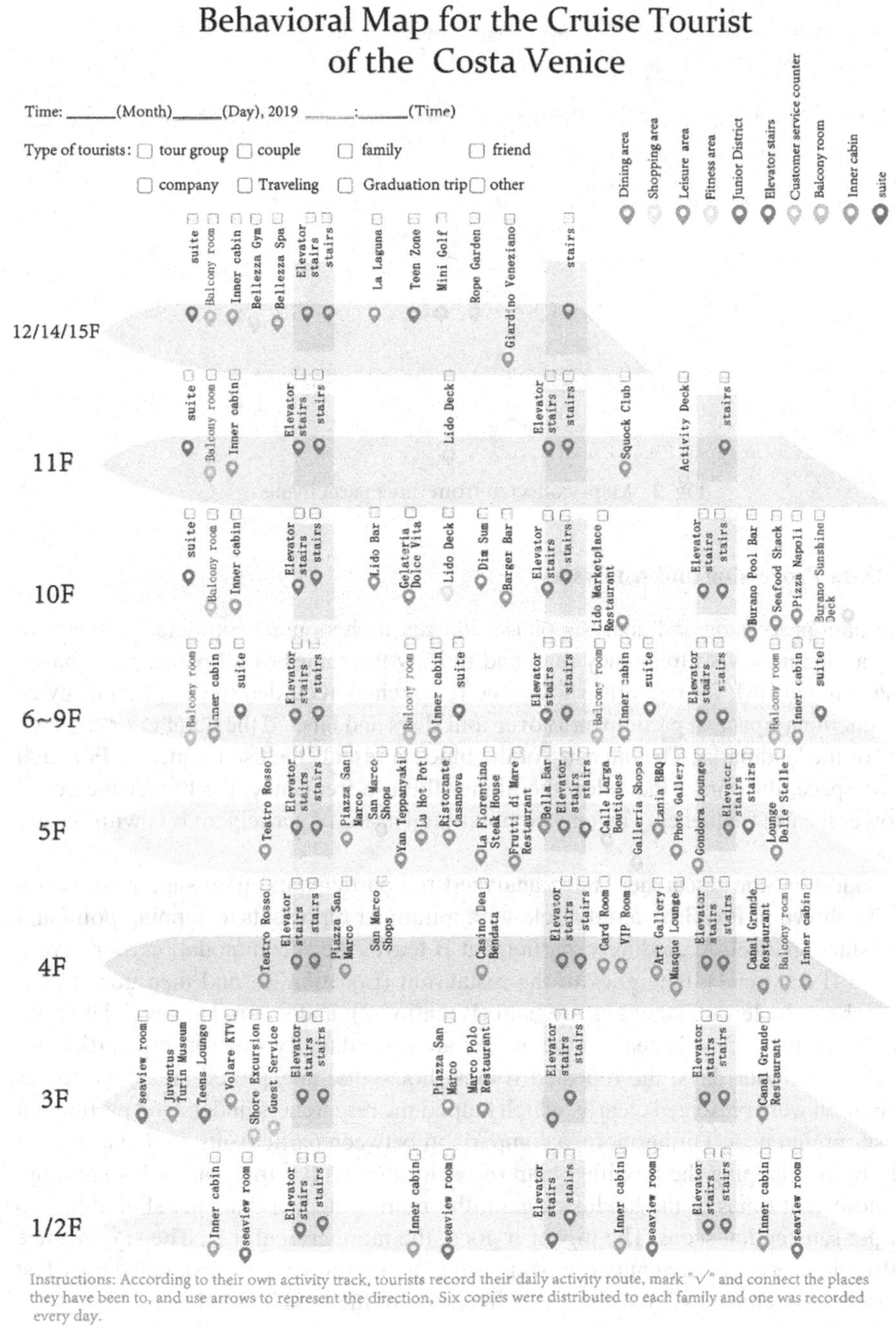

Behavioral Map for the Cruise Tourist of the Costa Venice

Fig. 1. BM for participants (translated from the original version). The deck of each floor is arranged from bottom to top, and different colors are used to distinguish the bases.

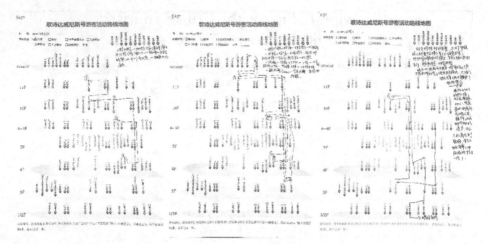

Fig. 2. Maps collected from three participants.

4.3 Data Processing and Analysis

In the data processing and analysis phase, the researchers used computer software to create a landmark visit frequency map and redraw the routes of all participants based on the original BM. As Fig. 3(a) shows, the researchers recorded the use frequency of each functional space of participant B over four days and labeled the number on the blue boxes of the landmarks. The intensity of the blue represents the use frequency. For each type of space, the darker the color, the higher the use frequency, the lighter the color, the lower the use frequency. The unused functional space of participant B is without any marks.

In addition, the researchers have analyzed the participants' path superposition on BM. As shown in Fig. 3(b), a red circle with a number represents a stopping point and its satisfaction level. Accordingly, participant B leaves the accommodation room every morning (Location 1), then goes to the restaurant (Location 3), and then goes to the space where the leisure activities are held (Location 2), and so on. However, he chose to go to the fitness area instead of the leisure area one day. By superposing participant B's routes for four days, the repeated routes, hot spots, unexplored areas, and other information were presented clearly, which helped the researchers understand participant B's user preferences. Furthermore, a comparison between participants could be applied easily by overlapping the modified map of each other with a transparent background. The more visit times of the landmark gain, the more welcome that spot should be. So does the satisfaction score. The higher it goes, the more critical it is. The significance of differences across different participants' satisfaction score will be evaluated by T-test statistic measurement to determine the differences of the satisfaction level.

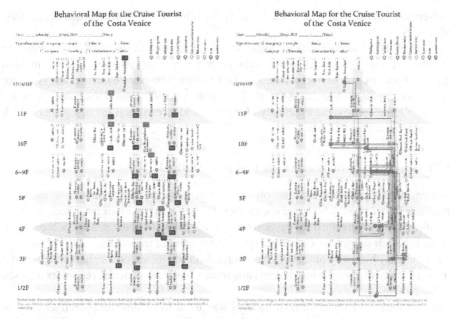

Fig. 3. (a) Left:The spatial usage frequency on the BM. (b) Right: The superposition of routes on the BM.

5 Discussion: Behavioral Mapping as a Patch of the User Research Method

5.1 Limitations of Behavioral Mapping Method

Although the BM method has many advantages, it has shortcomings in obtaining the attitudes and viewpoints of a user, also the meaning of their Behavior (Sommer and Sommer 2002). Besides, most BM is recorded from a third-party perspective, thus trained observers are required. Also, there are more challenges in the study of users' behavior in complex spaces, such as an extended recording period, a large number of recorded objects, and the distribution of recording locations. In such cases, the use of BM method requires researchers to invest a lot of time and energy in recording.

There are also shortcomings in the reliability of data collected by BM and the accuracy of data interpretation. For example, the design of the observation period, time point, and observation time directly affect the quality of the data collected. In terms of data interpretation, the recorded data only describes the user's behavior in a series of time nodes and thus cannot be used to conclude that the behavior is persistent. However, in data interpretation, it is easy to generate results based on this uncertain information, thus affecting researchers' judgment. For the above reasons, the BM method usually needs to be cooperated with other user research methods to gain more comprehensive and in-depth insights into user behavior and improve the accuracy of research results. For example, the accuracy of the collected data can be enhanced by repeated experiments or applied the Interobserver Agreement to investigate the consistency among independent observers around each recording site to verify the reliability of the recorded data (Doris et al. 2009).

However, BM has not formed a systematic combination of methods with other user research methods. The insufficient correlation between research methods may make it difficult for researchers to obtain a data set that comprehensively describes the behavior of the target objects. In addition, since the BM method is mostly used to explore behavior patterns in large spaces, the efficiency of this method is essential for researchers, such as saving time and workforce.

5.2 Triangulation with Other User Research Methods

Observational Method. Scientific observation is a purposeful, planned, systematic, and repeatable method, which can obtain data directly without other intermediate parts. It requires researchers to directly observe the object with their senses according to specific research purposes. Therefore, researchers can get information close to the real status of target objects by observing their natural status. Researchers also often with the help of a variety of modern instruments and means, such as cameras, recorders to assist the observation.

In this study, the observational method was used to obtain the passenger flow of each functional space in a specific period of time. The researchers recorded the original passenger flow data of six functional areas at different periods and used the Excel spreadsheet tool to sort out the data. The researchers drew a line chart of the passenger flow in each of the six functional spaces at different periods of times of each day, as well as the total passenger flow in each pointed area over several consecutive days. Based on this data, the hotspot graph of Chinese tourists' activity preference was formed, which shows the demand degree of Chinese tourists for various functional spaces of cruise ships.

In-Depth Interview. An in-depth interview is a kind of unstructured, direct, and personal method, which is suitable for understanding complex and abstract problems. It is mainly used to acquire a deeper understanding of issues for exploratory research. The in-depth interview can exchange information freely, and observe participants' reactions instantly. However, the number of interviews is usually limited since the in-depth interview method requires a lot of time.

In this research, the purpose of the in-depth interview was to investigate the user's recognition of the usage of the cruise ship's overall functional space, identify specific functional settings of each section, and have a deeper understanding of the preferences of Chinese tourists.

5.3 Triangulation of the Three User Research Methods

The heat map formed based on the observational method results reflects the overall preference of Chinese tourists in the cruise space, which is objective and quantitative. The results of in-depth interviews reflect Chinese tourists' personal experience and feeling, which is subjective and qualitative. These two methods are the most common user research methods. The Behavioral Map method, though not common, has played a crucial role in this study. It reflects a small sample of individual Chinese tourists and is objective and quantitative. As Fig. 4 shows, these three research methods are complementary and mutually verified to have a comprehensive and in-depth understanding of the target group.

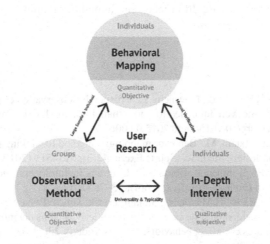

Fig. 4. Trangulation of the three user study methods

In addition, in order to verify the accuracy of the data obtained by the BM method, the researchers asked the participants about the time and place they visited during the in-depth interviews, and compared the data with the maps drawn by the participants. Also, at the end of the interview, participants were asked whether the BM method was intuitive and played the role of navigation, and the result is positive. Therefore, it is believed that the BM method is an excellent supplement to the observation method and in-depth interview method.

6 Conclusion

Behavioral mapping is a new method proposed by the research team based on cruise user research. The researchers first made maps according to the spaces of the target cruise ship and then distributed maps to three different types of tourists each day, and collected all the maps on the last day. Tourists were asked to mark the places they have visited each day on a map, which allows researchers to figure out the daily movements and behavior patterns of each type of tourists. This method requires the cooperation of the tourists, and the final result is a simple and intuitive presentation of the tourist route. According to the survey results obtained by the BM method, combined with the observation method and in-depth interview method, the logical relationship between the different functional spaces of the cruise ship can be further analyzed, so as to guide the design of cruise space that is more aligned with the lifestyle of the Chinese tourists.

In conclusion, user-centered cruise space design is a relatively new field, and appropriate measurement tools are needed. The Behavioral Mapping method used in this study provided a rich source of data to observe Chinese tourists' behaviors. It also shows the advantage of easy operation, friendly use, which offers researchers a powerful tool to save time and energy in user research. However, due to its defects of individuality and one-sidedness, it needs to be used with other research methods. The Behavior Mapping method can be used as a patch for different research methods to make the research results more comprehensive and accurate.

References

Berney, S.C., Rose, J.W., Bernhardt, J., Denehy, L.: Prospective observation of physical activity in critically ill patients who were intubated for more than 48 hours. J. Crit. Care, **30**(4), 658–663 (2015). https://dx.doi.org/10.1016/j.jcrc.2015.03.006

Cosco, N.I., Moore, R.C., Islam, M.: Behavior mapping: a method for linking preschool physical activity and outdoor design. Med. Sci. Sports Exerc. **42**(3), 513–519 (2010)

Choi, Y.S., Bosch, S.J.: Environmental affordances: designing for family presence and involvement in patient care. Health Environ. Res. Des. J. **6**(4), 53–75 (2013)

Milke, Doris L., Beck, Charles H.M., Danes, Stefani, Leask, James: Behavioral mapping of residents' activity in five residential style care centers for elderly persons diagnosed with dementia: small differences in sites can affect behaviors. J. Hous. Elderly **23**(4), 335–367 (2009). https://doi.org/10.1080/02763890903327135

Hanington, B., Martin, B.: Behavioural mapping (Chap. 6). In: Hanington, B., Martin (Eds.), Universal Methods of Design: 100 Ways to Research Complex Problems, Develop Innovative Ideas and Design Effective Solutions, pp. 18–19: Beverley, MA: Rockport (2012)

Healy, S., Manganelli, J., Rosopa, P.J., Brooks, J.O.: An exploration of the nightstand and over-the-bed table in an inpatient rehabilitation hospital. Health Environ. Res. Des. J. **8**(2), 43–55 (2015). https://doi.org/10.1177/1937586714565612

Hokstad, A., et al.: Hospital differences in motor activity early after stroke: a comparison of 11 Norwegian stroke units. J. Stroke Cerebrovasc. Dis. **24**(6), 1333–1340 (2015). https://doi.org/10.1016/j.jstrokecerebrovasdis.2015.02.009

Joseph, A., Hamilton, D.K.: The pebble projects: coordinated evidence-based case studies. Build. Res. Inf. **36**(2), 12–145 (2008)

Kramer, S.F., Cumming, T., Churilov, L., Bernhardt, J.: Measuring activity levels at an acute stroke ward: Comparing observations to a device. Biomed. Res. Int. **2013**, 460–482 (2013)

Lay, S., et al.: Is early rehabilitation a myth? Physical inactivity in the first week after myocardial infarction and stroke. Disabil. Rehabil. 1–7 (2015). https://dx.doi.org/10.3109/09638288.2015.1106598

Ngesan, M.R., Zubir, S.S.: Place identity of nightie urban public park in Shah Alam and Putrajaya. Procedia Soc. Behav. Sci. **170**, 452–462 (2015)

Bayramzadeh, S., Joseph, A., San, D., et al.: The impact of operating room layout on circulating nurse's work patterns and flow disruptions: a behavioral mapping study. HERD: Health Environ. Res. Des. J. **11**(3), 124–138 (2018)

Sommer, R., Sommer, B.: A Practical Guide to Behavioral Research: Tools and Techniques. Oxford Press, New York (2002)

Scott, M.M.: A powerful theory and a para- dox: ecological psychologists after Barker. Environ. Behav. **37**(3), 295–329 (2005)

Zadeh, R., Shepley, M., Williams, G., Chung, S.: The impact of windows and daylight on acute-care nurses' physiological, psychological, and behavioral health. Health Environ. Res. Des. J. **7**(4), 35–61 (2014)

Zubair, A., Barus, L.S., Soemabrata, J.: Passenger behavioral mapping and station facilities design at commuter line train station (case: tangerang station, Indonesia). Int. J. **16**(58), 151–156 (2019)

Zifferblatt, S.: Architecture and human behavior: toward increased understanding of a functional relationship. Educ. Technol. **12**(8), 54–57 (1972). www.jstor.org/stable/44418593. Accessed 21 Apr 2020

A Study of Sightseeing Illustration Map Design

Wai Kit Ng[1(✉)] and Jing Cao[1,2]

[1] Graduate School of Creative Industry Design, National Taiwan University of Arts,
New Taipei City, Taiwan
[2] School of Media and Design, Hangzhou Dianzi University, Hangzhou, China

Abstract. Maps have gained popularity with the advent of the latest technological advances.The Maps have been transformed from the traditional paper maps into digitized ones, which have become aninseparable part of our daily lives. This study focuses on the sightseeing illustration maps. Specifically, the study endeavors to analyze the experiences of the local and the foreign tourists.This exploratory study is based on a comparison of the two groups of visitors and to discuss their experiences. The results of the study would contribute to our understanding of the preferences of the tourists and the impact of the novel digitized sightseeing illustration maps. The results illustrate that the subjects were significantly influenced by the pretty visual design, the educational backgrounds, the age, and the knowledge of the localities which would be helpful in the sightseeing illustration map designing and recommend certain areas for potential research in the future.

Keywords: Cognitive differences · Sightseeing illustration maps · Consumer behavior

1 Introduction

International tourist arrivals grew 5% in 2018 to reach the 1.4 billion mark. This figure was reached two years ahead of UNWTO forecast. At the same time, export earnings generated by tourism have grown to USD 1.7 trillion [17].

Since they are constantly used by tourists as route finders. Similarly, tourist maps can also be under-stood as "icons" of destination space, particularly as they are taken home as souvenirs or mementos [7].

Indeed, many tourist centers are still offering paper tourist maps despite the popularity of mobile maps. It is argued that tourist maps play a significant role in guiding tourists and marking destinations [4]. Different from the space description in traditional maps, modern illustration map s have many breakthroughs in creativity [11]. The purpose of this study is to understand the different versions and understandings of the tourists. We attempted to explore what factors affected tourists' perception of map vision. The framework provides the experience effect of tourists on sightseeing map is an important factor in the development of maps design industries, and can also be used as the standard of design feasibility.

P.-L. P. Rau (Ed.): HCII 2021, LNCS 12773, pp. 80–89, 2021.
https://doi.org/10.1007/978-3-030-77080-8_8

2   Literature Review

2.1   Sightseeing Illustration Map

Tourist maps are essential resources for visitors to an unfamiliar city because they visually highlight landmarks and other points of interest [5].The contents of the city images so far studied, which are refer-able to physical forms, can conveniently be classified into five types of elements: paths, edges, districts, nodes, and landmarks [10]. The results also help to uncover market opportunities as well as to identify local elements and attractions that are most valued by visitors [1]. Illustration Map is used in express concrete and easily recognizable illustrations used on maps displays, To promote the legibility, fun and characteristics of map information transmission [13]. In addition to basic composition elements, colors, images, lines, text, etc. must be used flexibly to highlight themes and strengthen the spatial impression of area [17].

2.2   How Did the Visual Effects Tourists Behavior

Cognition is not abstract information processing implemented by the brain alone, but it draws on the resources of the brain, body, and environment (thus, we say that cognition is embodied and situated). Importantly, the environment also includes other cognitive agents—other humans and cognition makes use of them as well. Thus, cognition is distributed, potentially across many distinct individuals [16]. An enormous amount of quantitative information can be conveyed by graphs; our eye-brain system can summarize vast information quickly and extract salient features, but it is also capable of focusing on detail [8]. Cognitive theorists postulate that, in the mind, knowledge resides within sets of organized and interlinked mental schemata which can be activated by experience [6]. A tourist attraction may contain several popular views and landmarks. The most popular view or landmark is selected to represent the primary visual feature of a tourist attraction in a map [3]. The apparent clarity or "Legibility" of the cityscape.

By this we mean the ease with which its parrs can be recognized and can be organized into a coherent pattern/Just as this printed page, if it is legible, can be visually grasped as a related pattern of recognizable symbols [10]. All visitors interviewed explained their choice of direction in terms of visual elements. It might be an architectural feature, a work of art or some evidence of street entertainment [2]. Visual communication relies both on eyes that function and on a brain that interprets all the sensory information received. An active, curious mind remembers and uses visual messages in thoughtful and innovative ways. Knowing about the world and the images that it conveys will help you analyze pictures [15]. Attitude structure contains three elements: cognition, affect and conation, organized and internal consistency, Especially the centrality of the attitude. Attitudes, like most other consumer behaviors, are learned through learning. Consumers obtain and form attitudes directly or indirectly from their own experience and information from interpersonal and media [11]. Social influences determine much but not all of the behavioral variations in people. Two individuals subject to the same influences are not likely to have identical attitudes, although these attitudes will probably converge at more points than those of two strangers selected at random. Attitudes are really the product of social forces interacting with the individual's unique temperament and abilities [14].

3 Methodology

Sightseeing illustration map is a visual work combining information transmission, marketing communication and artistry, It is widely used in scenic spots. At present, there is a lack of understanding of the actual effect and design method of illustration map. This research investigates the sightseeing illustration map, Industry under the influence of cognitive and behavior. Figure 1 is the structure of this research.

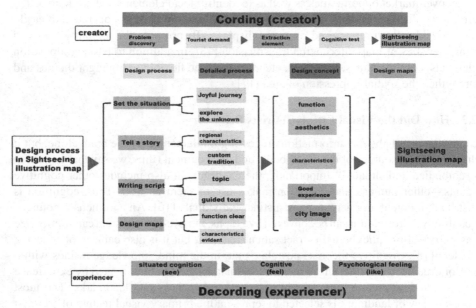

Fig. 1. Framework of the study adopted from [12]

3.1 Research Problem and Hypothesis

To validate the research data, multiple data sources were employed to strengthen the research reliability. These included interviews and questionnaire survey. With regard to the spatial structure, a closer and multi-point perspective was adopted differing from the existing Hong Kong illustration map, which was more in line with the actual sighting of the tourists. The architectural structures and the decorative details were separated from the regular and common tourist illustration maps.

Based on the interview survey the following conclusions could be made:

1) The number of travels affects the perception of the Hong Kong element. Around three trips would be more helpful for recognizing the cultural characteristics than from zero to two times.
2) Regarding the visualization of the map samples by the male and the female participants, the difference was hardly noticeable.
3) All participants expressed a great sense of visual perception and the aspiration to visit in the affirmative.

The following hypotheses could be proposed based on the results:

H1: Significantt differences could be noticed in the reading experiences and attitudes between the local and the foreign tourists.

H2: Visual perception has a significant impact on the visit aspirations.

H3: The number of trips would impact the susceptibility of the element of Hong Kong (Figs. 2 and 3).

3.2 Experimental Samples

Fig. 2. Sample –Tai Hang-Tin Hau   **Fig. 3.** Sample –Lei Yue Mum

As for the selection process of the research object, the products were chosen from a set of maps. Certain basic elements were used during the planning and creation stage, to enhance the guiding functions of the tourist illustration maps. The bountiful landscape reflected the diversity and humanistic approach to life of Hong Kong and enhanced the overall artistic and memorial value.

To understand the attitudes of the tourists towards the new sightseeing illustration maps, the data was assimilated using a questionnaire in October, 2020. In the case study concerning the foreign participants, the users were allowed to use the mobile scan QR code for watching maps on the HTML5 format (one page picture and one page explanation), Besides, poster maps (115.5 cm × 54.5 cm) were also displayed on the walls. The survey included the visitors to the exhibition in the University within New Taipei in Taiwan.

3.3 Questionnaire Design

In case of the local participants, only the HTML5 version maps were available. The selected solutions illustrated different possibilities for digital pictures and posters for the subjective evaluation of the interactivity of the user for quicker comparison. In the questionnaire, the background of the participants was also requested, including age, gender, nationality, number of times travelled in Hong Kong, and the reading media referred. The main questionnaire had a total of eight questions. The tourist participants had to indicate their feelings on a five-point scale, to evaluate the responses of the participants to each item. (see Table 1)

Table 1. Five points scale of participants attributes of the maps.

Descriptions
F1. Are you satisfied with the visual aesthetics of the map?   □1 □2 □3 □4 □5
F2. Do you think the attractions on the map are clearly recognized?   □1 □2 □3 □4 □5
F3. Do you think the road directions clear on this map?   □1 □2 □3 □4 □5
F4. Can you feel the elements of Hong Kong through visual presentation?   □1 □2 □3 □4 □5
F5. Do you think this work is difficult to read?   □1 □2 □3 □4 □5
F6. There be any intention to actually visit after reading the map?   □1 □2 □3 □4 □5
F7. The inconvenience of using electronic map in travel   □A □B
F8.   Is this work a good design?   □1 □2 □3 □4 □5

4   Results and Discussions

This study analyzed participants reading experience towards the maps. Questionnaire passed the cronbach alpha test scores .788. Reliability increases after deletion of questions; the topic design still meets the reliability requirements (Table 2).

Table 2. Correlation coefficients between each experience-factor

Variable	F1	F2	F3	F4	F5	Overall
F1	1	.404**	.503**	.592**	.358 *	.710**
F2	.404**	1	.555	.433**	.280 *	.331*
F3	.503**	.555**	**	.542**	.420**	.497**
F4	.592**	.433**	1	1	.228	.562**
F5	.358*	.280*	.542**	.228	1	.303*
Overall	710 **	.331*	.420** .497**	.562**	.303*	1

*p < .05. **p < .01.

Correlation coefficients were calculated in this study to examine the relationship every pair of items of maps experience. In the subject of the main questions F1 to the F5, visual aesthetics of the map, clearly recognized of the map, road directions clear on this map, elements of Hong Kong through visual presentation and Overall had significant correlated. As can be seen in Table 1, Visit aspiration is significantly correlated with that of elements of Hong Kong through visual presentation (Table 3).

Table 3. Analysis of variance for maps experience by educational background

Variables	Works	F	M	Scheffe's post hoc
F1	Secondary School	10.148**	4.36	1 > 3
	University		4.10	2 > 3
	Graduate		3.21	
F2	Secondary School	4.5*	4.00	
	University		3.95	2 > 3
	Graduate		3.26	
F3	Secondary School	9.464**	4.55	1 > 2
	University		3.55	
	Graduate		3.05	1 > 3
F4	Secondary School	3.512*	4.09	
	University		3.10	
	Graduate		3.00	
F5	Secondary School	4.321*	4.91	1 > 3
	University		4.10	
	Graduate		3.95	

$*p < .05.$ $**p < .01.$

An analysis of one-way ANOVA shows that the educational background was resulting in significant differences with all five items. The differences were between Secondary School group and university group, Secondary School group and graduate group.

Results of Scheffe's post hoc analysis suggested that in three groups scores of University and Graduate educational Background participants were significantly less than that of Secondary School group. This result should be related to the fact that Secondary School group most of them are local residents and over 40 years old. That reasons why they had a higher standard of than other groups.

From this, we can conclude that the evaluation of the map is related to the understanding of Hong Kong.

Table 4. Chi-square tests for inconvenience in using the map by age

Independent Variable			Items				x^2 (df)	
			1	2	3	4		
Age Young		f	9	8	4	3	13.117*	
		%	18.0%	16.0%	8.0%	6.0.%	(6)	
midlife		f	2	4	1	7		
		%	4.0%	8.0%	2.0%	14.0%		
mature adult-hood		f	2	2	0	8		
		%	4.0%	4.0%	0.0%	16.0%		
Total		f	13	14	5	18	50	50
		%	26.0%	28.0%	10.0%	36.0%	100.0%	100.0%

Chi-square test was manipulated in this study to examine the association between three group of age variables dependent variable of maps experiences. Table 4 shows that three sets of Chi-square score were higher than the critical values suggesting that inconvenience in using the map in significant differences in their experience.

Figure 4 shows that as regards the most of the young participants selected item 1 "Easy to go wrong", The second was item 2 "too complicated". However, in the results of the mature adulthood group participants most selected item4 "other". Finally, the midlife group also participants most selected item4 "other", The second was item 2 "too complicated".

The multiple regression model overall with all five items produced F = 10.705, suggesting a statistically significant (p < .001). Taking of five design items of Overall as independent variables, can properly predict the overall evaluation score, five predictors produced R = .741, R2 = .549. As can be seen in Table 5, have significant correlation in the Visual β (t = 4.067, p < .001). visual is most important factor in participants overall evaluation of the map design.

5 Summary

The sample group consisted of the local and the foreign tourists. The groups were constructed in such a manner that valid comparisons could be drawn between the two groups. Based on the statistical analysis of the data, certain important findings have been discussed below.

(1) The local and the foreign tourists: With regard to the hypothesis H1, 'There are significant differences in the reading experiences and attitudes between the local and the foreign tourists', the statistical data indicated no significant differences in the experience of the map readings. However, in case of the Hong Kong element

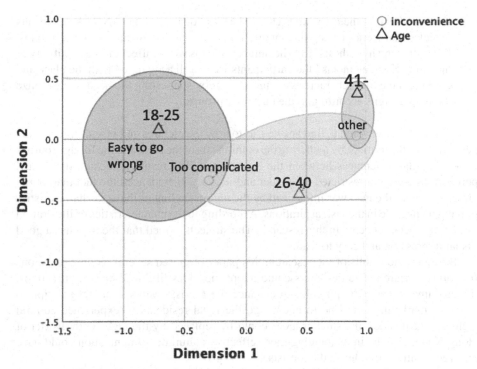

Fig. 4. The relation between inconvenience in using the map and age

Table 5. Multiple regression analyses for five design items to predict preference of the maps

Independent variable	Predictor variable	B	r	β	t
Overall, of the maps	Visual	.485	.710***	.544	4.067***
	Recognition	.049	.331**	.053	.428
	Clarity	.113	.497***	.148	1.051
	HKelement	.117	.562***	.178	1.308
	Difficulty	.017	.303*	.021	.181
	$R = .741$ $R^2 = .549$ $F = 10.705$***				

*$p < 0.05$ **$p < 0.01$ ***$p < 0.001$

perception, certain differences were felt, while the susceptibility of the foreign tourists was lower than that of the local tourists.

(2) The attraction effect: Regarding the hypothesis H2, 'Visual perception has a significant impact on visit aspirations', we could observe that all the participants had positive feelings about the visuals which were quite similar. Through an evaluation to explore how the visuals affected the visit aspirations towards the maps, the outcome illustrated that there was a close relation to the Hong Kong element perception.

(3) Factors affecting the map experience: The Chi-square and the ANOVA test results revealed that there were significant differences in age and educational backgrounds. Regarding the hypothesis H3 'The number of trips would affect the susceptibility of the Hong Kong elements,' the participants were well acquainted with the place and could feel the local characteristics more, thereby suggesting the need to customize the map designs according to the customer groups.

This study endeavored to find answers as to how the pictures and visuals affected the perception of the tourists, by gathering the opinions from the local and the foreign tourists to identify the differences between the two groups. The results indicated that visual perceptions were not restricted by gender and identity. It evidenced the fact that tourist effectiveness of the map was influenced by the understanding of the area characteristics playing a vital role in the visit aspirations. According to the characteristics of illustration map, it has been reflected in this research. The subjects agreed that the map has a good visual impression and easy to read.

Based on the results of the research, this paper proposed several recommendations for future research. The designers could adopt measures like well-known attractions, the city images, and Uniqlo color to enhance the tourist desires. Another solution to this city marketing could be to encourage the local residents to experience cultural sightseeing. Besides, the smaller areas could be replaced by the eighteenth district of Hong Kong. Finally, implementing more effective cultural communication could have ensured positive attitudes in the tourists.

References

1. Ana, R., Filipa, B., Ana, C.: Motivation-based cluster analysis of international tourists visiting a World Heritage City: the case of Porto. Portugal J. Destination Mark. Manage. 49–60 (2018)
2. Cédric, C.: How can the use of a mobile application change the course of a sightseeing tour? A question of pace, gaze and information processing. Tourist Stud. 202020(1), 49–74 (2019)
3. Chao, H.L., Jyun, Y.C., Shun, S.H., Yun, H.C.: Automatic tourist attraction and representative icon determination for tourist map generation. Inf. Vis. 13(1), 18–28 (2014)
4. Chu, Y.Y., Ching, Y.H., Wen, H.C., Yu, S.W., Yung, J.C.: Investigating the choice between a mobile map and a paper tourist map in an urban traveling task. Inf. Soc. Res. 111–150 (2019)
5. Floraine, G., Maneesh, A., Robert, W.S., Mark, P.: Automatic Generation of Tourist Maps , SIGGRAPH (2008)
6. Gregory, C.R.: The cognitive psychology of knowledge: basic research findings and educational implications. Austalian J. Educ. 35(2), 131–153 (1991)
7. Ignacio, F.: Tourist maps as diagrams of destination space. Space and Culture 14(4), 398–414 (2011)
8. Chambers, J.M.: Graphical Methods for Data Analysis. CRC Press, Boca Raton p. 5 (1983)
9. Kristen, A.L.: Emotions emerge from more basic psychological ingredients: a modern psychological constructionist model. Emotion Rev. 5(4), 356–368 (2013)
10. Kevin, L: The Image of The City. The MIT Press, Cambridge (1960)
11. Liao, S.L.: Consumer Behavior: Concepts and application, Fcmc.TW (2007)
12. Lin, R.: Transforming Taiwan aboriginal cultural features into modern product design: a case study of a cross-cultural product design model. Int. J. Des. 1(2), 45–53 (2007)
13. Ng, W.K.: A Study on Design Aesthetics and Creation of Sightseeing Illustration Map, Department of Visual Communication Design Chaoyang University of Technology (2020)

14. Michael, J.B.: Marketing: Critical Perspectives on Business and Management. Routledge, Abingdon on Thames, p. 37 (2001)
15. Paul, M.L.: Visual Communication: Images with Messages, Wadsworth Publishing Company, California (1994)
16. Semin, G.R., Smith, E.R.: Socially situated cognition in perspective. Soc. Cogn. **31**(2), 125–146 (2013)
17. Wang, K.L.: Cultural Landscapes Reappearance of City in Taiwan--The Depicting Process of a Map Called Jiantan Landscape's Illusion, Design Research Annual, 135–166 (2019)
18. World Tourism Organization: International tourism continues to outpace the global economy, p. 2 (2019)

Brand Construction of Chinese Traditional Handicrafts in the We-Media era—A Case Study of "Rushanming", a Ru Ware Brand

Shuang Ou[1]([⊠]), Minghong Shi[2,3]([⊠]), Xin Wen[2]([⊠]), and Rungtai Lin[3]([⊠])

[1] Academy of Arts and Design, Tsinghua University, Beijing, People's Republic of China
43313431@qq.com
[2] Shenzhen Technology University, Shenzhen, Guangdong, People's Republic of China
2425035228@qq.com
[3] Graduate School of Creative Industry Design, National Taiwan University of Arts, New Taipei City, Taiwan
rtlin@mail.ntua.edu.tw

Abstract. There are various traditional handicraft works in China and they have formed different kinds of brands after years of inheritance and development. Because of the market economy, traditional handicrafts gradually transform from rather stable family inheritance to less stable enterprise inheritance. The construction of enterprise brands is beneficial to the inheritance and sustainable development of traditional handicrafts. By applying a data analysis method, this thesis takes "Rushanming", a Ru Ware brand as an example to summarize and analyze the experience in constructing Chinese traditional handicraft brands in the we-media era.

Keywords: We-media · Traditional handicrafts · Ru ware · Brand value

1 Introduction

In the information age, the original model of social operation has changed greatly and the existing frame constructed within regional markets is disintegrated rapidly, which then make the development of traditional handicraft enterprises trapped in a dilemma–they have difficulty in inheriting skills and showing their culture value, and their markets shrink greatly. Therefore, at present when regional characteristics and traditional culture connotations are disappearing rapidly, brand construction of traditional handicrafts needs to start immediately.

A brand can improve enterprises' potential productivity and market competitiveness, so building a brand is the only way for an enterprise to develop. Throughout the development history of the market economy, the interactivity between market economic structure and brand construction is reinforced gradually. Through brand construction, handicraft works are able to enter the mainstream consumer market and achieve success.

© Springer Nature Switzerland AG 2021
P.-L. P. Rau (Ed.): HCII 2021, LNCS 12773, pp. 90–102, 2021.
https://doi.org/10.1007/978-3-030-77080-8_9

Under the background of globalization, multiple brands, aggressive brand extension and complex sub-brands appear in the market, which promote the market to be more prosperous and energetic [1]. With product demands becoming increasingly diversified and market competition getting fierce, the brand can help traditional handicraft enterprises decide on their market positioning and promote the upgrade and update of traditional handicrafts. What's more, it can also accelerate the permeation of traditional handicraft works into other industries so as to expand business and realize business innovation.

2 Literature Review

Each historic brand survives from a long period of handicraft competition and business competition, so it contains immeasurable brand, economy and culture value. Nowadays, brand development is facing new problems at the new stage. How can it be solved and promote the sustainable inheritance and development of brands? The role of brand construction in the market economy has been widely recognized by world-leading scholars.

Raggio, R., Leone, R. believed that a brand is the large asset of many companies. To prove this, they worked on a project of brand value evaluation, which discussed brand evaluation, brand value and brand application in practice [11]. Based on this research, they also highlighted that enterprises always pursue applicable value of brands so as to maximize brand value. Meanwhile, they further discussed the measures to develop brand value, aiming to shed light on future practitioners [12]. Sharma, P., Mishra, S.S., Sengupta, R.N. pointed out that brand positioning was a key factor that led to enterprises' differentiation advantages. They believed that a company should formulate a brand development program by understanding itself and the surrounding environment, investing in brand identity and image-building plans, and brand reviewing. It showed the importance of brand positioning in enterprises' brand construction. As for the importance of brand positioning [13]. Santos, F., Burghausen, M., Balmer, J.M.T also held the same opinion. They expounded the concept of traditional brand positioning and explained how the bidirectional function between products and enterprise brands reinforced each other, which were of great strategic importance to middle and small-sized enterprises, as they could not only give repositionings to their products or brands, but also get some meaningful enlightenment from products' brand inheritance and construction of enterprise inheritance [14]. Zhang et al. proposed that the interaction between creativity performance and brand authenticity could actively promote consumers' brand experience, and improved management experience on how to promote the sustainable inheritance of traditional brands through investigating specific cases [15].

Ou et al. pointed out that since the social environment, consumption trends, and consumption structure in contemporary China have changed greatly, promoting the transboundary integration and development of traditional handicrafts with innovative design philosophies could convey the culture value of traditional handicrafts better and construct aesthetic concepts of the contemporary era [10]. In Advances in Chinese Brand Management, Heine, K., Gutsatz, M. mentioned that Chinese economic orientations were gradually changing from low-end manufacturing to high-end and creative industry, from capital-driven to innovation-driven, and from "made in China" to "designed in China". They also believed that brand construction was becoming a key factor driving economic transition in China [5]. Meanwhile, Indian scholars Yadav, R., Mahara, T. considered that medium and small-sized enterprises played a crucial role in Indian economy, and pointed out if they wanted to flourish in the competition, innovation is extremely important [17].

In Brand value, Arvidsson, A. discussed the essence of brand value. They believed that in social productive process, consumers created symbolic and emotional wealth around the brand in their daily communication and interaction, which clarified the role of consumers in brand creation [4]. Meanwhile, Masè, S. and Cohen-Cheminet, G. put forward that product innovation was not only limited to technology-driven and market-push strategies, but also included design-driven innovation. They further highlighted that this kind of innovation should be associated with the meaning conveyed by the brand and business model of the enterprise [8]. Furthermore, from the consumer perspective, Schroeder, J.E., Borgerson, J.L., Wu, Z. revealed how consumers created the meaning of culture heritage brands from consumers' point of view. They reckoned that Chinese enterprises are ready to stand out in the new-style brand competition using their abundant culture heritages, rather than relying on cheap mass production. Finally, they came up with the idea of applying Chinese characteristics and cultural knowledge to establish brand value, in order to tackle cultural tensions, instead of avoiding them [6]. Nedergaard, N., Gyrd-Jones, R. also agreed that innovation is a key factor in driving brand growth. They discussed how enterprise brands made contributions in leading and driving innovation, and further investigated the significance of brand leadership in innovation management and approaches of brand innovation [9]. Kim, Y.K., Sullivan, P. proposed that when formulating brand strategies in the market, consumers' eagerness to positive experience were necessary to show the true themselves, or cooperatively create designs or thoughts on the brand. In addition, a model of emotional brand strategies should be sketched [7].

Urde, M. put forward the idea of making a brand into a strategy resource, and regarding invisible value and symbols as resources are the necessary step of brand positioning. He advocated exploring brand mission based on the model of brand positioning, constructing the brand by establishing vivid pictures and forming internal brand identity. He also advised to form the brand by bidirectionally combining the rationality with emotions, and simultaneously constructed a brand hexagon. All in all, the core of brand-driven organization is to establish "value" with strategic foundation so as to finally create product awareness, products, brand association, and consumer loyalty to the brand [16].

3 Methodology

Through comprehensive literature review, this paper collects and analyzes scholars' opinions worldwide on brand construction, and then draw the conclusion that brand construction is of great significance to enterprise transformation and development. Meanwhile, this paper describes the scale, development situation and brand construction of 310 Chinese traditional handicraft companies through questionnaires, which lay a solid foundation for further discussion on this topic. Through in-depth data analysis on 310 traditional handicraft companies, this research summarizes the problems of companies in brand construction and clarifies the directions to solve these problems. Finally, through case studies, the value of constructing traditional handicraft brands is analyzed, valuable experience in constructing traditional handicraft brands is summarized.

4 Construction Situation and a Case Study of Chinese Traditional Handicraft Brands

4.1 Construction Situation of Chinese Traditional Handicraft Brands

Guided by market economy, traditional handicraft workshops transform from household production into enterprise operation. Therefore, on one hand, the brand becomes the core asset of enterprise development; on the other hand, it improves the stability of inheriting intangible cultural heritages. The investigation of the 310 domestic Chinese traditional handicraft enterprises shows that most traditional handicraft enterprises are small and micro businesses with the scale of small and medium-sized enterprises [18]. Based on the overall trend of the result, the further the enterprises develop and the bigger scale they have, the stronger brand awareness they will have (see Fig. 1).

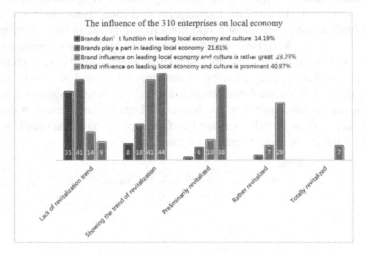

Fig. 1. Annual sales and brand awareness of the enterprises

As shown in the chart above, the 310 enterprises show five different development trends. Roughly 40.97% of the interviewed enterprises believe that their enterprise brands have a prominent influence on leading local economy and culture; 23.23% of them believe that they have a great influence on leading local economy and culture; 21.61% of them believe that they have some influence, and only 14.19% of them do not think they play a role in leading local economy and culture (see Fig. 2). This data shows that traditional handicraft enterprises have a correct understanding of the importance of the brand, and they have rather strong cultural confidence.

However, compared with high brand self-cognition and self-assessment, the proportion of brand promotion in enterprise total investment is not high. Most enterprises

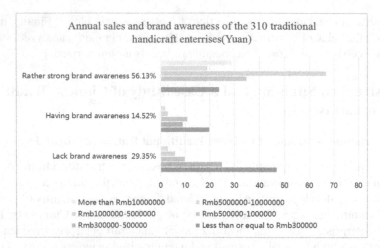

Fig. 2. Brand influence of the 310 enterprises on local economy

claim their investment proportions in the following order: heritage production → brand promotion → product distribution → enterprise operation. However, based on their practical operation, their investment proportions in reality are in the following order: heritage production → enterprise operation → product distribution → brand promotion (see Fig. 3). According to the descriptive statistical analysis to the 310 enterprises, it can be concluded that the average value of enterprises in brand promotion is 56, the maximum value is 129, the minimum value is 11, and the standard deviation is 24 (see Table 1). This means that most enterprises are short of investment in brand promotion. In other words, although most traditional handicraft enterprises have had brand awareness, they are still at the initial stage of brand construction.

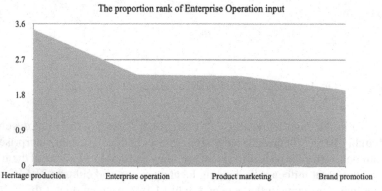

Fig. 3. The proportion rank of enterprise operation input

Based on the marketing channel, regular customers and fans are main consumer groups of traditional enterprises at present, contributing 82.26% of the sales. Their main

Table 1. The descriptive statistical analysis results of 310 enterprises' total points, average value, maximum value, minimum value and standard deviation in heritage production, enterprise operation, product marketing and brand promotion.

	Total points	Heritage production	Enterprise operation	Product marketing	Brand promotion
Average	280	89	52	82	56
Maximum	444	144	99	115	129
Minimum	159	38	21	58	11
Standard deviation	60	21	17	13	24

consumption channels are from exclusive shops, Circle of Friends and government procurement, which shows that these enterprises have not broken through the acquaintance sales model, and they fail to expand their consumer circle through brand communication. It reflects that these enterprises pay too much attention to traditional sales territories, and show weak awareness of applying brands to resource integration for expanding their consumption fields [18].

A brand is intangible asset, including four aspects: brand awareness, approved quality, brand association and brand loyalty [1]. It can improve enterprises' potential market competitiveness. Constructing a brand is the only way for the enterprises to develop. At the age of consumerism where information technology and smart devices are popular, public consumption views are increasingly mature; consumers clearly show their appeal to product brands to increasingly pursue for various high-efficiency and high-quality experience. Consumption is gradually upgrading vertically and differentiating horizontally. When selecting traditional handicraft works, consumers become active instead of being passive, and they change from perceptual consumers to rational ones. As a result, on the way of building a brand, enterprises need to clarify their position and find out the most appropriate direction for their brands to develop.

According to the research, handicraft works within RMB300 are the most popular in the market, while works of handicraft brands are usually worth more than RMB 500 belonging to high-end products [19]. Handicraft brands over-emphasize their own value but lack of efficient communication with the market, which leads to a psychological gap between the producers and buyers. Among interviewed enterprises, a large number of them are modern enterprises separated from family firms or enterprise brands transformed from master brands. In particular, the brand "Rushangming" originating from the producing area of Ru Ware products in Henan province is of great innovative and exemplary significance. The next section will briefly analyze this case, as an initial exploration of the universal law of constructing traditional handicraft brands in We-Media age.

4.2 A Case Study

In the 1980s, Ru-kiln reproduced again. Since 2015, the government has begun to pay attention to protecting, inheriting, innovating and developing Ru Ware culture. Thus, Ru Ware industry gradually shows the trend of developing from technical recovery phase to industrial renaissance stage. However, the industry chain in the ceramic production field in Ruzhou is still weak [2]. The increasing number of foreign and domestic enterprises which imitate Ru Ware not only brings opportunities to the popularization of Ru Ware culture, but also brings challenges to the formation of Ru Ware industry in the source area at the initial stage of its renaissance [3].

The Initial Establishment and Creative Product Design of "Rushanming". In the international and transregional competition of Ru Ware products, the brand "Rushanming" originating from Ruzhou comes out of Ruzhou and explores a characteristic development road through seeking for cross-border cooperation and integrating resources of good quality. The brand was first established by Li Keming, the nephew of Li Tinghuai who is a great master of Ru Ware. It consists of only 3 to 5 core members, but it is a closed loop including R&D, production and processing, brand promotion, and marketing. Meanwhile, three product systems mutually supported by household items, High-end handmade and custom-made cultural and creative products are constructed to serve for diverse living needs from the perspective of industrialization.

Li Keming entered the Ru Ware enterprise as an administrator. What he had to face at first was marketing. After some market research, he realized that master ware products with high price and low practice were hard to sell. After further studying in Academy of Arts and Design, Tsinghua University, he further understood the history of Ru Ware and found out the problems in this industrial development. He began to design and research practical Ru Ware products and establish Ru Ware brands.

The process of brand construction consists of product research, manufacturing, logistical support, brand communication, channel allocation, and other links. It involves complicated and systematic engineering. The implementation of brand strategies mainly focuses on designing visual images, brand association, and establishing the deep relationship between consumers and brands. Each of these tasks is carried out under the guidance of brand identity and brand positioning [1]. Therefore, Li Keming thought deeply about brand positioning during practice and dealt with all kinds of difficulties.

In 2016, Li Keming started his first trial: the development model of integrating Ru Ware with wine. He designed the series products "Ru-liquor", registering and establishing the brand "Rushanming" in January 2017. He drew inspiration from Ru Ware treasures of the Northern Song Dynasty in overseas museums, reproduced bottles "Yuhuchun" into winebottles, cooperated with master winemakers in Henan, and finally pushed out "Ru-liquor" series products. After consumers drink the wine, they can keep the bottles as a vase to display flowers instead of throwing them away, which achieves the effect of $1 + 1 > 2$ (see Fig. 4).

When reproducing bottles "Yuhuchun", he applied 3D printing techniques to best restore the original works. Because white wine is easy to permeate and volatilize, he tested again and again and finally overcame technical problems caused by cracked pieces of Ru Ware products. Finally, he obtained the opportunity to cooperate with his partners

Fig. 4. "Ru-liquor"winebottles work as vases

with his sincerity and excellent business plan, which promoted "Ru-liquor" products to become real products from creative ideas.

Guided by these thoughts, Li Keming designed "a box of Ru pigs", "moneyed mouses" (see Fig. 5), "moonlight treasure rabbits" (see Fig. 6) and other series products in succession. They combined with seasonable festivals and consumption situations, based on which he carried out marketing promotion. For the favorable appearances, high elegance, moderate prices, practical value and story lines, these products win him a large group of fans, good reputation and favorable market returns.

Fig. 5. Moneyed mouse   **Fig. 6.** Moonlight treasure rabbit

Market Research and Consumer Connection. Market research is the precondition of product positioning and product design. The popularization of internet and e-commerce simplifies the process of market research and reduces research costs. When releasing the products above, Li Keming didn't adopt the traditional sales models of offline exclusive shops. On the contrary, he used internet crowdfunding, which means to raise project fund from netizen in the way of group buying + forward purchasing. In this way, the sponsors make use of communication features of internet to show their creativity to the public and aim at net friends' support and attention. The model can help them efficiently understand consumers' real demands and market sizes, so they can carry out production with the raised project fund in advance, which not only saves costs, but also avoids oversupply. At the same time, crowdfunding overturns the traditional operation model, and puts the construction of brand images ahead of product production and consumption.

Let's take the series products "Ru-liquor" as an example. The series initiated 6 times of crowdfunding through Jingdong, Yitiao and other crowdfunding platforms, and they showed an ever-increasing trend from sales, page views and supporters. Especially, between the two crowdfundings in March and June in 2020, a three-month gap, their supporters and the amount of raised money reach a new height. (see Table 2)

During the two crowdfundings, "Rushanming" continued its internet publication, took the initiative with the brand and connected human emotions with its protects to guide mass consumption. Based on the sales of Ru-liquor series products, its brand stickness starts to show preliminarily. Meanwhile, after continuous communication with the brand, consumers begin to show interest in the brand and become its supporters. They generalize the products and the consumer group is further expanded.

Table 2. Crowdfunding situation of Ru-liquor series products during 2016–2020.

Ru-liquor's crowdfunding	2016	2017	2018	2019	March, 2020	June, 2020
The number of supporters	447	381	309	566	765	1123
Target amount (Yuan)	50000	50000	30000	150000	100000	50000
The amount of total raised fund (Yuan)	121893	307839	170511	784060	842042	1469764
Achieving rate	244%	615%	568%	523%	842%	2939%

The Construction of Its Brand Image and Brand Broadcasting. In establishing and spreading the brand "Rushanming", it paid attention to exploring the historical and cultural value of Ru Ware, linking up the ancient to the modern and putting forth new ideas. In view of the rarity and long history of ru ware products handed down, it is of great significance to explore the aesthetics of the northern song dynasty to meet rising consumer demand in the new era. After the TV show whose story background is the northern song dynasty became popular, Li keming seized the opportunity and pushed out the same kind of Ru Ware products, which yields twice the result with half the effort. besides, he made a fuss about the character "Ru" in Ruzhou and Ru Ware so as to produce a series of associations and homonymous meanings and leave repetitious and overlapped impressions on consumers.

The background of "Rushanming" is based on Northern Song Dynasty. With the advertisement "Ru" Ware products are so excellent that they can describe the scenery of the sun shining after rain, the remarkable color of Ru Ware products can be described. Moreover, in order to enhance consumers' visual impression, Li Keming applied azure and white, unique to Ru Ware, to be the dominant tone in designing product packaging. Through the analysis of the stroke structure of the printed font in the Song Dynasty, Li keming adopts the unique font with sharp end and vertical strokes in regular script, and designs it on this basis to enhance the smoothness of the font edges and corners and make the font full of modern sense (see Fig. 7).

Fig. 7. Product packages of "Rushanming"

"Rushanming" fully applies we-media platforms–publishing its WeChat Official account and opening an online flagship store to be the windows of advertising and selling its products. Hence, when we browse web pages, we can easily sense the detailed introduction including creative inspiration, craft process, product connotations and other aspects. Meanwhile, Li Keming also works as an internet celebrity and sells his products on network platforms. Surprisingly, his single performance once got 74000 praises with 17000 interactive comments. Through getting involved in different training and social activities, he introduces his products and ideas to the public to enhance his influence within and out of the industry. Besides, he also actively participates in industrial dialogues and research activities to expand his interpersonal resources and increase opportunities of cross-border cooperation. Moreover, when interviewed by official media and we-media, he seizes opportunities to advertise the production techniques of Ru Ware products to the public, aesthetics of the Northern Song Dynasty and his creative design concepts, which wins him a large group of fans.

5   Research Results and Discussion

Based on the case study of traditional handicraft brands, the construction model of traditional handicraft brands is generated in forms of Urde's hexagon model.The left side of the hexagon is perceptual and emotional factors and the right side is rational and

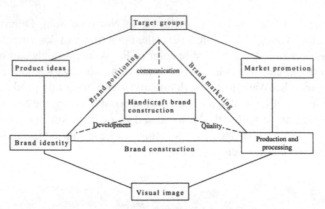

Fig. 8. Construction model of traditional handicraft brands (collected and drawn by authors of the research)

fixed factors. (see Fig. 8) The traditional craft brands should be able to spread, develop and quality-orient. Spreading corresponds to the target group of the brand. The first visual identity of the brand must be widely recognized and sustainable development, the quality of products should be guaranteed in production and processing. Brand positioning must be accurate, which including brand identity, product creative positioning, and target audience positioning. The construction of the brand includes identification of its logo and the overall visual image, and the choice of production and processing. Brand marketing includes identifying target groups precisely, making appropriate promotion strategy for the market and providing support for production and processing. Finally, these factors will create the value of traditional handicraft brands.

Brand value is also reflected on consumers trust and support. Through continuous interaction with consumers, consumers will become loyal customers for the brand, which will produce public effects and repurchases. When consumption behavior occurs, in addition to the performance of the product, their ideas and popularity as well as the influence on social relationships are also taken into the consideration range. The brand effect of "Rushanming" appears after nearly 4 years of operation. It can be seen that the innovation of traditional handicrafts should be based on traditions.The contemporary value of traditional handicrafts will be created only if traditional cultural value is explored deeply, the brand should keep up to the date so as to make modern technology and media for our own use. In the initial stage of the brand "Rushanming", small teamwork and we-media marketing have enabled the market rapidly expand with low-cost brand communication. The approach has explored a new road for the development of traditional handicraft enterprises and. It is worth learning from other enterprises.

There is very few Ru Ware products of the Northern Song Dynasty were handed down and the public impressions and understanding of them are only limited to history books and museums. However, the brand "Rushanming" makes the public close to Ru Ware to understand its short but prosperous peak, and arouses their yearn for the elegant tastes of people in Song Dynasty. Meanwhile, through representing treasures of the Northern Song Dynasty with technology and lower costs while ensuring quality, "Rushanming" makes the elegant tastes in Song Dynasty close to the ordinary people in modern times.

Through creatively transforming traditions, "Rushanming" broadens the application of Ru Ware products in contemporary life, and successfully builds the image of Ru Ware products among mass consumers, especially among young people.

6 Conclusion and Suggestions

Craftsmen innovate their works on the basis of inheriting traditions. If they can be mass-produced, they will become products and then enter the consumption market and maximize the value of products by obtaining economic returns, which will feed back handicrafts in return. Brands can help the rapid transformation of innovations into products, so that the value of crafts manship can be finally reflected.

It can been seen from the case of "Rushanming" that in the process of brand building, the key points are to clarify the corporate brand development positioning and the interaction with consumers as brands create the communication between enterprises and market centering based on consumer demands, allowing consumers to gradually build trust and form emotionally dependence, which is also a process of brand construction and maturity.

Internet economy is a new economic form, which greatly reduces the cost of brand building and communication, especially for small and micro traditional handicraft companies. Therefore, traditional handicraft enterprises should make full use of the convenience brought by the rapid development of internet and seize the opportunity of cross-border cooperation to expand their brand effect through sharing communication or distribution channels.

References

1. Aaker, D.A., Joachimsthaler, E.: Brand Leadership, 1st edn. Free Press, New York (2000)
2. Zhuping, Z.: Problem research on the inheritance and innovation development of traditional cultural industries in internet era–taking Ru ware as an example. J. Chengdu Inst. Public Adm. (1), 81–85 (2019)
3. Junpu, Z.: Discussing the definition and classification of Ru ware from the perspective of protecting cultural ecological diversity. Beauty Times Creativity (3), 58–60 (2020)
4. Adam Arvidsson, A.: Brand value. Brand Manage. **13**, 188–192 (2006)
5. Heine, K., Gutsatz, M.: Luxury brand building in china: eight case studies and eight lessons learned. In: Balmer, John M T., Chen, Weifeng (eds.) Advances in Chinese Brand Management. JBMAC, pp. 109–132. Palgrave Macmillan UK, London (2017). https://doi.org/10. 1057/978-1-352-00011-5_5
6. Schroeder, J., Borgerson, J., Wu, Z.: A brand culture approach to Chinese cultural heritage brands. Brand Manage. **22**, 261–279 (2015)
7. Kim, Y.-K., Sullivan, P.: Emotional branding speaks to consumers' heart: the case of fashion brands. Fashion Textiles **6**(1), 1–16 (2019). https://doi.org/10.1186/s40691-018-0164-y
8. Masè, S., Cohen-Cheminet, G.: Repetto, a Paris-based craft enterprise growing into a global brand: design-driven innovation and meaning strategy. In: Jin, B., Cedrola, E. (eds.) Product Innovation in the Global Fashion Industry. Palgrave Studies in Practice: Global Fashion Brand Management, pp. 113–142. Palgrave Pivot, New York (2018)
9. Nedergaard, N., Gyrd-Jones, R.: Sustainable brand-based innovation: the role of corporate brands in driving sustainable innovation. Brand Manage. **20**, 762–778 (2013)

10. Ou, S., Shi, M., Deng, W., Lin, R.: Research on the development path of "new technology" and "traditionalization" of Chinese embroidery. In: Rau, Pei-Luen Patrick. (ed.) HCII 2020. LNCS, vol. 12192, pp. 396–407. Springer, Cham (2020). https://doi.org/10.1007/978-3-030-49788-0_29
11. Raggio, R., Leone, R.: Drivers of brand value, estimation of brand value in practice and use of brand valuation: introduction to the special issue. Brand Manage. **17**, 1–5 (2009)
12. Raggio, R., Leone, R.: Chasing brand value: fully leveraging brand equity to maximise brand value. Brand Manage. **16**, 248–263 (2009)
13. Sharma, P., Mishra, S.S., Sengupta, R.N.: An abstract: brand orientation as antecedent to brand value: construct redefinition and conceptual model. In: Krey, N., Rossi, P. (eds.) Boundary Blurred: A Seamless Customer Experience in Virtual and Real Spaces. AMSAC 2018. Developments in Marketing Science: Proceedings of the Academy of Marketing Science, p. 251. Springer, Cham (2018). https://doi.org/10.1007/978-3-319-99181-8_79
14. Santos, F., Burghausen, M., Balmer, J.: Heritage branding orientation: The case of Ach. Brito and the dynamics between corporate and product heritage brands. Brand Manage **23**, 67–88 (2016)
15. Zhang, S.-N., Li, Y.-Q., Liu, C.-H., Ruan, W.-Q.: A study on China's time-honored catering brands: Achieving new inheritance of traditional brands. Retail. Consum. Serv. **58**, 102290 (2021)
16. Urde, M.: Brand orientation: a mindset for building brands into strategic resources. Mark Marketing Manage. **15**(1), 117–133 (1999)
17. Yadav, R., Mahara, T.: Preliminary study of e-commerce adoption in indian handicraft SME: a case study. In: Pant, M., Ray, K., Sharma, T.K., Rawat, S., Bandyopadhyay, A. (eds.) Soft Computing: Theories and Applications. AISC, vol. 584, pp. 515–523. Springer, Singapore (2018). https://doi.org/10.1007/978-981-10-5699-4_48
18. The Research of Targets, Standards and Strategies in Prospering Chinese Traditional Handicrafts in Ministry of Education' s Fund Programme on Human Social Science Research in 2019
19. Editorial Department of Chinese Handicraft, The Investigation Report of Chinese Handicraft Ecology, Chinese Handicraft (2018). https://www.sohu.com/a/231412804_289194

The Interweaving of Memory and Recollection: A Case Study of Memorial House "Qiyun Residence"

Yikang Sun[1]([envelope]) and Jianping Huang[2]

[1] School of Fine Arts, Nanjing Normal University, Nanjing, Jiangsu, P.R. China
sunyikang120110@hotmail.com
[2] Graduate School of Creative Industry Design, National Taiwan University of Arts,
New Taipei City, Taiwan
50516059@qq.com

Abstract. As a place to relax and rest after work, the "family residence" is also an important space for gathering family affection and centripetal force which is full of "Memory" and "Recollection". Everyone has their ideas about the style and planning of the "family residence". Looking back at history, designers have had many successful cases in their overall planning and design of their home space, but there are relatively few designs focusing on "monumental". This article takes a family memorial "Qiyun Residence" as an example. Based on Maslow's hierarchy of needs theory, and the relationship between "Technology", "Humanity" and "Cultural Creativity", it analyzes the origin and main idea of "Qiyun Residence". The results show that "Qiyun Residence" presents a style that can be said "Simplicity without simple, and Simple without boring." It's connecting the owner's precious "Memory" and "Recollection", and this design method can also inspire similar designs.

Keywords: Qiyun Residence · Family memorial house · Human-centered design · Memory and recollection · Emotional and experience design

1 Introduction

The purpose of the design is to improve the quality of human life and enhance the cultural level of society. At the beginning of the 20th century, design reformers represented by Bauhaus have been guiding the new direction for modern design and driving the world's modern design movement by its concept that combines craftsmanship with industry and pursues "rational" in design thinking in the 20th century. The evolution of design over the past 100 years, behind the changeable styles, has never stopped thinking and questioning the essence and connotation of design. These changes mainly involve the following aspects: (1) From Function to Feeling, (2) From Creating to Sensing. How to make "design" and "human" get along more harmoniously, and how to make design more in line with "humanity" is an important prerequisite for the sound and sustainable development of design, and it is also one of the main topics of future research.

© Springer Nature Switzerland AG 2021
P.-L. P. Rau (Ed.): HCII 2021, LNCS 12773, pp. 103–115, 2021.
https://doi.org/10.1007/978-3-030-77080-8_10

There are many types of residential-related designs, which can be roughly divided into two categories: (1) A building used to shelter from wind and rain and provide living and working space; (2) Various supplies for life and work scenes. To some extent, the design of "family space" can be regarded as the beginning of all designs. With the development of technology and the maturity of construction skills, today's buildings can meet the requirements of "functional level". When the "function" has satisfied, with good "feelings and experience" to become the people further goal, which is consistent with Maslow's hierarchy of needs [8]. The reason why "home" allows us to relax, apart from letting us temporarily put aside the turmoil of work and social interaction, the more important thing is that the meaning of "home" gives us comfort.

This article takes a family memorial house "Qiyun Residence" as an example to introduce the origin and execution of its design. From the overall plan to the detailed scrutiny, it examines how it uses creative ingenuity to achieve "commemorative" purposes. This design pattern can provide a reference for similar designs.

2   Framework and Design Mode of Family Memorial Hall

In addition to providing comfortable living space, "family residence" is also an important carrier of feelings and memories among family members. Compared with the planning and design of "functional" spaces, the design of "commemorative" spaces is more difficult. For individuals or families, the design of the memorial is still a relatively novel topic. Different from the design of general family space, the planning and design of commemorative houses not only need to consider the size and layout of the existing living space but also need to think about how to create a suitable space to achieve the purpose of "memorial".

2.1   The Role and Significance of Family Memorial Hall

In the past, family memorial halls mainly appeared in the form of ancestral shrine, or the ancestral home from childhood was used as a relatively fixed place of memorial. Many people can still count the various things that lived in the ancestral house when they were young. These "memory" and "recollection" are extremely beautiful.

Generally speaking, for families, the memorial space and field are represented by "Ancestral Hall". The patriarchal clan system and family system are the foundation of old Chinese society, and the concept of a clan is deeply rooted in people's minds. Ancestral halls are used for many purposes. They are not only a place for "ancestor worship". Later generations will also use the ancestral hall for activities when they handle weddings, funerals, birthdays, and ceremonies. Besides, family members sometimes use the ancestral hall as a meeting place to discuss important affairs within the family.

With the development of society, especially in the process of urbanization, the way of living has also changed. Not every family has the conditions to reserve a place dedicated to ancestors, but the cherishment of "family affection" and the remembrance and gratitude for the good "memories" of the past have always existed in life. Maybe people don't have so much space to build family ancestral halls, but the nostalgia for ancestors and the interaction between family members have never stopped. For example, many people put photos of relatives in their wallets and carry them with them, or put family and family photos at home.

As far as "Qiyun Residence" is concerned, all family members have comfortable residences and can satisfy their daily needs. They also hope to have a relatively independent and fixed space to recall their ancestors, gather family affection, and inherit family traditions. "Qiyun Residence" came into being under this background.

2.2 The Family Memorial Hall Reflects the Changing Laws of Human Needs

The demand for family memorial halls may not be as large as that of general functional spaces, focusing on people's "spiritual level" needs. Modern design is committed to solving the relationship between "people" and "product", so that "product" can satisfy "people's needs" to the maximum [7]. According to Maslow's hierarchy of needs theory, "needs" belong to the lower level, satisfying physiological and physical needs, and are instinctive responses. When the physiological needs are met, the pursuit is the desire of the psychological level, that is, the stage of gradually rising from "needs" to "wants" and even "desires". This is also in line with the development law and direction of modern design: (1) From Function to Feeling; (2) From Use to User; and (3) From High-tech to High-touch.

In the past, economic development was not sufficiently developed, and the national income was low. Due to conditions, many families often lived in a small house for several generations. Although it is crowded and not convenient enough, having a place to live is the greatest satisfaction. Even so, many families still maximize the use of space, even if it's just a corner, they still hang pictures of their ancestors and family members as a memorial. It shows that people's cherishment of family affection will not be reduced due to insufficient material conditions, but rather spiritual sustenance.

With the development of society, people's income continues to increase and the quality of life has been greatly improved. Daily necessities have more than just practical functions. The object-oriented design concept has gradually shifted to focus on consumers' "desires" and has become a human-centered design trend [3]. What consumers are after is no longer just to satisfy their physiological needs, but to affirm themselves at the spiritual level. Under such economic conditions, "wants" often exceeds "needs", which is also the trend of future design development. For designers, to understand the difference between "needs" and "wants", only when the design is "wants" or "needs" can they be more favored by Qualia designs. The concept and starting point of "Qiyun Residence" is from the level of spiritual needs and exerts its value.

2.3 Inspiration from "Sensitive Technology", "Human-Centered Design", and "Cultural Creativity"

The design of the family memorial hall should take "memorial" as the main axis while taking into account both use and experience. Therefore, after clarifying the design goals, designers need to learn to make good use of technology to satisfy human nature and use creative ingenuity and cultural creativity to shape the required design. Therefore, it is necessary to clarify the relationship between "technology", "humanity", "cultural creativity" and "design".

"Sensitive Technology" and "Human-Centered Design" Promote the Benign and Sustainable Development of Modern Design

Since the 20th century, the evolution of modern design can be roughly divided into five stages, from the original "Design for Function" to the current "Design for Feeling" [2], it is also in line with Maslow's hierarchy of needs theory. In recent years, industrial design represented by smart products, holding high the banner of "function" and "technology", has indeed brought convenience to life. However, it also allows modern people's lives to be kidnapped by "technology" and affected by the ubiquitous "intelligence". Although the concept of "human-centered design" is gradually being valued, there is still a lot of room for improvement, and there are many problems that need to be resolved urgently.

In the field of architecture and interior design, many houses use "technology" as their selling point, connecting various electrical appliances and lighting equipment in the home to the network system and integrating them into various ecosystems. Users only need to use the App to control it at any time. From a technical perspective, there is no difficulty, but is it really necessary? Let the air conditioner started working half an hour in advance, so that users can immediately feel the coolness when they return home. Although such "functional design" claims to be human-oriented, does it allow users to be kidnapped by technology? Since work and social interaction have consumed a lot of people's energy, they actually hope that the "residence" can be simpler and not too complicated, so that they can feel warm and caring. Therefore, it would be better if the design can consider human nature and human feelings more!

The planning and design of "Qiyun Residence" basically followed the spatial pattern of the original house, and did not make a very large renovation, but used movable partitions such as movable partitions to activate the space. Since "Qiyun Residence" focuses on commemoration, it does not introduce too many "technological" objects in the design. Everything is presented merely, and leaving more space for users to feel and experience themselves. This reflects: technology is the foundation of design; and humanity is the beginning of design (Figs. 1 and 2).

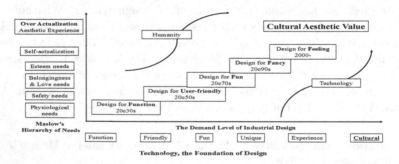

Fig. 1. Technology, the foundation of design [5].

"Cultural Creativity" Fully Realize the Cultural Connotation and Humanistic Feelings of Modern Design

What cultural creativity needs to meet is the needs of the user's psychological level, that

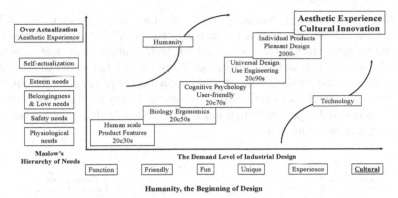

Fig. 2. Humanity, the beginning of design [5].

is, how the designer will give the user a product (design) through culture and creativity to give them a spiritual touch. In recent years, the cultural and creative industry has received much attention from many parties, to inherit and carry forward the inherent culture. Under the thinking of cultural and creative industries and in the consumption process of modern culture, products gradually focus on the development of spiritual value and promote their development of artistic creation of fine culture [1]. Lin argues that the core of the cultural and creative industry is "craftsmanship (culture)" forming "business (industry)" through "creation (design)". And "culture" is a type of life, it is formed by the "life propositions" of a group of people, which nurtures a "taste of life" and is recognized by more people to form a "style of life" [6].

What is the beauty of "crafts" when technology makes mass manufacturing commonplace? In the technological age, what is "beauty"? What "form" is the appropriate technical aesthetics? This issue has been debated throughout the development of the design history, but there is still no conclusion, only a relatively compromise solution. Perhaps, a solution can be found from a "cultural" perspective.

Compared with Hi-tech's 3C industry, Lin proposed that cultural creativity is Hi-touch's 4C industry (Fig. 3). The first "C" stands for "Cultural", and cultural and creative products are creations based on daily life's culture; the second "C" refers to "Collective", and cultural and creative products are selected from cultural relics that can represent cultural characteristics. Converting the daily necessities created; the third "C" means "Cheerful", cultural and creative products are pleasing to the eye and delightful; finally, the fourth "C" refers to "Creative", cultural and creative products integrate with the daily life's culture creative goods [5]. In short: culture is the root of design.

In the design of "Qiyun Residence", the photo walls that can be seen everywhere, as well as the photos of the ancestral house and the genealogical tree, all reflect its "memorial" purpose. These photos not only record good memories of the past, but also connect the bonds between family members. It is like the culture which playing the role of inheritance and continuation. Therefore, culture is the root of the design.

Looking to the future, designers will play the roles of "interpreters of technology, leaders of humanity, creators of sensibility, and creators of taste." Specifically, the creators of taste are "Hi-touch" that touches consumption, cultural and creative products are "Hi-touch" that expresses humanity, industrial products are pursuing "Hi-tech" of physical nature; cultural and creative products are "Hi-touch" that appeals to sensibility the demand for industrial products is rational "Hi-tech"; cultural and creative products focus on the story of life, and industrial products pursue the rationality of production. Therefore, cultural and creative products often have a touching story behind them. These stories can enrich the connotation of life, express the quality of "High-tech" through cultural creativity, and appeal for "High-touch". The taste of culture and aesthetics is a perfect interpretation of the economy.

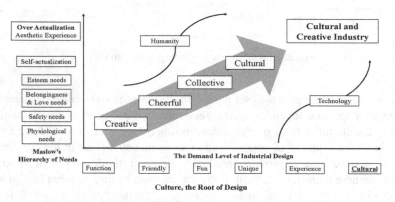

Fig. 3. Culture, the root of design [5].

3   Analysis of the Conception and Design Idea of "Qiyun Residence"

3.1   The Origin and Design Idea of "Qiyun Residence"

"Qiyun Residence" was established by Li's descendants to commemorate the ancestral house (Fig. 4). It was named "QiYun" to commemorate father Mr. Li Shiqi and mother Ms. Lin Suyun. The idea of "Qiyun Residence" was inspired by Kuotuan and his wife Yunching. After then, the five sisters named Chingmei, Shumei, Hsienmei, Lingmei, and Chuanmei has formed a planning team and cooperated and communicated fully. At the same time, an executive team composed of son-in-law Sitsung, Yungchen, Rungtai, Hsiensheng, and Yingchieh, will display their expertise and show creativity. Thanks to the concerted efforts, the Li family's descendants were finally able to drink water and think about the source, advocating the ancestral virtues, pursuing the ancestors carefully, inheriting the ancestors, and carrying forward the family tradition. For many years, it has always been the wish of Kuotuan and Yunching to promote charity in the name of parents. The completion of "Qiyun Residence" allowed them to fulfill their long-cherished wish (Fig. 5).

Generally speaking, the design of the memorial hall is more than thinking about the past, but lacking in modern style. It's relative lacks "Design for User-Friendly". The design concept of "Qiyun Residence" can be summarized in the following 3 points:

Fig. 4. The ancestral house of Li family which full of "memory" and "recollection".

Fig. 5. "Qiyun residence" that continues "memory" and "recollection".

(1) **Convey Nostalgia, and Shape the Modern Elegance.**
The design of "Qiyun Residence" adopts a modern and fashionable style, and in terms of connotation, it makes people think about ancient feelings;
(2) **Minimalist design and perfect function.**
The "Qiyun Residence" shows the memorial hall. The minimalist design can also be created, and it can be both functional and fashionable;
(3) **Cultural creativity and national aesthetics.**
The "Qiyun Residence" hopes to promote the task of local "cultural characteristics" and "national aesthetics", and achieve the goal of reproducing "Taiwanese elegance".

3.2 "Memory" and "Recollection" Trigger the Nostalgic Feeling of "Qiyun Residence"

Different from mass-manufactured industrial products, the "temperature" of hand-made products is unmatched by technology products. Such products can even touch consumers' hearts. Such products can also reflect certain living customs and traditional culture. In addition to providing basic functions, it is also symbolic and storytelling. Similarly, the design of a commemorative space is more for remembrance and gratitude, and secondly is to meet the needs of "function".

In the process of urbanization, many old houses and ancestral homes have to face the situation that they must be demolished, except for those recognized as national, county, and city monuments and are better protected. For many people, perhaps the buildings are not suitable for use due to their age, but every brick, tile, grass, and tree is full of good memories of the past. Even if it is rebuilt in situ, it is completely referenced to the structure and details of the original building, but there is always something less.

Fig. 6. The genealogical tree: a symbol of family prosperity and cohesion.

On the wall of "Qiyun Residence", a large number of photos of family members are displayed with concise and warm words. This all tells the thoughts of every family member for the most affection. This is the best expression of the "sense of ritual". The genealogical tree is more symbolic of the bonds between family members (Fig. 6).

In the specific design process, based on past research, the designer summarized the steps of cultural product design into the following four stages: (1) set a scenario, (2) tell a story, (3) write a script, and (4) design a project [4], and used this idea in the design of "Qiyun Residence". Specifically, the positioning of "Qiyun Residence" is to recall parents, gather relatives, and inherit family traditions. Therefore, all designs are developed around the above positioning: for example, old photos tell stories of the past and become an important element of decorating the space. Sorting out so many old photos is a journey of nostalgia. To better highlight the meaning of the commemorative space, the designer strives for simplicity in the choice of style and form, highlighting the connotation of "Qiyun Residence" in the simplest form.

3.3 "Style" and "Function" Reshape the Modern Elegance of "Qiyun Residence"

"Design" emphasizes making adequate planning and arrangements according to the purpose. However, "design" itself is a process of constant change and diligence, and nonsensical actions often bring unexpected "surprises". In the process of conception, design, and construction of "Qiyun Residence", due to little poor communication, a design style with "Mondrian" style as the main axis was formed, which became one of the characteristics of "Qiyun Residence".

"Qiyun Residence", from the establishment of the overall style to the careful consideration of details, allows the simple design to be creative, functional, and stylish at the same time. The main functional zones of "Qiyun Residence" are roughly composed of four parts: "The Meeting Room", "The Reading Room", "The Modern Leisure Space (The Bar)", and "The Family Activity Space and Information Room". Table 1 shows the style of "Qiyun Residence" and its positioning of different areas. The following selects the different functional areas of "Qiyun Residence" to introduce its design concept and creative ingenuity.

Table 1. The style of "Qiyun residence" and its positioning of different areas.

Area	Functional Positioning	Style
The Meeting Room	A place for family members to discuss matters.	The design style of "Qiyun Residence" follows a simple route.
The Reading Room	Collect the books, photos and videos of all family members.	Due to the designer's personal preferences and the "small accidents" during the construction process, the "Mondrian" style was formed and became one of the characteristics.
The Modern Leisure Space (The Bar)	Leisure, communication, and relaxation.	
The Family Activity Space and Information Room	Family activities or personal memorial space.	

The Meeting Room

The meeting room has a large space and can be used for family gatherings or business discussions (Fig. 7). Different from the general design, the meeting room has abandoned the traditional screen and replaced it with a large TV and a large whiteboard. The whiteboard also functions like a curtain. In addition to the long table in the middle of the room, some corner tables can be lifted on the surrounding walls. Usually, the tabletop can be pasted with action photos or decorated with paintings; when not in use, it can be closed and will not occupy space (Fig. 8). On the wall on the side of the conference room, the photos of the former residence and the genealogical tree are displayed. The picture of the ancestral house is placed at the bottom, indicating that the ancestral house was preceded by the current residence, which expresses the meaning of gratitude; and then the genealogical tree is stacked in the ancestral house to express the inheritance and convey the meaning of the family's advancement.

Fig. 7. The meeting room.

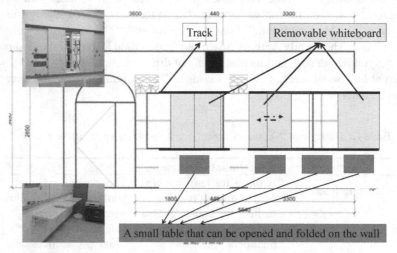

Fig. 8. A removable whiteboard and wall tables that can be stored.

The Reading Room

The reading room can not only display photos and albums of family activities in different periods, but also display various achievements of family members (Fig. 9). In this space, family members can not only Memory the past and appreciate family affection, but also look forward to the future and learn from each other. The design of the bookcase divides the space with vertical and horizontal lines. The several constitutions of the chain washing and the elegant tones can further highlight the symbolic significance of the space as a collection of books. Part of the partition edges is decorated with bright colors so that this quiet and leisure space will not appear too dull and monotonous.

The Modern Leisure Space (The Bar)

The design of the bar counter becomes an interface that combines both "thinking about the nostalgia" and "fashion", and it is also an interface for transforming public space into a family activity space (Fig. 10). Unlike other functional areas, the bar counter provides simple drinks and small points, allowing family members to communicate and interact more easily, and the design style tends to be modern and lively. The original design originally echoed the minimalist style of the whiteboard like the "meeting room", composed of simple square skateboards; during the construction process, a "Mondrian"

Fig. 9. The reading room.

style was accidentally formed. Judging from the final effect, it has become one of the characteristics of "Qiyun Residence": simplicity without simple, and simple without boring.

Fig. 10. The bar.

The Family Activity Space and Information Room

Family activity space can not only hold family gatherings, but also provide personal meditation (Fig. 11). For this special space, the "movable door wall" is designed to block out daily life spaces such as toilets and kitchens. Since the design of "Qiyun Residence" has not made major changes to the original room type, the design of the "movable door wall" not only meets the requirements of "function", but also facilitates concrete implementation. Similar to the design of "reading room", "meeting room" and other spaces, while continuing the simple and geometric "Mondrian" style, it also partially decorates the walls with bright colors, making the entire space simple and lively. On the wall, stories and photos among family members are displayed with different themes. It is these precious "Memories" and "Recollections" that constitute the connotation of "Qiyun Residence". These details make "Qiyun Residence" unique to each family member.

Living space is limited, but emotions are unlimited. In the family information room, because the walls are limited, the performance of family members is unlimited. The existing mounting plates are all symbolic and can be replaced at any time. With modern cloud technology, the data about each member of the family can be placed in the cloud and can be viewed at any time.

Fig. 11. The family activity space and information room. (According to different themes, a large number of photos of family members are displayed on the wall).

4 Summary

In recent years, public welfare issues have received increasing attention. Whether it is the official policy orientation or the spontaneous behavior of the people, the development of the society is more benign and warmer. In addition to commemorating parents and inheriting family traditions, "Qiyun Residence" also hopes to become a "cultural education center" in the future. It plans to provide various services including book lending and holding public welfare lectures to promote local characteristics. And further, promote Taiwanese culture.

The planning of "Qiyun Residence", whether for daily use, further experience, or even extended to serve the community, its "function" and "experience" are organically combined with ease. "Qiyun Residence" provides a possible way for the planning of the same type of family memorial space. As long as the connotation is sufficient, the simplest way, the purest form, no matter the size of the space, anyone can create a memorial field belonging to their own family. It is true that "Qiyun Residence" has just been completed, and many details will be continuously improved and supplemented during use. If the space is not large enough, people can also choose a room or even a corner of the home and remodel it slightly. The continuation of the tradition does not need to be too fancy external. As long as we are a caring person and be grateful all the time, anything can become more meaningful and warmer.

Currently, we are not short of useful "High-tech" designs, but we look forward to more "High-touch" designs. For the owner, an object or a house, because of the "Memory" and the touching "Recollection" it carries, perhaps their functions are outdated or even the entity no longer exists, but they have irreplaceable value and meaning. For everyone in the Li family, "Qiyun Residence" is unique, not only the "first one" but also the "only one". Although it is only for other people, because there is no common memory, recollection, and perception, "Qiyun Residence" may not be of reference value for them. However, the design concepts contained in "Qiyun Residence" and the experience accumulated in the implementation process can provide a certain reference for the same type of design.

From the perspective of cultural creativity, the above design concepts are also one of the cores needed for future design. The current design content is very extensive, ranging from architecture to small daily necessities. Although there are many functions, it is very complicated. While competing with powerful functions, whether it conforms to humanity is often ignored. When the homogeneity of design becomes more and more obvious, only designs that move people's hearts and conform to humanity can win the favor of consumers.

Acknowledgements. The author wishes to thank Mr. Kuntuan Lee, General Manager of Duan Kwei Machinery Enterprise, and special thanks to the Li Family for authorizing the photos of "Qiyun Residence". Their thoughts on "Qiyun Residence" also provided great inspiration for the author. At the same time, the author would like to thank Prof. Rungtai Lin for directing this study.

References

1. Chen, H.Y.: A study on the information design of cultural product (Unpublished Master's Thesis). Ming Chuan University, Taipei (2009). (in Chinese). https://hdl.handle.net/11296/wab5md. Accessed 10 Sept 2020
2. Chien, C.W., Lin, C.L., Lin, R.: A study of five "F" in product design. In: Shih, Y., Liang, S.M. (eds.) Bridging Research and Good Practices towards Patients Welfare: 2014 Proceedings of the 4th International Conference on Healthcare Ergonomics and Patient Safety (HEPS), p. 409. CRC Press, Boca Raton (2014)
3. Chuang, M.C., Chen, Y.T., Chang, Y.J.: A study on tactile image of products: using handless cups as a case study. J. Kansei **1**(1), 28–45 (2013). (in Chinese)
4. Lin, R.: Transforming Taiwan aboriginal cultural features into modern product design: a case study of a cross- cultural product design model. Int. J. Des. **1**(2), 45–53 (2007)
5. Lin, R., Kreifeldt, J.G.: Do Not Touch: Dialogue Between Design Technology and Humanity Art. National Taiwan University of Arts, New Taipei (2014).(in Chinese)
6. Lin, R.: Preface: the essence and research of cultural and creative industry. J. Des. **16**(4), i–iv (2011). (in Chinese). https://doi.org/10.6381/JD.201112.0002
7. Maslow, A.H.: A theory of human motivation. Psychol. Rev. **50**(4), 370–396 (1943). https://doi.org/10.1037/h0054346
8. Maslow, A.H.: The farther reaches of human nature. J. Transpersonal Psychol. **1**(1), 1–9 (1969)

Disseminating Intangible Cultural Heritage Through Gamified Learning Experiences and Service Design

Yunpeng Xiang[1], Jingzhi Wang[1], Jing Fa[1], Naixiao Gu[1], and Cheng-Hung Lo[2(✉)]

[1] Suzhou Education Association of Exchange, Suzhou Education Bureau, 198 Gongyuan Road, Suzhou, China
[2] Xi'an Jiaotong-Liverpool University, 111 Renai Road, Suzhou, China
CH.Lo@xjtlu.edu.cn

Abstract. In recent years, preservations of Intangible Cultural Heritages (ICHs) have become an important agenda for the relevant researchers, designers, and technologists. Traditional archiving methods to protect and promote ICHs may include the recognitions from governments or representative bodies and the textual or video documentaries. However, barriers between the general public and ICH still exist and affect the accessibility of ICH at both conceptual and practical levels. In this case, an educational approach for disseminating ICH values can be considered. Initiated with this thought, we set out to develop an educational project that embeds ICH knowledge in a gamified learning experience. The project is framed as a service design practice to ensure an integrated consideration of the involved stakeholders, including the collaborated local sector, designers, educators, and target learners. The development process results in a game-based education program that serve both the interests of the promoting bodies and the intended audience. In the program, we incorporate three types of ICH originated locally in Suzhou. This paper, therefore, presents the design rationale and development process of this educational program, as well as the result of an initial test with a group of participants. The participants' feedback demonstrates the advantages of SD approach in developing a user-centered and service-oriented public course. Meanwhile, the playful elements put the included ICHs into their social context and incentivized the participants in learning the related cultural knowledge.

Keywords: Service design · Cross-cultural education · Intangible Cultural Heritage · Education for international understanding · Game-based learning

1 Introduction

Intangible Cultural Heritage (ICH) is an important asset of regional cultures. Many countries have established its own list of ICH under the definition and guidance provided by UNESCO. Much effort has been put in extending the list with the research evidence on the historical origin, development process, survival status, and inheritance. This might somehow drive the work towards more to specializing ICH as a subject area that is only

© Springer Nature Switzerland AG 2021
P.-L. P. Rau (Ed.): HCII 2021, LNCS 12773, pp. 116–128, 2021.
https://doi.org/10.1007/978-3-030-77080-8_11

accessible to certain groups of cultural experts. China has also established a hierarchical list of ICHs according to their scales of impact. As of 2014, China has more than 3,145 national ICHs. In Jiangsu Province, for example, there are more than 4,500 ICHs recognized at provincial, municipal and county levels. With such a vast and diverse set of ICH resources, China's ICH program presents tremendous opportunities, as well as challenges. Suzhou, where our project was carried out, is a historical city in China. It alone has thousands of multi-level ICHs manifesting the rich local cultures. To properly protect and promote ICHs, local sectors and practitioners incorporated various methods, e.g., creating ICH websites, building ICH museums and heritage bases, generating publications and documentaries, holding festivals, competitions, and exhibitions...etc. However, issues still exist in the dissemination approaches, audience size, inheriting talents, and so on (Li 2019). ICH-related education plans have been in the discussion about tackling those issues. Qiao (2018) put forward three major scenarios of current ICH education in China, which includes combining university education and ICH for cultural integration and practice, incorporating local communities for cultural popularization and training, and infusing primary and secondary education with relevant cultural experience. Since 2009, Suzhou government has promoted and implemented curriculums for international understanding in primary and secondary schools. Suzhou Education Association of Exchange (SEAIE), the initiator of this project, has played a key role in managing the exchanges between local and international education bodies, as well as exploring education models for international understanding. SEAIE aims to create more opportunities for local and international students to communicate and collaborate. It has developed activities and programs connecting students of different nationalities. As the business grows, it needs to integrate more local resources for developing educational projects related to cross-cultural communication.

UNESCO (2003) emphasized that the sustainability of ICHs should consider their identification, preservation, protection, promotion and transmission (particularly through formal and non-formal education) so as to achieve their revitalization. Educational strategies for inheriting and promoting ICH should be diversified, rather than focusing on retaining their original inheriting models (Michela et al. 2014). In other words, ICH-related education includes not only cultivating new talents in mastering ICH skills, but also disseminating ICH-related knowledge to general public. A great example is the I-Treasure project supported by the European Community (Cozzani et al. 2016). The project has been initiated to build a platform that meets the demands of different types of ICH learners. The general public can thus learn about typical ICHs through this platform. The platform is well intended and packed with rich content. However, the platform could be less attended by those who do not live in the specific cultural environment of certain ICHs. The users may not earn a profound understanding to the values of the included ICHs (Fairchid and Helaine 2009). This implies that the related cultural environments could play a vital role in generating further interest in ICH knowledge and skills. Besides, as globalization accelerates, it is increasingly obvious that countries need to balance the promotion of their own culture with an international perspective to ensure the people from aboard could appreciate local cultures in a more effective sense. UNESCO has encouraged adapting a more international education approach to achieve the mutual understanding and respect between different cultures, countries and peoples. However, current educational models are still in the need of a suitable environment for ICH to become more accessible and even applicable in modern context (Peng et al. 2017). The same notion has been taken into account in this project. We set out to develop an educational project that embeds ICH knowledge in a gamified learning experience. And it is expected that this activity-driven and experiential learning approach can help reduce the cultural barriers such as languages and backgrounds and engage the learners in a playful way.

2   Design Process and Method

The project was developed in line with SEAIE's mission in the education for international understanding. To accommodate the shared interests among the involved stakeholders such as the learners with different cultural profiles, we adopted Service Design (SD) strategies to develop the project. It is also possible to explore the possibility of building logic bridge to connect ICH historic contexts with modern life (Chen 2018). The following sections will discuss how we carried out this public education project that embedded ICHs in a gamified learning experience.

2.1   Exploring the Interests of Target Learners

The target audience was set by SEAIE as the students between the ages of 7 and 18 who studied in the local and international schools in Suzhou. The first step of this SD project was to elicit the potential learners' needs and preferences. In the collaboration with a local international school, the project team managed to recruit a total of 65 students for a brief survey on their experience about Suzhou's culture and their interests in culturally-related activities.

According to the students' feedback, 83% of them reported having general ideas on Suzhou's culture (Fig. 1). 56% of them have attended some local culture-relevant courses and activities (Fig. 2). They are aware of some classic and popular ICH projects in Suzhou but do not have a deep understanding of them. Less than 40% of them gave positive feedback to their previous experience on culture-related courses (Fig. 3). When exploring further the factors affecting their responses to the previous experience, one can see from Fig. 4 and 5 the importance of curriculum design, language barriers, and field experience. Nonetheless, they showed the interest in a more interactive and challenging experience such as games and puzzles (Fig. 6). We also interviewed teachers who had delivered culture-related classes. They agreed with the potential flaws in traditional culture courses. For example, students were hardly engaged with the teaching materials. And it was difficult to promote active learning with the limited peer interactions in the classes.

2.2   Constructing the User Journey Map

We adopt the concept of game-based learning in developing this educational project. Deterding (2011) pointed out that game-based learning was to use gaming elements in non-gaming environments, and this approach needed to consider the importance of

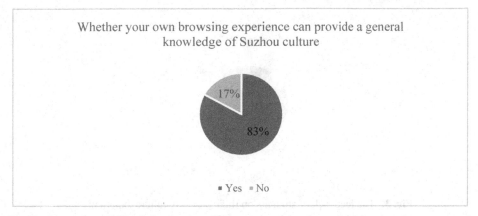

Fig. 1. More than half of the local and international students thought they have a rough understanding on Suzhou and its culture based on their past experience

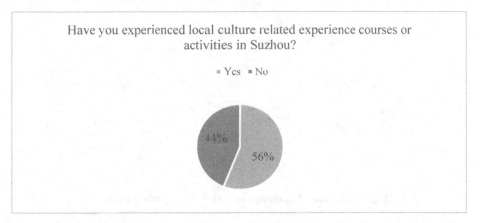

Fig. 2. More than half of the local and international students has been involved some local cultural courses and activities in Suzhou

learners' experience and perception during the entire process (Huotari and Hamari 2012). In the method of game-based learning, it will cultivate better learning attitude, increase student motivation, and help achieve better learning outcomes (Kapp 2012; Perrotta et al. 2013; Michelle et al. 2020). We deem the learning process as the user journey in our proposed course. To satisfy the requirements of learnability and playability, the project integrated the elements of role playing, puzzles, ICH knowledge, self-study, and collaborative activities. The user journey is described as follows.

At the beginning, the participants are given a background story to read and assigned with specific roles in the story. In each chapter of the story, an ICH concept with one related cultural idea is introduced as the thematic elements. After the participants finish reading the prepared story, they make their ways to that ICH's local sites, which can be a workshop, studio, exhibition venue, museum, traditional theater... to name but a few. At the local site, they experience the ICH's practice either by observing or following along

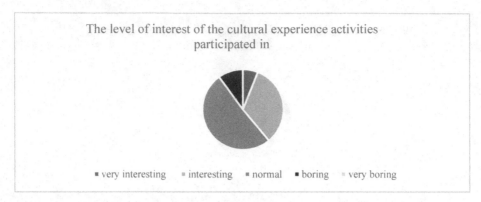

Fig. 3. The feedback of students on their previous culture experience

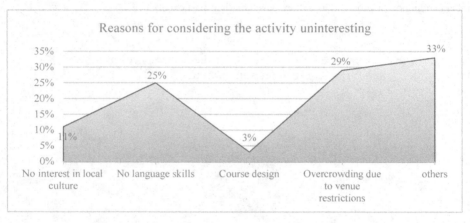

Fig. 4. Reasons of students caused the activity unattractive

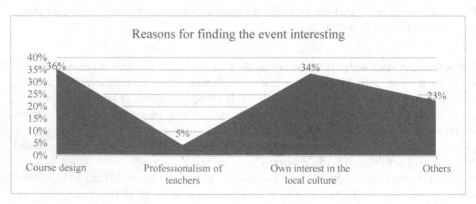

Fig. 5. Reasons of students caused the activity attractive

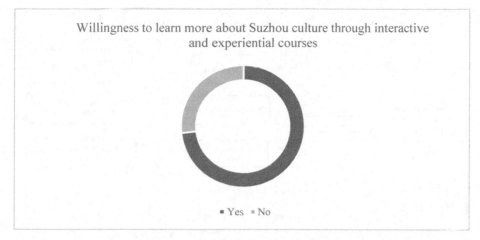

Fig. 6. Students' willingness to attend interesting and interactive courses for Suzhou culture

the designed exercises. As the journey map (Fig. 7) shows, several touching points have been set to enrich the users' emotional experience. First of all, the participants will receive the material package (the Game box), including a set of character cards. The cards contain character images, which were newly-illustrated in this project, and character descriptions. We intend to hand over tangible and accessible materials to the participants and entice them at the very beginning. After receiving the package, they compose and get into the characters along the storyline. Then, all participants need to complete the study module, which may appear to be a daunting but necessary task to some participants. Once they move to the following stages of the more dynamic ICH experience and puzzle-solving activities, we expect that their emotional states can be elevated with the increased levels of active participation and peer interaction.

2.3 Designing the Curriculum with Detailed Materials

The user journey map gave us a general framework for planning the learning progress and trajectories in the proposed course. Based on the framework, we set out to develop detailed elements in course, including character design, experience design (both ICH experience and study module) and course material design.

Characters
In the part of storyline, we adapted a classic Chinese writing, *Six Chapters of a Floating Life*, written by Fu Shen in Qing Dynasty. The new novel documents the life stories among six characters, who lived in and around Suzhou. In addition to the dynamics among the characters, the novel contains a rich description of the culture and lifestyle of Suzhou. As shown in Table 1, there are 6 characters with illustrated personalities. They may have different thoughts and opinions on the shared agendas. The participants get a chance to consider the conflicts between the characters through the role-playing activities.

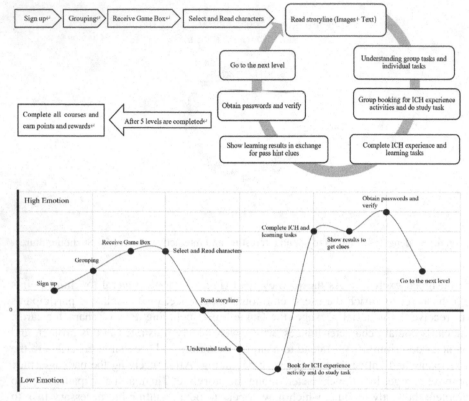

Fig. 7. Designed journey map with emotional points

ICH Experience

Both of the ICH experience and study module were designed along the plots of the story. The aim of the study module is to establish a general understanding of the cultural and social background of the included ICHs. Taking the Suzhou embroidery, a recognized local ICH, as an example: It firstly tells the story about the origins of Suzhou embroidery, as well as the comparison between the traditional and industrialized embroidery approaches. After that, the participants will do simple exercises to practice Suzhou embroidery. When the participants visit the Suzhou embroidery base, they are invited for a task to tell the difference between the embroidery products made by the traditional techniques and those made by the machines. It is expecting that in this way, the participants' impression on Suzhou embroidery will be strengthened. Finally, they need to solve the puzzle that gives away the password leading to the next activity. The puzzle is related to the main cultural or social ideas presented in both the story and the ICH experience. If the participants, who come with diverse cultural backgrounds, can solve the puzzle, it probably indicates the effectiveness of integrating ICH experience, cultural appreciation, and international understanding within this activity (Table 2).

Table 1. Characters designed in this project

Name	Yue Shen	Qian Chen	Zhuo Shen	Yun Tang	Le Shen	Yi Tang
Age	15–21	15–21	11–17	10–16	6–12	7–13
Personality	Traditional literati, conservative and stable, not easy to be changed, self-restraint and restraint, have a huge sense of mission to maintain the family	Traditional women, virtuous and capable, respect their husbands, love their children, depend on them for their lives, and follow the traditional rituals, thus slightly losing their individuality	Intelligent and studious, spontaneous, curious about new things, but still not easy to give up what is there; gentle and affectionate, kind but also weak	Gentle and intelligent, with the softness and beauty of a traditional woman, loving life and having an interest in life, but also with a lively and cheerful mind, curious about new things	Lively and naive, with a certain rebellious spirit, the pursuit of the concept of "gender equality", but with the social incompatibility of the time, so sharp personality, it is difficult to be truly happy	Bold and brash, loves his sister, is charitable and chivalrous, aspires to a world outside of the dynasty and aspires to sail far away
Group	Traditional		Middle		Vanward	

Course Materials

To ensure the curriculum conducted smoothly and playfully, the teaching materials have been designed systematically. It contains two main parts, including the Game Box for the students who will participate in this curriculum and Tip Cards for the ICH inheritors and teachers that may host both ICH experience and study modules (Fig. 8). With the Tip Cards, the ICH inheritors and teachers could follow the guidance of curriculum designers to finish the ICH experience and study modules. In the Game Box, there are six character envelops and five groups of props for sections, respectively. The contents of materials in each character envelop are different with others, which are designed according to the settings of each character. For example, when a participant chooses one character, his character cards will only show his own story and ideas to help him understand himself in the story (Fig. 9), but if he prefers to achieve more information about the entire story and reasons on development, he need communicate and discuss with others. In this way, the connection between the six participants in the roles of characters would be strengthened. Meanwhile, the link between the participants and ICHs is also designed in this Game Box. After participants finished reading stories, they would move to ICH experience. In their envelops, some packages of ICH are introduced to them, such as in the section of Chinese tea, there are multi materials are designed to tell the story of Chinese tea (Fig. 10). With all these well-designed materials, the whole process of this 'ICH education Game' could be controlled by the course designers as much as possible, while their efficiency still need testing in the future.

Table 2. Storyline and main structure of the project

	First Chapter	Second Chapter	Third Chapter	Fourth Chapter	Fifth Chapter
ICH	-	Suzhou Food	Tea	Suzhou embroidery	-
Task	Complete the Idiom Crossword Card	Tasting and solving puzzles in the restaurant	Experience the tea making process, experience the comparison of Chinese and Western tea seats, and complete the decryption in the tea room	Complete the embroidery sample and solve the puzzle in the Suzhou embroidery showroom	Performing the Lantern Scene
Culture	Suzhou food culture: "No food from time to time, no food out of date"	The symbiotic relationship between human and nature	Clash of civilizations is the desire to communicate and learn from each other: to learn from each other's strengths, there is no high or low civilization.	1. Experience the function of Chinese subtle emotional expression and craft products 2. The display and collision of female self-awareness	1. The clash of ideas, struggles and compromises on the road to gender equality 2. The meaning of "home" in Chinese style
Study module	Chinese	Critical Thinking	World History	Economics and Mathematics	Literature
Password (in Chinese)	不时不食	和合共生	和而不同	各美其美	

Fig. 8. Images of tip cards (left) and the game box (right)

Fig. 9. Images of different character cards in the game box

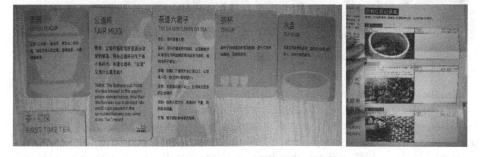

Fig. 10. Images of Chinese tea package in the game box

3   Initial Test

To verify how the proposed course works in practice, we recruited a total of 12 participants for an initial test. There were 6 students from a local primary school and the other 6 students from an international school. They were led to complete the five chapters of the story by acting the assigned characters. They had also gone through the tasks specified in the study modules, the ICH experience, and the puzzles (Fig. 11). After experiencing the course, we ran an unstructured interview to gather their feedbacks. The result shows that most of them are satisfied with the playfulness of the service. The emotional responses also correspond well to what was articulated in the user journey map. For example, once they got their own character cards, they were very keen to know their own personalities and discuss actively with the others to understand their relationships. When they succeeded the task of finding the password of each section, their emotional state reached to a high level. However, when the study module was deemed as too challenging or boring, they also lost the concentration on the following parts.

Fig. 11. Images in the test period

On the other hand, most of the participants responded that they could understand the cultural and social knowledge conveyed in the course. They were glad to share all the learnt knowledge with their friends and relatives. For example, at the section of Chinese tea culture, after they learnt about the history, tea making process, and the appreciation of Chinese and Western tea, they were excited with sharing with their parents and friends about the tea sets and tea-making skills. Some of their parents gave spontaneous feedback on seeing the children changing views on local ICHs, related Chinese culture, and the possible connections with other countries' cultures. This supports our initiative of achieving international understanding and cross-cultural communication through a gamified project.

In addition to the feedback from the participants, we were interested in the reactions of ICH inheritors and teaching staff to the course. We ran a focus-group session with these stakeholders. In the session, they commented on the complexity of this course and the challenges of running it with limited manpower. The reason why this course appears to be engaging is the small set of participants. The trade-off was that the content and message might not be disseminated to a larger number of audiences at the same time, as the ICH workers suggested. Moreover, the quality of the course delivery relies heavily on the host's ICH experience, cultural literacy, and language proficiency.

4 Conclusion

The project is a novel attempt to combine ICHs, internationalized education, and gamified learning experience into a cultural course. We have framed it as a Service Design project and considered the target learners' needs and preferences. Starting from constructing an overall user journey, we established the key touching points and incorporated local ICH sites in the proposed course. We then devised the corresponding role-playing activities, ICH exercises, study modules, and puzzle games to engage the participants in various ways. The result of the initial test reflects the advantages of SD approach in developing a user-centered and service-oriented public course. Meanwhile, the playful elements put the included ICHs into their social context and incentivized the participants in learning the related cultural knowledge.

Although this ongoing project received positive feedback from the participants in the initial test, we are aware of the rooms for improvement by scrutinizing the critiques. For example, the ICH practitioners and the teachers involved in this project gave advice on reducing the complexity and considering the flexibility in terms of human resources. These suggestions will be taken on board in the future iterations of development. One possible aspect of simplifying the participants' task is re-design the study modules, which appeared to be less effective in engaging the participants. Other improvements could include developing auxiliary programs to train seed teachers and building a rewarding system to recognize the participation and achievement of learners.

References

Chinese State-Level Non-Material Cultural Heritage List. https://www.ihchina.cn/project.html

UNESCO Headquarters, Paris: Convention for the Safeguarding of the Intangible Cultural Heritage. UNESCO Org. (2003)

Fairchild Ruggles, D., Silverman, H.: From tangible to intangible heritage. In: Helaine Silverman, D., Ruggles, Fairchild (eds.) Intangible Heritage Embodied, pp. 1–14. Springer, New York (2009). https://doi.org/10.1007/978-1-4419-0072-2_1

Chen, J.: Redesign of immaterial culture heritage of Lotus Lantern based on service design thinking. Hundred Sch. Arts **162**(3), 235–239 (2018)

Deterding, S., Dixon, D., Khaled, R., et al.: From game design elements to gamefulness: defining gamification. In: Proceedings of the 15th International Academic MindTrek Conference: Envisioning Future Media Environment. ACM (2011)

Huotari, K., Hamari, J.: Defining gamification: a service marketing perspective. In: Proceeding of the 16th International Academic MindTrek Conference (2012)

Kapp, K.M.: The Gamification of Learning and Instruction: Game-Based Methods and Strategies for Training and Education. Pfeiffer, San Francisco, CA (2012)

Perrotta, C., Featherstone, G., Aston, H., Houghton, E.: Game-Based Learning: Latest Evidence and Future Directions (NFER Research Program: Innovation in Education). NFER, Slough (2013)

Michelle, T., et al.: The agency effect: the impact of student agency on learning, emotions, and problem-solving behaviors in a game-based learning environment. Comput. Educ. **147**, 103781 (2020)

Cozzani, G., Pozzi, F., Dagnino, F.M., Katos, A.V., Katsouli, E.F.: Innovative technologies for intangible cultural heritage education and preservation: the case of i-Treasures. Pers. Ubiquit. Comput. **21**(2), 253–265 (2016). https://doi.org/10.1007/s00779-016-0991-z

Li: Research on the Innovation Model of Jiangsu Intangible Cultural Heritage in the Context of Big Data - Taking Suzhou Intangible Cultural Heritage as an Example. Shen Hua, pp. 88–89 (2019)

Ott, M., Dagnino, F.M., Pozzi, F.: Intangible cultural heritage: towards collaborative planning of educational interventions. Comput. Hum. Behav. **51**(10), 1314–1319 (2015)

Peng, Tan, Li: A new exploration of intangible cultural heritage education of Ying Shi - a study of Ying Shi art as vernacular teaching material at Ying Xi Middle School, Yingde Guangdong Landscape Archit. **39**(5), 15–19 (2017)

Qiao, Dong, Su: The Chinese Experience: Diverse Intangible Cultural Heritage Practices. Jiangxi Fine Arts Press (2018)

Exploring the Creation of Substandard Stones in Fuzhou Shoushan Stone

Xi Xu[1,2]([✉]) [iD]

[1] Khon Kaen University, 123 Mittraphap Road, Khon Kaen 40002, Thailand
[2] Fuzhou University of International Studies and Trade, 28, Yuhuan Road, Shouzhan New District, Changle 350202, Fujian Province, China

Abstract. The purpose of this study is that Shoushan stone, as one of the four traditional 'seal stones' in China, has been named as the 'National Stone' of China" by the Chinese Association of Treasures and Jade Stones. With the depletion of mineral resources, people's requirements for Shoushan stone carving are becoming more and more stringent, Shoushan stone with transparent texture, delicate moist and brilliant color is no longer easy to get; while the majority of consumers still find it difficult to accept a large number of so-called 'inferior' Shoushan stones, fine and coarse, bright and dark, adulterated quartz, black spots, etc. but no one wants to talk about them. The imminent problem of how to obtain creative resources for a new generation of young artists who do not have good stone resources has caught my attention. This study aims to create new forms of artistic expression by carving or re-creating 'inferior' Shoushan stones and their carvings, to break the stereotype of Shoushan stone art among consumers, to re-establish the beauty of other forms of Shoushan stone, and to attract more people to understand and pay attention to Shoushan stone art. The development and application of Shoushan stone secondary stones are studied, and their development contents are analyzed and discussed by applying theories of artistic innovation and design psychology, so as to elaborate how to better utilize the ingenious carving of secondary stones in Shoushan stone stones to enhance their artistic value.

Keywords: Shoushan stone · Sub-stone · Cultural creation · Art

1 Introduction

Shoushan stone is a precious stone unique to Fuzhou, with its crystal quality, moist, colorful, natural color and distinctive color boundary, with the characteristics of rarity, humanity and appreciation, loved by people at home and abroad, mainly located in the northern suburbs of Fuzhou City, Jinan District and Lianjiang County, Luoyuan County, the junction of the 'Golden Triangle' area. Shoushan stone, as one of the 'Four Great Seal Stones' of China, has been named the 'National Stone of China' by the China National Treasure and Jade Stone Association. As the mineral resources of Shoushan stone continue to be depleted, the requirements for Shoushan stone carvings have become more and more stringent. Shoushan stone with transparent texture, warm and delicate, rich and colorful is not easy to get. The average consumer still likes good Shoushan stone,

© Springer Nature Switzerland AG 2021
P.-L. P. Rau (Ed.): HCII 2021, LNCS 12773, pp. 129–138, 2021.
https://doi.org/10.1007/978-3-030-77080-8_12

and no one is interested in the so-called 'inferior' Shoushan stone, which has a large number of fine and coarse, bright and dark, mixed with quartz and black dots. Because of the driving factors of the market environment, the problem of resource conservation of Shoushan stone needs to be solved urgently. In Shoushan stone carving, materials are becoming more and more scarce at this time how to use a grain of gold sand, a little black point, a piece of quartz, a crack, to create the beauty of ingenuity; how to apply the natural beauty of Shoushan stone, to create the beauty of art; how to carry out the so-called 'inferior' Shoushan stone creation, is now the problem to be studied is also extremely It is also of great practical significance. In recent years, we have found that many young artists use a lot of stones that are not considered by others in their creations, and these unpromising stones have shown different artistic effects through their secondary creations. Their secondary creation of Shoushan stone corners is, first of all, a kind of reuse and resource-saving artistic behavior, which is a kind of protection and an appeal for the depleting Shoushan stone resources; secondly, for the corners or 'second-rate' stones, their own relatively low price advantage creates the possibility of its creation, which is not bound to It is not bound to the careful creation of 'good'stones, but has a huge space for artistic expression.

One is that the material is large, suitable for carving some large works, and then he can take it or leave it, so he can play freely. According to the distribution of color and then according to the artist's ideas to take the trick how to take, three is cheap, especially for the young generation of artists to create the cost of how much is very critical. Then there is the 'second stone', which gives a sense of simplicity and authenticity, a sense of authenticity that will bring out our innermost and most realistic feelings in a very straightforward way. In the process of carving, a lot of scratches,.chisel marks and saw marks of the carving knife are preserved on the stone. This way of creation is more expressive than the traditional way. In my opinion, the material of stone is defined by people and the market, they are all given by nature. Can the artistic value be proportional to the value of the material? [1] This is also worthy of our deep thinking.

Due to the driving factors of the market environment, the problem about the resource conservation of Shoushan stone needs to be solved urgently. In this study, we plan to reuse and recreate the "second-rate" Shoushan stone or the corners of Shoushan stone carved with Shoushan stone, to make the corners of Shoushan stone present unique artistic value again by using clever carving techniques, to create new artistic expressions, to break the stereotypes of consumers about Shoushan stone art, and to re-establish the art of Shoushan stone in other forms. We try to explore whether the material value is higher than the artistic creation value.

The production of Shoushan stone is documented in the 'Stone Observation Book': 'The stone has complexes, water marks, and sand partitions, so the stone is solved by first looking at its reasoning, and then measuring its complexes, so avoiding water marks and chiseling sand partitions to solve it [2].' This focus has influenced the Shoushan stone industry for more than 300 years, but in today's lack of Shoushan stone resources, we should not still follow the wind, we should think about it, get a stone or "chisel sand partition to solve it"? What is more important is how to apply the art to the material.

The stone has a mind, but it can not express, and as the highest living creature, it is we put on a certain dress, put on a certain action, add a certain expression, give a certain thought, at this time, you are talking with the stone, spiritual communication, maybe you have a heart, love, to convey what the stone wants to say; maybe you misunderstood the stone's original intention. We all try to imitate nature and express nature, but often it is counterproductive and destroys nature, so it is not real.

Shoushan stone is rich in color, fine texture, has a high aesthetic value, can not be painted with artificial colors when processing, but will destroy the natural beauty of Shoushan stone, as Han Fei said in "Han Fei Zi - solve the old": "and the jade, not decorated with five colors, Sui Hou's pearl, not decorated with silver and yellow, its quality to the beauty, things are not enough to decorate." Therefore, in the creation process of Shoushan stone carving, the natural color and luster are retained, and the beauty is expressed more naturally.

Eumie Mourne is an artistic masterpiece by French sculptor Auguste Rodin that fully reflects Rodin's thoughts and artistic characteristics. Rodin shaped this figure sitting on bended knees with a bowed head and dried and sagging breasts, seemingly immersed in the memory of past events, while mourning for the present aging and dying [4]. When some people saw it, they covered their faces and walked away because of the unpleasant sight. When Rodin learned of this, he explained his opinion about beauty and ugliness: "Ordinary people always think that whatever is considered ugly in reality is not material for art - they want to forbid us from expressing what offends them by making them feel displeased and disgusted in nature. This is their great mistake. What is generally considered 'ugly' in nature, can become very beautiful in art......" Rodin believes that in art, as long as it is true and has personality, it is beautiful; only what is false or impersonal is in art The ugliness of art. He believed that for the artist, 'everything is beautiful - because he is constantly walking within the light of the real [5].'

In the creation of Shoushan stone carving, most people think that 'top quality' can create good works, while 'second quality' cannot create good works, or affect the creation. In fact, if a good stone is good, if it is placed in the hands of a short-sighted 'craftsman' who only knows how to imitate, it will destroy the dual beauty of nature and art; the so-called 'ugly'stone, dressed up with the word skillful The so-called 'ugly'stone, dressed up with the word 'ingenuity' to give it ideas, to evoke dialogue between people's hearts, to create a good product, that is not more beautiful, more ingenious?

2   Research on the Creation of Substandard Stones

With the lack of mineral resources of Shoushan stone, in recent years I have used a lot of these stones that others do not see in my creations, and these stones that are not looked at, there are articles to do, I have the following experience of the use of Shoushan stone 'secondary stone'. First, the richness of color, texture is particularly 'secondary stone' a major feature, is impossible to imitate artificial. Life is colorful, art is diversified, using colorful Shoushan stone to express the natural things, is not more accurate, more in line with nature, more close to life, to create endless passion and imagination space. Second, the material is large. When creating a piece of Tianhuang stone, you will not dare to make a knife on the left and a knife on the right, and you will be limited to use thin art and high relief carving techniques when creating a good stone. "creation, you can play freely. Thirdly, there are many varieties and different shapes. When creating, there are endless subjects to carve, and unexpected effects, and you can find the best way to create according to different shapes, without receiving the influence of material price, and can better express the creator's intention [3]. Figure 1 shows the incense holder created by using 'second-rate stones'.

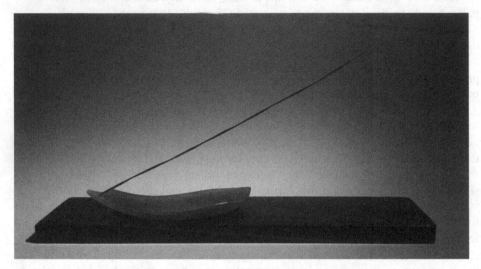

Fig. 1. Wire incense holder made of sub-stone

There is a type of Shoushan stone that people call 'sipingshi', but 'sipingshi' is not the name of the stone produced in the veins, but refers to the ancient Shoushan stone blocks and their carvings buried in the soil of the Guang Ying Yuan site outside Shoushan Village. The carvings, which were obtained by excavation, were quite rare. At that time, ancient monks collected Shoushan stones in large quantities and had a rich collection of stones, and had workshops and warehouses to supervise the production of tribute stones. It was destroyed by fire during the Northern Song and Jin dynasties (1125–1140), leaving a large amount of remnants of Shoushan stone mining application for about 250 years. After the temple was abolished, the stone was buried in the soil after fire moxibustion, and after nearly a thousand years of soil and moisture erosion, the fire and soil erosion of the temple ping stone showed a variety of unique percolation, ancient meaning, in the simple and introverted history of the paste embodies the jade Zhen gold sound of the old gas, its aura stone rhyme is not new stone comparable. Most of the existing Terrapin stones retain the broken marks left by the stone removal, and do not have a fixed form and texture, one stone, one product, which cannot be duplicated, but people who have seen Terrapin stones are enchanted by its appearance as ancient jade, so it is highly treasured. This is what the ancients gave us these signs of character, and then this type of stone has its own different artistic language, although the material value of such a stone is not very high but history gives it a higher human value. In the creation of 'sipingshi', it is mostly combined with Zen and Buddhist themes. Because it is excavated in the temple itself, and after the experience of fire in the process of expression can use the Buddha's residual beauty to composition, this theme is to use the history of culture little by little, most of the ancient people to retain the traces of this, and then add some of their own ideas, according to these cuts and chisel marks to carve, history and art is the interpretation of this stone is a more perfect summary. The subject matter of Fig. 2 is based on the cultural meaning of "sipingshi".

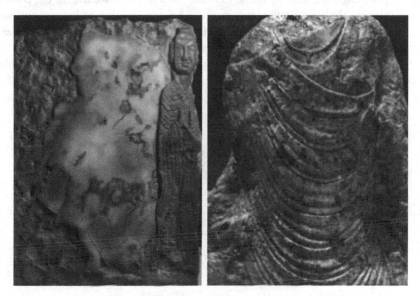

Fig. 2. Fragmented statue of Buddha created with "sipingshi

3   Exploration of the Peripheral Use of Substandard Stones

The secondary creation of Shoushan stone corners is, first of all, a kind of art behavior of reusing and saving resources, which is a kind of protection and an appeal to the depleting Shoushan stone resources; secondly, for the corners or "secondary products", its own advantage of relatively low price makes it possible to be created, and it is not bound to the careful creation of 'superior' stones, but its artistic expression can have a huge space to play, and create civilian artworks and daily necessities that are different from traditional Shoushan stone carvings. From the edge itself, it can create some works that are different from the mainstream form of traditional Shoushan stone carving, because of its smaller size, it can be made into earrings, pendants, keychains and other common decorative artworks, making the art of Shoushan stone no longer limited to the form of precious ornaments, and attracting more exposure and attention with its delicate form. From the aspect of artistic integration to creative research, the expression of Shoushan stone has many possibilities.

3.1   Combination of Stone with Lacquer Art and Painting

Shoushan stone and lacquerware are one of the three treasures of Fuzhou, and the fusion of the two is an innovation that few people have thought of and created. If the two can be skillfully combined, the art of Shoushan stone can be brought to its fullest. The combination of Shoushan stone and painting, with the natural stone pattern of Shoushan stone, and a little carving and modification, can create a flat painting with a body and texture. The following work is a combination of Shoushan stone and natural lacquer that I have done together with my friend to make a tea tray that we use for our daily tea drinking, inlaying the stone in the lacquer tire. The combination of lacquer and stone gives a sense of simplicity, and the combination of lacquer and stone gives a feeling of simplicity, and the use of different textures gives a sense of the vicissitudes. It can convey the sense of history falling off. For example, the tea plate made in Fig. 3 and the stone painting made in Fig. 4.

Fig. 3. A tea tray combining stone and lacquer art

Fig. 4. Stone painting

3.2 Stone and Wood

Through the years and the environment of the wood, more or less will have cracks produced, the common way to deal with that is to use the wood carved into a butterfly tenon embedded in the wood to prevent it from continuing to crack, Shoushan stone as a stone can be processed again, as a material to fix the cracks in the big board is also suitable, the combination of stone and wood, is also an innovative form. Use some natural some mechanism of rotten wood to design inlaid some Shoushan stone, after polishing to present a different feeling, the different colors of the stone and the unique visual effect of the collision of wood grain. For example, the table and the tea pot made in Fig. 5. These are all artworks created by using the combination of wood and stone, such works give people a sense of mood and antiquity (Fig. 6).

Fig. 5. Stone and wood collection of tables and tea pots

Fig. 6. A tea tray made of stone and wood

3.3 Stone and Life

With people's demand for quality of life, the clement of Shoushan stone is added to people's daily necessities, and the combination of material and material is used to create products that can be used in daily necessities. Shoushan stone can also be combined with metal to make jewelry, the hardness of Shoushan stone is easy to scratch and the combination of metal can play a protective role and can also have a very good visual effect such as earrings, rings, necklaces, pendants and other peripheral products. Such as Fig. 3, 4 and 5 is the creator using the edges of the production of necklaces pendants, etc., due to the unique material of Shoushan stone now has its own unique flavor in contrast with other jewelry. Figures 7 and 8 are created by the creators according to the daily life of some flat.

Fig. 7. Necklace and pendant made of Shoushan stone

Fig. 8. Vase made of Shoushan stone

This study also plans to open special Shoushan stone lectures and courses for the general consumer group, to popularize the knowledge of Shoushan stone as a national stone for more people, and through the form of creative workshops, through hand-crafted creation, more people can participate in the artistic creation of Shoushan stone, so that Shoushan stone culture becomes an affordable artwork for the general public. In the application of Shoushan stone 'inferior'to create cultural and creative peripheral, one is the economic reasons, it does not allow us to buy good stone to create; and another is that I studied art, personal aesthetic value orientation is different. Therefore, I like to create with 'good products', and I am more obsessed with creating spiritual works with 'inferior products'. For example, Fig. 9 shows tableware decorations made from ceramics in daily life.

Fig. 9. Tableware decoration made of Shoushan stone combined with ceramics.

3.4 "Sub-stone" Theme Creation

When my friend and I were creating the 'Wu Band When the Wind'series, we were inspired by the 'Eighty-seven Immortal Scrolls' painted by Wu Daozi, and we felt very shocked when we saw this painting. There is a kind of through the ancient and even the modern, very ancient taste, mystery, presentation is very strong, the performance of the figure I selectively carved ten characters as a representative, carved with the main character: the South Pole Heavenly Monarch, and each holding different offerings of jade maiden, the scene is grand, the characters have different forms, the background choose to connect with tiles, line carved with cloud patterns, and the front of the characters to form a unified, and later directly simplified, not carved character features, clothing pattern refinement, but have different manners to present. Instead, there are different manners presented, between seeming and unlikely. For example, Fig. 10.

Fig. 10. Ancient Chinese figures carved in Shoushan stone

4 Summary

In the era of aesthetic interests continue to improve and personal space of personal language is changing, in Shoushan stone carving most people are relying on the material, texture of the stone to create, try to ask whether the value of the material is higher than the value of artistic creation, or the value of artistic creation is higher than the value of the stone itself, this is a problem worthy of our consideration. The research study creators in the creation process to be combined with the consumer demand, the use of consumer demand such as a literary elegant appreciation, tea ware, incense ware, character ornaments, situational devices, to create different personalities for the user life aesthetic, highlighting the dominant thinking of people in the context, but also highlight the stone in the context of the thinking brought to people, but also to be able to attract the attention of consumers and like. The sense of interest and fun expressed in it is the re-expression of the sense of life, the understanding of the communication, the generalization of people, things and things in life [6], only through human interaction, contact, feeling each other's growth experience, joy, anger, sadness and happiness, through the communication of the heart of the natural creatures, feeling the ubiquity of life, and finally in the creation of skilled technical language, artistic language, to convey In the end, we will use our skillful technical language and artistic language to convey what we want to express deep in our hearts in order to create a work that can resonate with the audience.

References

1. Zhou, Y.: Analysis of factors influencing the price of Chinese artworks. Popular Lit. Art (13), 269–270 (2015)
2. Chen, Z.: Shoushan stone "thin meaning" carving aesthetic collect talk. Art 2, 81–85 (2013)
3. Liao, S.: A study of technicality and artistry in andrew wyeth's paintings. Hunan University of Science and Technology (2015)
4. Alastair, S.T.W.: Why should the thinker be called the mourner? (3), 47–54 (2019)
5. Sun, Q.: The Analysis of Literary Aesthetic Quality. Yunnan University Press (1994)
6. Manman: Study room literary and elegant appreciation of heavenly fate Wu belt when the wind Du sea in the poetic habitat. China New Era (6), 90–93 (2016)

Meet the Local Through Storytelling: A Design Framework for the Authenticity of Local Tourist Experience

Wenlin Zhang[✉]

College of Design and Innovation, Tongji University, 281 Fuxin Road, Shanghai, China
wenlindesign@hotmail.com

Abstract. Based on a rapid ethnographic investigation on a historical site in China, this study aims to formalize a storytelling design framework towards enhancing the authenticity of local tourist experience. Specifically, semi-structured interviews were conducted with residents, tourists, and tour guides. Participatory observation was adopted to understand storytelling behaviors. I analyzed the field data through affinity diagram in two rounds. The first round of analysis was guided by the theoretical lens of Staged Authenticity, bringing the emergence of the core category—the gap towards achieving the authenticity of local tourist experience. This gap is specified as six barriers: social exclusion; mental and physical cost; artificiality; cost-effectiveness; storytelling triggering; and local expertise diffusion. The second round of analysis was guided by the Storytelling Design Framework, resulting in an in-depth comparison of storytelling practices contributed by residents and tour guides. Based on these findings, I formulated the storytelling design framework for the authenticity of local tourist experience and prototyped an audio guide application to illustrate one potential use of the framework. This study enriches understandings of local tourist experience and provide groundings for design applications in tourism.

Keywords: Authenticity · Storytelling · Tourist experience · Local experience · Rapid Ethnography · Audio guide application

1 Introduction

The concept of authenticity—the sense of being true, real, and unique—has been widely discussed since its initial introduction to tourism research [1, 2]. Among these discussions, the actualization of authentic tourist experience drew considerable attention [3–5]. The perception of local experience, which refers to the experience of immersing in local culture and environments, is suggested to be an essential aspect for achieving authentic tourist experience. This is expressed by the stated desire for tourists to engage more intensively with the residents of tourist destinations [6], to involve more meaningful interactions with the local culture [7], and to gain more intimate sharing of personal tales than a standard telling of tourism facts [8].

© Springer Nature Switzerland AG 2021
P.-L. P. Rau (Ed.): HCII 2021, LNCS 12773, pp. 139–154, 2021.
https://doi.org/10.1007/978-3-030-77080-8_13

Storytelling involves the interpretation of folklore, legend, and social event. It is a potential medium to enhance the authenticity of local tourist experience. Several studies suggested the positive effect of storytelling on tourists' perception of locality, followed by an emerging strand of HCI studies to explore potential utilizations of digital storytelling in tourism and cultural heritage contexts [9–13]. For instance, Burova et al. [10] proposed a storytelling-based application to facilitate tourists' engagement with cultural exploration at the airport. They further evaluated that interactive storytelling is an effective method in persuading tourists to explore local products at the airport and thus promote tourists to explore local culture.

Despite fragmented solutions have been proposed to enhance local tourist experience, a fundamental framework to guide storytelling design towards the authenticity of local experience is scant. To address this issue, I conducted a field study to approach the first research objective—develop a contextualized understanding of culture-oriented tourism with a specific focus on residents' interpretation of locality and tourists' perception of authentic tourist experience. Based on these findings, a secondary objective is to formalize a design framework towards enhancing the authenticity of local tourist experience. This study adds to a contextualized understanding of culture-oriented tourism practices. The formulated design framework provides groundings for HCI researchers to develop solutions to better engage tourists in tourist destination and cultural heritage. The framework has implications for design practices toward enhancing the authenticity of local experience.

This paper is structured as follows. In Sect. 2, I frame the theoretical basis of Staged Authenticity and Storytelling Design Framework. Section 3 elaborates on the research methods and the process of data analysis. In Sect. 4, I report on the identified themes and barriers in current tourism practices. In Sect. 5, I formulate the design framework for the authenticity of local tourist experience, introduce its building blocks, and unfold rationales for deriving the framework. A potential utilization of the framework is demonstrated in Sect. 6 through an audio guide application. In Sect. 7, I situate this study in the field of HCI and tourism research to suggest implications, limitations, and avenues for future research.

2 Theoretical Framework

2.1 Authenticity of Local Tourist Experience

Authenticity is a widely discussed and debated concept in tourism research. Scholars discuss authenticity within typologies of objectivism, constructivism, and existentialism [4, 14–16], leaving a clear and unified definition controversial. The concept of authenticity I adopted in this study is from Sharpley [17], who defines authenticity in tourism context as traditional culture and its origin, in the sense of being true, real, and unique. Authenticity of local tourist experience has been widely discussed and empirically investigated since the concept of authenticity being introduced to tourism studies [1]. Evidence supports that the perception of local experience—the experience of immersing in local culture and environments—is an essential aspect for achieving authentic tourist experience. Specifically, researchers advocate that tourists are increasingly demanding authentic, experientially oriented opportunities involving more meaningful interactions with locals [3, 5, 18]. Arsenault and Trace [6] observed that tourists increasingly desire to engage more intensively with the residents. Paulauskaite et al. [7] explored the perceived authenticity of local experience on online accommodation rental platform Airbnb, investigating what influences customers to select Airbnb accommodations through qualitative interviews. The empirical findings indicated that interactions with hosts and with local culture were important to Airbnb users.

Tourists' anticipation for the authenticity of local experience can be theoretically sketched by Staged Authenticity—a well-acknowledged and disseminated theory to explain tourist attraction and tourist motivation [1, 2, 19]. MacCannell [1] introduced sociologist Erving Goffman's "dramaturgical perspective" to study tourism activities. According to Goffman [20], the social structure is assimilated to a large stage with the division of "The Front" and the "The Back". "The Front" refers to the place where performing actors, guests, and service staffs interact with each other. The actions for strangers or casual acquaintances are called frontstage behavior. "The Back" is the closed place for the performance of the frontstage, it can be described as a place with a close connection with truth, intimacy, and authenticity. MacCannell [1] suggested that in tourism development, local hosts present their culture (including themselves) as a commodity to tourists. This leads to the "staging" of the authenticity of the host's social life, which is a common phenomenon in modern tourism experiences. Deepak [21] further elaborated the positive effect of Stage Authenticity on ethno-tourism, that is, ethno-tourism with Stage Authenticity as its core will, to some extent, strengthen the authenticity of local culture and contribute to the reconstruction of traditional culture and ethnologic identity.

The present study adopts Staged Authenticity as a theoretical perspective to guide our empirical investigation for its revealed explanatory power on our research scope—HCI practices for enhancing local culture-oriented tourist experience. I take the example of Airbnb to illustrate the relevance between Staged Authenticity and HCI practices. The partial success of Airbnb owed to its strategy to enable interactions between tourist and house-owner towards co-creating culturally immersive tourism experience [7]. HCI practices involved in Airbnb, like mobile applications and online service platforms, play a key role in mediating Staged Authenticity between house-owner and accommodating tourists. For example, a house-owner may commercialize the local culture as room decorations and local stories being upload to the Airbnb platform. The tourists may order on the platform and share the uploaded stories, making the digital platform a mediator to satisfy Staged Authenticity.

2.2 Storytelling

Stories can be of many forms such as folklore, legend, and social event. Storytelling as a means of social interaction is a way to create and sustain human connections and relationships [22]. Scholars have brought the concept of storytelling to the discussion of tourist experience, and the positive effect of storytelling for enhancing touristic experience and local culture exploration has been implied. HCI research, adopting digital storytelling techniques on mobile devices to guide visitors to a historical site enhanced the identity of the destination through affecting tourists emotionally and increasing the receptiveness and precision of conveyed information [9]. Burova et al. [10] conducted a field evaluation of an airport application to demonstrate that interactive storytelling is an effective method in persuading travelers to explore local products at the airport and thus promote tourists' exploration of local culture.

Digital Storytelling revolves around the idea of combining storytelling with a variety of digital multimedia, such as images, audio, and video [23]. Many forms of digital storytelling were initiated and tested on digital devices, ranging from thc development of audio-based [24], photo-based [25], location-based [26], to analog-based [27] storytelling prototypes.

Researchers have also introduced digital storytelling to tourist destination and cultural heritage contexts. For instance, Chowdhury [11] presented Pintail—a mobile companion application for guided storytelling in travel contexts. It prompts the user to create, remix, and reflect on their own travel stories. Specifically, Pintail synthesizes contextual story prompts from online travel reviews and doodle-books to inspire story creation. These prompts can guide users to create atomic travel stories. Created stories are then printed on paper and become analog story-artifacts or conversation-starters. Lombardo and Damiano [9] presented and evaluated a storytelling-based application on mobile devices for an anthropomorphic guide to a historical site. They addressed the storytelling paradigm in a mobile context in the way of third-person "performed narration" of a single character in an augmented reality scenario.

However, several limitations of digital storytelling solutions in tourism might be sketched. First, existing solutions largely engage machine-generated stories, leaving human touch and personality largely uncharted attributes. Second, existing narratives tend to strip context-dependency, leaving a culturally immersive tourist experience uneasy to be satisfied. Third, existing applications largely require a lot of time and efforts to create the narrative, which is contradicted to tourists' anticipation for non-invasive, easy to access interactions in tourism contexts. The transmissibility and replicability of the stories are limited as well.

To examine how digital storytelling could better enhance tourist experience, I seek a guiding framework to understand the form, media, and structure of storytelling. This is approached by adopting the coding scheme of Storytelling Design Framework [22] as a guiding scheme of coding. The authors thematically coded the interview data on older adults to identify seven considering factors in the design of systems using storytelling as a motivational framework [22]. These factors are illustrated in Table 1.

Table 1. Storytelling design framework proposed by Chu et al. [22].

Factor	Definition
Audience	The explicit or implicit audience for the act of storytelling
Content	Storytelling requires that content is produced, whether it is based on a true story or it is a fictional one
Process	The process of telling a story involves a myriad of processes from the retrieval of details, structuring, summarization, mental organization and tracking, that are typically done automatically or unconsciously
Context	The social and physical environment for the storytelling process
Medium	The medium through which storytelling happens. Media used (e.g., phone, email) are solely for communication and storytelling
Trigger	This aspect relates to how storytelling is catalyzed
Intention	The purpose behind storytelling

3   Methods

3.1   Data Collection

I adopted qualitative methods to examine the perceived authenticity of local tourist experience by exploring the meaning and interpretations of individuals in real-world contexts [28]. Rapid Ethnography is a collection of ethnographic data elicitation methods intended to provide a reasonable understanding of users and their activities given significant time pressures and limited time in the field [29]. I employ this strategy to understand context-dependent tourist experiences and develop storytelling design prototypes for its time-efficiency and gained popularity in HCI research.

The empirical investigation was situated in Confucian Temple, a historical tourist attraction located in Nanjing, China. In December 2019, I conducted a one-day participant observation on-site, specifically focusing on real-world tour guide application usage behavior of varied users. I collected pictorial and documentary data on existing audio guide applications on-site and online. After that, I recruited 10 participants for interviews. These participants are referred to as P1 to P10, numbered by their date being interviewed. I conducted semi-structured interviews with 4 local residents (P1, P2, P3, P4) on-site, 4 tourists with enriched interest and experience in local tourist exploration online (P5, P6, P7, P8), and 2 full-time or part-time tour guides online (P9, P10). The duration of interviews ranged from 30–60 min. Virtually all interviews were audio-recorded with obtained informed consent, except for one local resident declined to be taped. In the latter case, the interviewer took lengthy written notes on-site. Interview questions are roughly divided into two parts. The first part aims to elicit interviewees' perception and interpretation of local events, stories, and tourist experiences. The second part deals with storytelling, aiming to understand the form, structure, and content of digital storytelling in tourism practices.

3.2   Data Analysis

I followed a two-step data analysis process to extract data on (1) perceptions and interpretations of authenticity towards local tourist experience and (2) storytelling practices in tourism context. In the first round of analysis, data were examined and extracted through the theoretical lens of Staged Authenticity [30], resulting in enriched quotes and narratives expressing the interpretation of authentic and local experience. In the second round of analysis, data were thematically coded based on Storytelling Design Framework [22]. The framework guided the researcher to elicit data in mapping storytelling practices. Affinity diagram—a tool that gathers large amounts of language data (ideas, opinions, issues) and organizes them into groupings based on their natural relationships—went hand in hand with the two-step data analysis process [31].

4   Results

As a result, three higher-order themes emerged from the first round of coding. We bound up with three essential terms adopted by the theory of Staged Authenticity—frontstage, backstage, and staged authenticity—to interpret the meaning of each identified theme (Fig. 1). Besides, a comparison between storytelling practices contributed by local residents and by tour guides were deduced from the second round of coding. The result is shown in Table 2.

Fig. 1. Themes emerged from the first round of analysis.

4.1 Frontstage: Barriers for Local Engagement

"The Front" refers to the place where actors, guests, and service staff interact with each other [20]. Pertaining to tourists' perception and interpretation of the frontstage in tourism, both locals and tourists cited the damage of commercialization to the authenticity of local culture-based tourist experience. Their perception of the damage can be described as the uniqueness of tourist experience and the fake local feature. To illustrate, P5 mentioned that "[I went to Slovenia] because so many domestic attractions have been commercialized. In fact, what you see is not the same as the original one anymore, even the locals themselves won't go there." P2 felt that "nowadays tourism attractions are almost the same, there are basically no cultural features, or to say the commercial [atmosphere] is too strong".

Despite existing tourist attractions did not fully support tourists' perception of locality, almost all interviewed tourists expressed their interest to explore the backstage of local attractions, labeled as "seeking local engagement" in present study. Seeking local engagement refers to the tourist is motivated to immerse, feel, and interact with local people, objects, events, and environments. As P5 put it, "I don't really enjoy planned tours. Sometimes, I actively search for local destinations on [a social media] to see where locals prefer to go." As is reported, P5 anticipated to experience local life, she thus went to a local restaurant (environment), having dinner inside the restaurant (events), and observed local people while gossiping (people). She felt she "indirectly involved in [local] activities (P5)", which is deemed as essential for her to "experience the organic life atmosphere that cannot be felt anywhere else (P5)."

However, several barriers were identified for tourists to seek local engagement. These barriers were categorized as F1. social exclusion, F2. mental and physical cost, and F3. artificiality. As P6 indicated, she "felt embarrassed to talk with local people" and "don't know where to start and end." She also felt "inappropriate and being excluded by local environments (P6)." The description expressed the feeling of social exclusion in local environments. P5 and P8 described their endeavor to search for, ask for, and go to places where local people prefer. They swift between multiple platforms to compare different places to plan the tour, indicating an increased mental and physical cost compared with a packaged tour. P10 exemplified the barriers to local engagement through audio guide applications in many tourist attractions. She put that "audio guide is a good way for strangers to experience the history of local attractions, but its narrative is very objective and it is not touching for me. It's more like a machine talking with me and made me hard to feels like the original." This description indicated the negative effect of the sense of artificiality on tourists' perception of authentic local experience.

To sum up, the interviewees perceived the damage of authenticity in local tourist attractions and expressed their interests and efforts to explore local engagement. However, the seek for local engagement, to a broad audience, confronted barriers of social exclusion, mental and physical cost, and artificiality.

4.2 Backstage: Barriers for Enabling Local Expertise

"The Back" is the closed place for the performance of the frontstage. It can be seen as a place closely connected with truth, intimacy, and authenticity [20]. As the researcher probed deeper into the backstage of local life, the concept of local expertise, defined as the interpretation of local history, events, and tourist attractions by local residents, emerged from the analysis. Local expertise is generally delivered and diffused through storytelling, mostly in a face-to-face manner. The interviewed residents explicitly or implicitly expressed their local expertise, including traditional lifestyle and local memories that are contextualized and being interpreted subjectively. For example, P1 told us his story about the spatial arrangement of a tourist attraction and shared his childhood memory of the attraction. Local expertise differs from the expertise provided by tour guides in that the narrative of local expertise generally involved the reflection of personal experience, making the information united with human touch and cultural context. Expertise provided by tour guides, however, involved more objective, professional, and accurate knowledge on tourist attractions. P9, equipped with part-time tour guide experience, indicated that "tour guides need to be authorized by obtaining the tour guide certificate. [...] It examines your professional knowledge about attractions, such as its culture, history, etc., and how to narrate your speech." An in-depth analysis between local expertise and tour guide expertise is presented in Sect. 4.4.

Despite all tourist participants expressed their interest in local lifestyle and culture, the value of local expertise, compared with tour guide expertise, has less been explored. P5 expressed her strong interest in the habits and culture of the locals and her desire to explore and experience these contents. She said "[The local house-owner] often talk to you about, for example, what festival is on Sunday and what event will be held at the center of the town. On Sunday, he said he would try to make a local specialty dish. When we were cooking, I went around [the chief] to observe what he was doing." P8 reported that "I prefer to search on [a social media] to look for restaurants and attractions local people recommended, because I feel that posts written with personal experience are more convincing. [...] When I went to eat, I intentionally or unintentionally noticed them [local people] talking about their life stories, that [the experience of eating] was quite unusual, as if I was diving into their daily routine secretly."

Some participants exemplified current solutions for delivering local expertise to tourists. P6 indicated that "The youth hostel owner is quite conversable, you will have a lot of opportunities to communicate with your buddies when you stay there. [...] He [the house-owner] will list some local features for you to find out. [...] I think this type of communication made me feels like immersing in local life." She also indicated that a customized tour equipped with a local tour guide is a good chance to communicate with the local. This type of tourism supports tourists to select to join a small-scale travel group, guided by a local with enriched knowledge about the city he lives in.

However, barriers to enable local expertise were also identified. The first barrier is B1. the cost-effectiveness of diffusing local expertise. P6 experienced customized tour to comment that "the reciprocal interaction between tour guides and tourists takes too much human cost and thus limited to be disseminated to a broader audience." P9, equipped with tour guide experience, described that she "often tell a similar story to a small group of audience for multiple times, which makes me feel a little bit exhausted."

The second barrier points to B2. the triggering of local storytelling. Interviews with residents revealed that the initiation of storytelling might be challenging. Most of the interviewed locals performed a rather conservative attitude in telling their personal experiences. Only after a long time of contextualization did they begin to narrate their personal stories. P1 said that he "has nothing to tell about local living" for his concern about the sensitive nature of the information he is delivering. Approximately 10 min later, he had been channeled into the interview scenario to share his story abundantly, ranging from the physical structure of the street, the interaction routine in neighbors, to his childhood memory on tourist attractions.

The third barrier concerns B3. the diffusion of local expertise. Existing solutions largely involve dyadic communication between the local and the tourist, meaning that the diffusion of local expertise was confined to a small-scale audience. Also, the delivery of local knowledge is often unrecorded, leaving the replay and share of the story impossible. P10 commented that "a digital platform helps a lot for the collection and sharing of stories".

In summary, although local expertise was highly anticipated by tourists, existing solutions have rarely delivered the resource of local expertise to a broad audience compared with the resource of audio guide expertise. Barriers for enabling local expertise were identified as the cost-effectiveness, the triggering of local storytelling, and the diffusion of local expertise.

4.3 Staged Authenticity: The Gap for Mediating Front and Backstage

In Sect. 4.1 and 4.2, I introduced two key conceptualizations emerged from data analysis—local engagement and local expertise. I also examined barriers for seeking local engagement and for enabling local expertise under the umbrella term frontstage and backstage, respectively. These barriers formulated the gap for mediating the frontstage and backstage towards the authenticity of local tourist experience.

Certain tourism solutions confronted some specific barriers, leaving other barriers underexplored. Casual acquaintance and conversation with local residents, according to

P6, confronted the barrier of B2. storytelling triggering and B3. local expertise diffusion, leaving some crucial aspects like F1. social exclusion and B1. cost-effectiveness unexplored. Existing audio guide applications in tourist attractions, according to P10, dealt with barriers like F1. social exclusion, F2. mental and physical cost, and B1. cost-effectiveness. However, B3. local expertise diffusion and F3. artificiality was less concerned. Customized tours, according to P9, dealt with the barrier of F1. social exclusion, B2. triggering storytelling, and B3. diffusing local expertise, but the B1. cost-effectiveness for tour guides is relatively low and the F2. mental and physical cost for tourists is relatively high.

4.4 Comparing Storytelling Practices Between Local Residents and Tour Guides

In this section, a comparison between storytelling practices contributed by residents and tour guides is formulated. The original Storytelling Design Framework [22] dedicated to the specific narrator of older adults deviates from our purpose to understand and compare various narrators. Hence, the current framework further incorporated the division of Narrator to analyze storytelling factors in more detail. The building blocks of the framework are illustrated in Table 2.

Table 2. A comparative analysis between storytelling practices contributed by local residents and tour guides

Narrator	Local residents	Tour guides
Audience	Family members; neighbors; remote friends	Tourist
Content	Historical events of the local place; personal experiences; political and social interpretations; family routines	Historical events of the local place; tourist instructions; consumption inducement
Process	Initiate by causal conversations or prompts; develop casually; end up casually	Tourist plan introduction; guide instructions; tourist attraction introduction; epilogue
Context	Daily routine; meeting with family and friends	Planned trip; package tour; customized tour
Medium	Face-to-face conversation	Face to face conversation; digital applications
Trigger	Nostalgic narratives	Reward; professional accomplishment
Intention	Maintaining human relationships; conveying lessons or messages; causal sharing	Job duty; the joy of sharing

5   A Storytelling Design Framework for the Authenticity of Local Tourist Experience

In Sect. 4, I identified the gap and barriers toward achieving Staged Authenticity in tourism contexts. The actualization of Staged Authenticity, in our specific context, depends on resolving the barriers for seeking local engagement and enabling local expertise. To bridge this gap, I formulated the storytelling design framework for the authenticity of local tourist experience (Fig. 2). The core idea is to adopt digital storytelling of local expertise as a virtual thread to link resident and tourist, labeled as "Meet the Local through Storytelling". The framework is deduced from the gap and barriers summarized above and built on the in-depth comparison between local storytelling and tour guide storytelling. The building blocks of the framework and the rationales for deriving them are illustrated as follows.

The framework consists of four essential modules: Narrator, Audience, Digital Storytelling, and Reward Mechanism.

- Narrator points to local residents equipped with local expertise and have the interest to share their personal experiences.
- Audience points to tourists with an interest in local engagement, which is represented as interactions with local people, objects, events, and environments.
- Digital Storytelling combines the human behavior of telling stories with a variety of digital multimedia, such as images, audio, and video. It is comprised of three key elements: Trigger, Mediator, and Interface. Trigger refers to the digital prompts for local residents to reflect and create their personal stories about local living. Mediator points to a tangible or digital platform for local residents to upload, adjust, and share their uploaded stories. An interface is also required for tourists to select and play stories uploaded by local residents.
- The role of the Reward Mechanism is to deliver the rewards provided by the audience to local residents who uploaded their stories. The rewards are in the form of money or virtual currency. As I pinpoint storytelling, specifying this module is beyond the scope of this study

I elaborate rationales on how the framework confront the identified barriers to achieve local experience.

- Pertaining to F1. social exclusion, the framework virtually connects local residents and tourists in a non-invasive manner, relieving tourists' embarrassment of being excluded by local environment.
- Pertaining to F2. mental and physical cost, our digital solution can be integrated into convenient devices like mobile phones, making the access of local engagement anytime, anywhere.
- Pertaining to F3. artificiality, the real, personal stories delivered through human voice contrast to the universal, machine tone delivered by existing audio guides.
- Pertaining to B1. cost-effectiveness, a local resident can deliver his story to a much broader audience digitally, relieving the physical effort to tell stories repeatedly.

- Pertaining to B2. storytelling triggering, digital prompts are suggested for local residents to reflect and create their stories, eliminating the embarrassment and worries for the invitation of the talk on-site.
- Pertaining to B3. local expertise diffusion, the proposed framework provides a digital solution to create, store, disseminate, and reward the uploaded stories, stepping towards a closed loop for local expertise diffusion.

6 A Storytelling-Based Audio Guide Application

An audio guide application interface was prototyped to demonstrate one potential use of the framework (Fig. 3). Residents can record and upload their own interpretations of the historical site through an intuitive, effortless interactive process. Travelers can select and play stories uploaded by residents. Narrating diversified local stories in residents' voice is suggested to enhance tourists' perception of local experience. The on-demand service offers payment for the resident once their story is played, and a local tourist experience can be better enhanced. There are three major use scenarios of the audio guide application (Fig. 4).

Upload and Download. Residents upload their own stories about local landmarks on the app. Tourists download the app and reserve a local tour.

Storytelling. When viewing landmarks, tourists can listen to the stories uploaded by locals through the application. In this way, landmarks can be deeply experienced and the interaction between the local and the tourist can be better enhanced non-invasively. Tourists can discard a specific narrative at any moment and jump to another one they prefer to explore content non-linearly.

Recall Anytime, Anywhere. After traveling, the app automatically generates video clips according to the pictures taken and the stories the tourist listened to.

7 Discussion

This study presented barriers for enhancing tourist's perception of locality and formalized a comparative analysis of storytelling practices between residents and tour guides. The idea of adopting digital storytelling as a mediator between residents and tourists was formulated as a prescriptive design framework, following an audio guide application prototype to illustrate one potential use of the framework. In this section, I situate the findings in HCI research, tourism research, and design practices. I further suggest limitations of this study and avenues for future research.

This study contributes to the field of HCI research in several aspects. First, the formulated storytelling design framework for the authenticity of local experience provides foundation for developing products, applications, and services in culture-oriented tourism contexts. The formulated framework extends to the guidelines proposed by Burova et al. [10] for a storytelling application for touristic public spaces. As the present study focused on the specific context of cultural exploration, our findings add to the

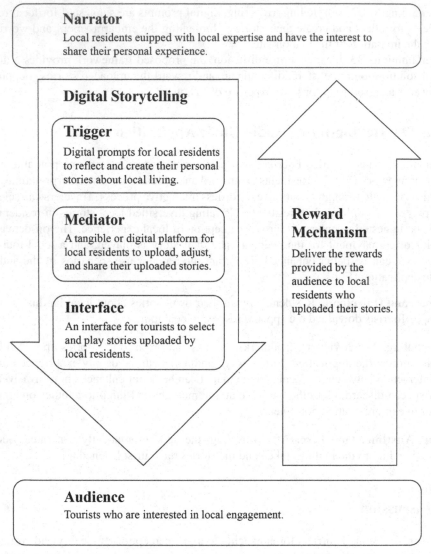

Narrator
Local residents euipped with local expertise and have the interest to share their personal experience.

Digital Storytelling

Trigger
Digital prompts for local residents to reflect and create their personal stories about local living.

Mediator
A tangible or digital platform for local residents to upload, adjust, and share their uploaded stories.

Interface
An interface for tourists to select and play stories uploaded by local residents.

Reward Mechanism
Deliver the rewards provided by the audience to local residents who uploaded their stories.

Audience
Tourists who are interested in local engagement.

Fig. 2. The storytelling design framework for the authenticity of local tourist experience.

proposed guideline of "facilitating cultural exploration" [10] through specifying a storytelling design framework to facilitate cultural exploration. Second, the identified barriers for achieving the authenticity of local experience provided an enriched and contextualized understanding of local culture-oriented tourism practices. This understanding provides empirical bases for HCI researchers to develop and refine existing solutions [9, 32] towards engaging tourists in a tourist destination and cultural heritage. Third, the in-depth comparison between storytelling practices presented by local residents and tour guides adds to existing discussions on digital storytelling solutions in the tourism sector.

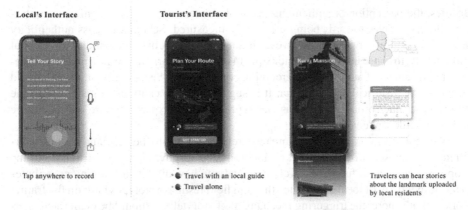

Fig. 3. Main interfaces of the audio guide application

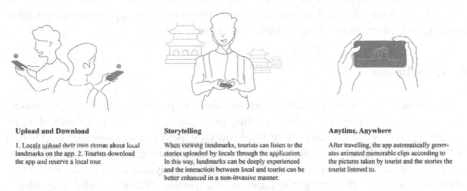

Upload and Download

1. Locals upload their own stories about local landmarks on the app. 2. Tourists download the app and reserve a local tour.

Storytelling

When viewing landmarks, tourists can listen to the stories uploaded by locals through the application. In this way, landmarks can be deeply experienced and the interaction between local and tourist can be better enhanced in a non-invasive manner.

Anytime, Anywhere

After travelling, the app automatically generates animated memorable clips according to the pictures taken by tourist and the stories the tourist listened to.

Fig. 4. Main use scenarios of the audio guide application

By suggesting local residents as potential narrators, rather than tour guide professionals, new pathways for digital storytelling might be unfolded.

Pertaining to tourism studies, the reported narratives and reflections of the interviewees are consistent with Paulauskaite et al. [7] who found that interactions with local culture contributed to user's perception of authentic experience while using Airbnb. The identified gap for achieving the authenticity of local experience in real context extends current understandings of authenticity in tourism practices.

For design practitioners, the audio guide application has practical value for commercialization in tourist attractions. The function of delivering local residents' stories to tourists through digital platforms can also be integrated into existing tourism solutions, such as Airbnb, to enhance tourists' perception of locality.

This study contends with several limitations. First, despite efforts to derive the framework with theoretical groundings and rationales, the evaluation of the framework has not been discussed. This is because the pivotal body of this paper concerns enriching understandings of local culture-oriented tourism to inform practitioners through proposing a design framework. Evaluation of the framework is out of the scope of the present study.

Besides, the perception of authenticity, central concept of this study, confronts controversial perspectives towards being effectively measured. Scholars discuss authenticity in tourism from multiple perspectives, ranging from objective authenticity, existential authenticity, to constructive authenticity [4, 14–16]. The blurry definition made the effective measurement of authenticity perception controversial. Second, this study was based on a specific historical site in China. It is suggested to be cautious in generalizing the findings to a much broader context as tourist practices and experiences are largely culture and context-dependent.

I suggest several avenues for future research. Pertaining to the field study, more diversified and representative cases can be incorporated for investigation. Pertaining to the proposed framework, future research may evaluate how digital storytelling contributes to the authenticity of local experience through field testing prototypes built on the framework. What's more, the triggering mechanism of storytelling should be elaborated. This is because our analysis found that local residents may be conservative in talking with others. Chu et al. [22] also indicated that older adults may have too little to talk about or too much to tell. Hence, it is crucial to develop and evaluate appropriate storytelling prompts to generate stories with suitable content and length.

8 Conclusion

In this paper, I presented a rapid ethnographic study to provide a contextualized understanding of local experience and storytelling practices in tourism. As a result, the gap towards achieving the authenticity of local tourist experience was identified, specified as barriers for seeking local engagement and for enabling local expertise. Six detailed barriers were identified as social exclusion; mental and physical cost; artificiality; cost-effectiveness; storytelling triggering; and local expertise diffusion. I also presented a comparison between local storytelling and audio guide storytelling. Based on the findings, the storytelling design framework for the authenticity of local tourist experience was formulated and illustrated by an audio guide application prototype. This study adds to a contextualized understanding of culture-oriented tourism practices. The formulated design framework provides the foundation for HCI researchers and practitioners to develop solutions to better engage tourists in tourist destination and cultural heritage.

References

1. MacCannell, D.: Staged authenticity: arrangements of social space in tourist settings. Am. J. Sociol. **79**, 589–603 (1973). https://doi.org/10.1086/225585
2. MacCannell, D.: The Tourist: A New Theory of the Leisure Class. University of California Press, California (1976)
3. Pine, J., Gilmore, J.: The Experience Economy. Harvard Business Review Press, Boston (2011)
4. Wang, N.: Rethinking authenticity in tourism experience. Ann. Tourism Res. **26**(2), 349–370 (1999). https://doi.org/10.1016/S0160-7383(98)00103-0
5. Tussyadiah, I., Pesonen, J.: Impacts of peer-to-peer accommodation use on travel patterns. J. Travel Res. **55**(8), 1022–1040 (2016). https://doi.org/10.1177/0047287515608505

6. Arsenault, N., Trace, G.: Defining tomorrow's tourism product: Packaging experiences. Canadian Tourism Commission, Vancouver (2004). https://doi.org/10.13140/RG.2.1.3264. 2406

7. Paulauskaite, D., Powell, R., Coca-Stefaniak, J., Morrison, A.: Living like a local: authentic tourism experiences and the sharing economy. Int. J. Tourism Res. **19**(6), 619–628 (2017). https://doi.org/10.1002/jtr.2134

8. Bryon, J.: Tour guides as storytellers—from selling to sharing. Scand. J. Hospitality Tourism **12**(1), 27–43 (2012). https://doi.org/10.1080/15022250.2012.656922

9. Lombardo, V., Damiano, R.: Storytelling on mobile devices for cultural heritage. New Rev. Hypermedia Multimedia **18**(1–2), 11–35 (2012). https://doi.org/10.1080/13614568.2012. 617846

10. Burova, A., et al.: Promoting local culture and enriching airport experiences through interactive storytelling. In: Proceedings of the 18th International Conference on Mobile and Ubiquitous Multimedia, MUM 2019, pp. 1–7. Association for Computing Machinery, New York (2019). https://doi.org/10.1145/3365610.3365640

11. Chowdhury, S.: Pintail: A travel companion for guided storytelling. Thesis. Massachusetts Institute of Technology, Cambridge, MA (2018)

12. Weiler, B., Ham, S.: Tour guides and interpretation. In: Weaver, D. (ed.) The Encyclopedia of Ecotourism, pp. 549–563. CABI Publishing, Wallingford, Oxfordshire (2001)

13. Bærenholdt, J., Haldrup, M., Urry, J.: Performing Tourist Places. Routledge Press, London (2017)

14. Kirillova, K., Lehto, X., Cai, L.: Existential authenticity and anxiety as outcomes: the tourist in the experience economy: existential authenticity and anxiety as outcomes. Int. J. Tourism Res. **19**(1), 13–26 (2017). https://doi.org/10.1002/jtr.2080

15. Kolar, T., Zabkar, V.: A consumer-based model of authenticity: An oxymoron or the foundation of cultural heritage marketing? Tourism Manage. **31**(5), 652–664 (2010). https://doi.org/10. 1016/j.tourman.2009.07.010

16. Steiner, C., Reisinger, Y.: Understanding existential authenticity. Ann. Tourism Res. **33**(2), 299–318 (2006). https://doi.org/10.1016/j.annals.2005.08.002

17. Sharpley, R.: Tourism, Tourists and Society. ELM Publications, Huntingdon, UK (1994)

18. Grayson, K., Martinec, R.: Consumer perception of iconicity and indexicality and their influence on assessments of authentic market offerings. J. Consum. Res. **31**, 296–311 (2004)

19. Taylor, J.: Authenticity and sincerity in tourism. Ann. Tourism Res. **28**(1), 7–26 (2001). https://doi.org/10.1016/S0160-7383(00)00004-9

20. Goffman, E.: The Presentation of Self in Everyday Life. Doubleday Publishing, New York (1959)

21. Deepak, C., Robert, H., Erill, S.: Staged authenticity and heritage tourism. Ann. Tourism Res. **30**(3), 702–719 (2003). https://doi.org/10.1016/s0160-7383(03)00044-6

22. Chu, S., Garcia, B., Quance, T., Geraci, L., Woltering, S., Quek, F.: Understanding storytelling as a design framework for cognitive support technologies for older adults. In: Proceedings of the International Symposium on Interactive Technology and Ageing Populations, ITAP 2016, pp. 24–33. Association for Computing Machinery, New York (2016). https://doi.org/ 10.1145/2996267.2996270

23. Robin, B.: The educational uses of digital storytelling. https://digitalliteracyintheclassroom. pbworks.com/f/Educ-Uses-DS.pdf. Accessed 11 Feb 2021

24. Chittaro, L., Zuliani, F.: Exploring audio storytelling in mobile exergames to affect the perception of physical exercise. In: Proceedings of the 7th International Conference on Pervasive Computing Technologies for Healthcare, PervasiveHealth 2013, pp. 1–8. ICST (Institute for Computer Sciences, Social-Informatics and Telecommunications Engineering), Brussels, BEL (2013). https://doi.org/10.4108/icst.pervasivehealth.2013.252016

25. Balabanović, M., Chu, L., Wolff, G.: Storytelling with digital photographs. In: Proceedings of the SIGCHI Conference on Human Factors in Computing Systems, CHI 2000, pp. 564–571. Association for Computing Machinery, New York (2000). https://doi.org/10.1145/332040.332505

26. Bentley, F., Basapur, S.: StoryPlace.Me: the path from studying elder communication to a public location-based video service. In: CHI '12 Extended Abstracts on Human Factors in Computing Systems, CHI EA 2012, pp. 777–792. Association for Computing Machinery, New York (2012). https://doi.org/10.1145/2212776.2212851

27. Brown, M.: Little Printer. https://medium.com/a-chair-in-a-room/little-printer-a-portrait-in-the-nude-4a5659ea731. Accessed 1 May 2021

28. Sandelowski, M.: Using qualitative research. Qual. Health Res. **14**(10), 1366–1386 (2016). https://doi.org/10.1177/1049732304269672

29. Millen, D.: Rapid ethnography: time deepening strategies for HCI field research. In: Proceedings of the 3rd Conference on Designing Interactive Systems: Processes, Practices, Methods, and Techniques, pp. 280–286. ACM, New York (2000). https://doi.org/10.1145/347642.347763

30. Creswell, J.: Research Design: Qualitative, Quantitative, and Mixed Methods Approaches, 4th edn. Sage Publications, Thousand Oaks, California (2014)

31. Martin, B., Hanington, B.: Affinity Diagramming, Universal Methods of Design: 100 Ways to Research Complex Problems, Develop Innovative Ideas, and Design Effective Solutions. Rockport Publishers, Beverly (2012)

32. Ardito, C., Buono, P., Costabile, M., Lanzilotti, R.: Enabling interactive exploration of cultural heritage: an experience of designing systems for mobile devices. Knowl. Technol. Policy **22**, 79–86 (2009). https://doi.org/10.1007/s12130-009-9079-7

The Living Inheritance and Protection of Intangible Cultural Heritage Lingnan Tide Embroidery in the Context of New Media

Shujun Zheng(✉)

Fuzhou University of International Studies and Trade, Hangzhou, People's Republic of China

Abstract. In the context of micro-communication, the communication of intangible cultural heritage is faced with the dilemma of lack of cultural characteristics, low utilization of resources, excessive commercialization and so on. At the same time, its "voice" is relatively weak, or "sound" will "sink" in the ocean of information. 5G era is coming, the way of communication based on WeChat, Weibo and other media will also produce subversive changes, intangible cultural heritage needs to seize the opportunity. From the perspective of new media, this paper analyzes the realistic predicament and internal reasons of the inheritance of intangible cultural heritage in the new era, and thinks that the communication strategy and development path of non-posthumous culture can be explored by using new media technology based on the thinking of "Internet +" to realize the innovation and demonstration promotion of non-posthumous culture inheritance mode; On the basis of "big data", the author constructs the communication chain of intangible cultural heritage, the mechanism of national folk communication and the diversified communication system. Using new media technology can greatly stimulate the endogenous motive force of the protection of non-heritage cultural tradition, enhance the influence of intangible cultural heritage, effectively promote the development and utilization of non-heritage culture, further expand the survival and cultural space of non-heritage, and enhance the soft power of Chinese culture.

Keywords: New media · 5G · Lingnan tide embroidery · Non-posthumous culture dissemination · Alive state inheritance and protection

1 Introduction

On October 15,2014, General Secretary Xi Jinping's important speech at the forum on literary and artistic work profoundly expounded the important role of Chinese culture in the rejuvenation of the Chinese nation. In his speech, he pointed out: "without the prosperity of Chinese culture, there will be no great rejuvenation of the Chinese nation." China is a country with profound and extensive cultural resources. Chinese culture is a cultural garden composed of 56 national cultures. [1] the external dissemination, internal inheritance and further development of intangible cultural heritage, from a macro point

© Springer Nature Switzerland AG 2021
P.-L. P. Rau (Ed.): HCII 2021, LNCS 12773, pp. 155–167, 2021.
https://doi.org/10.1007/978-3-030-77080-8_14

of view, it not only highlights the richness and diversity of our national culture, but also the practical need to enhance our cultural soft power in the international community.

With the iteration and transformation of new media technology, Cell phones and PAD as carriers of the new media transmission has also undergone earth-shaking changes, Micro-communication based on WeChat, Weibo and other new media platforms has become an important medium for the dissemination of intangible cultural heritage. In order to explore the research progress and achievements of the present stage in the dissemination of intangible cultural heritage by micro-communication, During the advanced retrieval of CNKI database, the author uses "WeChat" and "non-posthumous", "Weibo" and "non-posthumous" and "self-media" as the title of the article, A total of 11 papers were retrieved in all journals, Most of them are non-legacy universities, non-legacy government application strategies, Only 17 articles are related to the theme of intangible cultural heritage using micro-communication. Visible, Although the micro-communication mode with WeChat, Weibo and so on as the medium is more and more influential, But how to use micro-communication to better spread intangible cultural heritage, At this stage, the research is relatively small. So how to better use micro-communication for intangible cultural heritage dissemination? This paper analyzes the development of intangible cultural heritage communication and the dilemma of intangible cultural heritage communication in the context of micro-communication, In the context of micropropagation, How to use WeChat, Weibo and other new media platforms to spread intangible cultural heritage reasonably and effectively, To provide reference value for the subsequent dissemination of intangible cultural heritage.

2   Background to the Study

(i)   **Status analysis.** As the hidden wealth passed down from generation to generation in the process of human existence and development, intangible cultural heritage contains rich historical, cultural, economic and even scientific values, and plays an important role in the survival and development of human beings.In October 2018, the General Office of the CPC Central Committee and the General Office of the State Council issued "Some Opinions on Strengthening the Reform of the Protection and Utilization of Cultural relics", pointing out that it is necessary to strengthen the excavation, interpretation, dissemination and utilization of non-legacy resources, so that the unique advantages of cultural relics resources become the profound nourishment of the majestic power of building Chinese Dream together. We must attach importance to the protection and inheritance of non-heritage, and make non-heritage innovative development. Since ancient times, the two places have been prosperous and rich in culture. Lingnan has distinct location advantages, which has a strong demonstration and driving effect for the traditional culture to go out, realize the living state inheritance and innovation communication, and will certainly provide a new way of thinking for the overall protection and ecological inheritance of non-heritage culture. And help to innovate the communication path and enhance the communication effect. In the 13th Five-Year Plan, it is particularly pointed out that we should actively seize new opportunities for the implementation of the national strategy, give full play to the location advantages of "Belt and Road" and "Yangtze River Economic Belt", and transform the advantages of cultural resources into new strengths of the prosperity of local cultural industries. Efforts will be made to enhance cultural soft power. In the context of new media, let non-legacy catch the "Internet +" express, and use Internet thinking to promote the innovation and development of Ling South Africa.

(ii) **Existing studies.** Guangdong Province has a very rich Lingnan characteristics of intangible cultural heritage resources, in the first batch of national intangible cultural heritage list published by the State Council in 2006, Guangdong Province was selected 29 (a total of 518 items). Guangdong embroidery (including wide embroidery and Chao embroidery) was identified as the first batch of national intangible cultural heritage. Guangdong embroidery has unique Lingnan cultural characteristics, and Suzhou embroidery, Xiang embroidery, Shu embroidery together known as China's "four famous embroidery". Chaozhou embroidery is referred to as Chaozhou embroidery, which is a branch of Guangdong embroidery. Although Chao embroidery has a history of more than 1000 years, its cultural connotation and artistic form are the long history and cultural precipitation of Chaoshan people from ancient times to present. Chao embroidery art has extremely rich cultural connotation, its cultural and artistic value far exceeds Chao embroidery art itself, showing a complete Chaoshan art system, modeling system and color system.

(iii) After the implementation of the Law of the People's Republic of China on Intangible Cultural Heritage, Guangdong Province issued a supporting local law—the Regulations of Guangdong Province on Intangible Cultural Heritage adopted on July 29, 2011. This regulation reflects the importance attached by Guangdong Province to the protection and inheritance of intangible cultural heritage, and also provides local laws, regulations and policies for the protection and inheritance of intangible cultural heritage in the province. Guangdong Province and all cities under its jurisdiction attach great importance to the protection of intangible cultural heritage, and set up a number of intangible cultural heritage museums and exhibition halls in some key protection categories and birthplaces, and identified some intangible cultural inheritors. For example, Chaozhou City set up Chaozhou Xiangqiao District Chao embroidery Research Institute, set up a similar intangible cultural heritage protection center, by Chao embroidery inheritor Li Shuying master responsible for imparting Chao embroidery technology, Promote the improvement of Guangdong Province intangible cultural heritage protection mechanism.

3 Development of the Dissemination of Intangible Cultural Heritage

In the process of inheritance and continuation of intangible cultural heritage, intangible cultural heritage has always had its special status. With the continuous development of modern science and technology, the dissemination of intangible cultural heritage has gradually changed from its most primitive "oral instruction" to the way of communication through the latest media. In the course of this historical change, some of the traditional intangible cultural heritage is gradually dispelled, while some of them are modernized and reconstructed, which is endless in the new field of communication. [2] this paper divides modern communication into "traditional media period" and "new media period".

(i) **Period of traditional media.** From the early days of the founding of New China to the 1980s and 1990s, the dissemination of intangible cultural heritage was mainly in the traditional way, that is, television, big screen, radio, newspapers and other main carriers and channels of communication. The intangible cultural heritage of this period is basically spread by the mainstream media, from the selection of topics to the final presentation to the audience by a professional team to complete, accuracy, authority is beyond doubt, One of the most representative is about intangible cultural heritage related topics of film and television works. At the beginning of the founding of New China, the dissemination of intangible cultural heritage in China entered a prosperous stage, and the film and television works with Lingnan Chao embroidery dress as the subject matter were loved by the audience. These film and television works show the local customs of Lingnan region, with remarkable national characteristics, rich regional color and other characteristics, leaving a profound impression on the history of intangible cultural heritage dissemination. After the reform and opening up, the number of films on intangible cultural heritage in China has increased sharply. In this period, minority films have explored the cultural history, ethnic customs, cultural connotations and unique national characteristics of ethnic minorities, and many excellent works have emerged. It is these national theme films that show the national customs, cultural history, spiritual features and other aspects of Lingnan in front of the audience, so that the ordinary people can deeply understand the cultural tradition of Lingnan, and play a unique role in promoting the spread of Lingnan Chaoxiu culture.

However, in the current new media era, the traditional media is more focused on the living conditions, national customs, beautiful natural environment and religious beliefs in Lingnan area. The content reports concerning the cultural connotation of Lingnan area are generally not discussed in depth or relatively old, which leads to superficial or one-sided understanding and even misunderstanding. The rich cultural value of Lingnan Chao embroidery has not been excavated, and the cultural background has not been fully utilized. With the passage of time, the result is that the intangible cultural heritage of Lingnan is diluted and even replaced.

(i) **New Media Period.** Since the 21st century, Internet technology has begun to develop rapidly, and new media have emerged. At this stage, it is mainly based on mobile Internet technology to mobile phone, PAD as the carrier of new media, its manifestations are mobile TV, Internet TV, WeChat, Weibo, APP and so on. In all kinds of new media media, the micro-communication mode represented by WeChat and Weibo has become an important channel for the dissemination of intangible cultural heritage. With the continuous updating and iteration of the version, its main characteristics, such as openness, interactivity, sociality, timeliness and so on, make its communication power and influence more and more powerful.

1. the main body of intangible cultural heritage communication has undergone fundamental changes. In the Internet age, a photo or a small video, with a

paragraph of text can spread a thing. Everyone's position is equal, the right to speak is no longer the exclusive right of "official" and "authority", and the public can express their views [4].

In the past, the main body of intangible cultural heritage dissemination was government departments or mainstream media, but based on the functions of Weibo and WeChat, a large number of self-media people continue to emerge, ordinary people also have the right to publish information, people's "identity" has changed. Not only the receiver of information but also active as the publisher of information in the process of minority culture communication, people can obtain all kinds of Lingnan Chaoxiu culture information in the network, but also can publish and reprint the information related to Lingnan Chaoxiu culture to the Internet.

2. break the bottleneck of intangible cultural heritage dissemination. As long as the mobile phone is turned on APP, people can communicate with friends and strangers around the world to achieve one-to-one, one-to-many, many-to-many communication, without any geographical restrictions, as long as there is a network to achieve information acquisition and communication. For the dissemination of intangible cultural heritage, the regional bottleneck in the period of traditional media has been broken, and Lingnan Chao embroidery culture can be spread to all parts of the world through micro-communication. WeChat pays attention to acquaintance social, WeChat group communication between people is very frequent, has a high relationship intensity and trust. This group has strong social mobilization ability, in the process of intangible cultural heritage dissemination, we can make full use of this kind of group with high relationship intensity and strong trust to transmit information to more people in a short period of time.

3. improve the efficiency of intangible cultural heritage dissemination. With the help of mobile devices such as mobile phones and tablets, the information can be transmitted quickly in the first time and to people around the world in a very short time. Micro-propagation does not require cumbersome background, through mobile devices can be obtained at any time, any place to process and publish intangible cultural heritage information, its dissemination of information process is extremely simple and convenient, only need to upload text, voice, pictures or video and release, Get rid of a series of traditional media production process. Therefore, making full use of the simple advantages of micro-dissemination can spread Lingnan Chao embroidery culture to all over the world in the shortest time, which is of great benefit to the dissemination of Lingnan Chao embroidery cultural heritage and the promotion of Lingnan Chao embroidery cultural activities.

4 The Practical Dilemma of the Protection and Inheritance of Intangible Cultural Heritage

It covers many fields, such as traditional skills, folklore, folk literature, traditional music and so on. It has the characteristics of vitality, diversity and uniqueness. In the Ming

Dynasty, the folk activities in Chaozhou area were active, which promoted Chao embroidery to have a wide development market. Due to the increase of Chao embroidery personnel, Chao embroidery absorbed the advantages of local embroidery, and actively innovated and broke through in order to survive and develop more. In the Qing Dynasty, Chao embroidery played an important role in Chaozhou handicraft products. Chao embroidery has learned and absorbed other famous embroidery techniques in its continuous development, so that after the late Qing Dynasty, especially in the early days of the founding of the Republic of China to New China, Chao embroidery was deeply influenced by Gu embroidery, once also known as "Gu embroidery", until 1962. Chao embroidery was not named. Tide embroidery through learning and absorption of Gu embroidery, Suzhou embroidery, wide embroidery and other excellent embroidery skills, enriched the performance and skill level. Especially by mastering the performance and application of light in embroidery, the lifelike degree of embroidery image is improved. After the founding of New China, Chaozhou City integrated 13 Xiuzhuang and set up trade unions. After 60 years, mass commercialization gradually replaced the traditional small production mode of embroidery without batch. After the Cultural Revolution, the tide embroidery ushered in the peak of creative development. In the 21st century, people of social insight called for the rescue and protection of the lost tide embroidery skills, and included the tide embroidery in the first batch of national intangible cultural heritage list.

In recent years, the protection of Lingnan Chao embroidery has made great progress, but with the acceleration of urbanization and the change of social culture, the living environment of Lingnan Chao embroidery has gradually deteriorated. Taking the inheritance of non-heritage as an example, most of the inheritors of Lingnan Chao embroidery have been nearly ancient and rare in recent years, Xu Lijuan, the oldest national inheritor of lake drama, has been 86 years old, and some old rap artists have died. In recent years, Huzhou has issued a series of policy documents, such as "Measures for the Identification and Management of Non-Legacy Inheritors in Huzhou City", "Measures for the Application of Intangible Cultural Heritage Projects" and "General Guide for the Protection and Inheritance of Intangible Cultural Heritage", which provide policy support and institutional guarantee for strengthening the work of non-legacy protection. To do a good job of non-posthumous inheritance and protection, we should deeply plant folk, take root in local areas, approach villagers, and protect the geographical space and cultural field of non-posthumous living state inheritance as a whole.

During the period of the prevalence of traditional media, ethnic minorities can also spread their own culture at a specific time and place through their own arrangement, but this is only aimed at the audience groups in a specific region. After leaving a specific region, their programs will be in a state of reception because of regional and environmental problems. With the development and transformation of new media, the overall communication environment has changed greatly, and time and space are no longer obstacles to cultural communication. However, in order to cater to the special reasons of audience usage habits, increasing public attention and stimulating regional economic development, minority culture has emerged the following problems in the process of communication:

1. The cultural status of ethnic minorities is in a weak position. One of the important characteristics of the globalization of cultural communication is the squeeze of strong

culture on weak culture. As a weak culture, minority culture will have to bear great pressure from foreign strong culture in the process of communication. [5] in the face of the strong impact of mainstream culture and the continuous "squeezing" of the development space of minority culture, the trend of convergence of minority culture is becoming more and more obvious. People tend to enjoy the pleasure of "fast food" culture, so that minority culture is relatively weak. Although new media such as Weibo and WeChat provide a new platform for the dissemination of minority culture so that it can make more voices, minority culture is gradually disappearing under the impact of strong culture and the influence of "fast food" culture. Even some minority cultures are facing the dilemma of disappearing in the process of communication. For example: due to the impact of a large number of modern information, among the Miao people in Qiandongnan Prefecture, Miao songs are replaced by pop songs, and even many Miao people can no longer speak their own national language and know little about their own national history. [6] in the process of communication, the author studies and finds that the public numbers of Weibo and WeChat are relatively short of the content of the essence of national culture. In the face of the impact of mainstream culture and the influence of information explosion, Weibo, WeChat and other new media generally focus on event content, rarely dig into its cultural connotation, resulting in its lack of appeal in the process of communication, the audience is only quick browsing is difficult to impress, some WeChat public numbers even update lag or do not pay attention to the event, resulting in the content lost communication vitality.

2. minority cultural characteristics are missing. According to the cultural scholar Egerther, "the proximity of the region has always been a prerequisite for human groups to carry out and maintain communication links, and a fixed place of residence is necessary for the development of more frequent and meaningful communication." [7] different cultural backgrounds will produce different minority cultures. Because of the influence of geographical location, historical conditions and humanistic environment, ethnic minorities shape their unique cultural temperament, and their cultural background is relatively fixed. At the same time, minority cultures will spread in specific regions and groups of people. Different national cultures have different national characteristics, which should be expressed in the process of communication. [8] but in the new media environment, minority culture is gradually dispelled and assimilated in the process of communication, which makes the national characteristics not outstanding. In today's fast-paced society, few people will calm down to see the relevant content of minority culture, in order to meet the consumer habits (needs) of the audience, leading to the release of minority cultural content tend to surface. Even distorting minority history and historical figures for the eye, resulting in fundamental changes in the nature of minority culture. In this case, the communication of minority culture loses its own essence, and it is difficult for people to understand its connotation, which eventually leads to the gradual disappearance of the unique connotation of minority culture.

3. low utilization rate of cultural resources. In the process of spreading minority culture, whether people publish content through Weibo public number, WeChat public number, WeChat group or WeChat circle of friends, the fundamental purpose of which is mostly to improve their level of attention. In order to achieve this goal,

people will choose the content with high attention and interest. But in the process of micro-dissemination, the method adopted by communicators is to "follow the trend" and spread whatever content is of high concern recently. In order to cater to the audience's usage habits, people focus more on these places when reporting on ethnic minorities. People pay more attention to ethnic minorities, such as pastoral scenery, special costumes and unique buildings, and only stay in shallow presentation, rarely involving specific, detailed national cultural background and historical origin, few people really understand the connotation of minority culture, only one-sided reception of minority culture content, A large number of ethnic minority culture essence content is ignored, abandoned, eventually leading to the public will gradually solidify the understanding of minority culture.

4. business color is too strong. Compared with the traditional media, the entry threshold of micro-communication practitioners is relatively low and the quality is uneven. At the same time, many WeChat new media communicators are self-financing and cater to the audience in pursuit of high attention. Tend to cater to marketization, minority culture is only its tool as a customer. Under the impact of the development of modern society, some regions focus on local economic industries. Under the operation of commercial mode, the media appear as "additional products" of economic activities. The main purpose of communicators is to promote the development of local tourism, attract more tourists' attention and further improve the local economic level. In this environment, the original intention of minority culture communication has changed fundamentally, which is no longer a simple cultural communication, but an auxiliary advertising platform for the local development of tourism industry. On the other hand, in order to achieve the desired communication effect, the traditional art of ethnic minorities will be commercialized "processing transformation", such as the remodeling of national heroes, the change of classical songs, the ridicule of national culture, etc., the traditional art is divorced from the essence of national culture and overcommercialized.

5 An Effective Path to Inheritance and Dissemination of Non-posthumous Living State in the New Media Context

Ways to Promote the Effective Dissemination of Minority Cultures. It originates from life and belongs to life, deeply rooted in the fertile soil of traditional culture, bearing the memory of human civilization. Only in the specific cultural space or the paradigm of religious ritual, the non-posthumous communication context and communication field will slowly form. "Some Opinions on Strengthening the Reform of the Conservation and Utilization of Cultural Relics" clearly put forward that it is necessary to "make good use of traditional and emerging media, widely disseminate the cultural essence and the value of the times contained in cultural relics, and better construct the [8] of Chinese spirit, Chinese value and Chinese strength.

In the context of new media, we should take non-legacy as the breakthrough point and research theme, actively grasp the law of network communication, actively explore the way of non-legacy inheritance, and constantly strengthen the inheritance and propaganda of non-legacy culture through various media platforms. [9] good results have been achieved in extending the inheritance mode and spiritual connotation of non-posthumous culture by using the advantages of technology to expand the inheritance scope of non-posthumous culture and deepen the communication space.

In a word, the social micro-communication mode plays an increasingly important role in the process of information dissemination, and its communication power and influence can promote the effective communication of non-posthumous. From the long-term consideration of how to use WeChat, Weibo and other new media micro-communication platform to effectively spread non-legacy culture, this paper believes that we can start from the following points:

1. deep excavation Lingnan Chao embroidery culture essence. General Secretary Xi Jinping pointed out in the report of the Nineteenth National Congress of the Party: "deeply excavate the ideology, humanistic spirit and moral norms contained in the excellent Chinese traditional culture, and combine the requirements of the times to inherit and innovate, so that the Chinese culture can show its permanent charm and style of the times." In the process of spreading Lingnan Chao embroidery culture, due to the lack of understanding and understanding of the cultural connotation of Lingnan Chao embroidery, most of the reports on Lingnan Chao embroidery culture remain in a relatively shallow layer. Only choose the content with the highest attention to spread the report, but lack of cultural background, national tradition, national customs and other essence of the deep excavation. Therefore, when using micro-communication to spread Lingnan Chao embroidery culture, we should not stay on the surface. It is necessary to dig and develop the historical story and cultural background behind Lingnan Chao embroidery culture, increase the depth and breadth of cultural information, provide unique cultural understanding for the audience, and avoid misunderstanding of Lingnan Chao embroidery culture.

2. on the basis of "big data" to build Lingnan Chao embroidery cultural communication chain. First of all, it is necessary for Lingnan Chao embroidery culture communication workers to change their thinking, establish big data consciousness, and look at Lingnan Chao embroidery culture communication work from the "vertex" of big data. Big data can provide a strong data base for Lingnan Chao embroidery culture communication. Second, speed up the integration of traditional media with new media such as Weibo and WeChat. Traditional media are influenced by their mode of communication and are limited in time and space. It is difficult to respond in the first time. However, in the new communication environment, such restrictions should be broken. Traditional media and new media should accelerate integration, form influence more quickly, integrate related resources, strengthen the use of official micro-level and official blog, and enable the audience to participate more actively. At the same time, open the barriers of "big screen" and "big screen", "big screen" spread "small screen" way, So that Lingnan Chao embroidery culture communication in technical means, communication mode and other aspects have been effectively promoted, so that Lingnan Chao embroidery culture communication more diversified, forming a "big data" based on the three-dimensional communication matrix.

3. construction of Lingnan Chao embroidery folk communication mechanism. At the same time, we should pay attention to the construction of our own folk communication mechanism and the cultivation of cultural communicators with new media

professional literacy. For Lingnan Chao embroidery culture, the local people are the most affected by their own national culture, they know their own national culture tradition and heritage, are born Lingnan Chao embroidery culture disseminator, expression and inheritor, and they are also the most direct contact with the outside world, this paper believes that it should be inspired as a national pride, this link into a link with other ethnic groups, through them to spread the Chao embroidery culture, should establish the Lingnan Chao embroidery culture communication practitioners and the inter-ethnic people communication mechanism, regular training in new media business knowledge, Let them understand the relevant operation mode of learning micro-communication, effectively combine Lingnan Chao embroidery traditional culture with new media communication, rather than just staying in "commercialization" application. Cultivate a group of people familiar with Lingnan Chao embroidery culture and professional technology, make it become the spokesman of Lingnan Chao embroidery culture.

4. build a diversified communication system. In the process of spreading Lingnan Chao embroidery culture, we should actively cooperate with various communication channels. The relevant comments of the official media have a particularly significant impact on the communication effect of Lingnan Chao embroidery culture. We should pay attention to strengthening the cooperation between Lingnan region and official Weibo, WeChat public number, establishing diversified and multi-level communication system, and improving the overall communication effect. Excavate the content of Lingnan Chao embroidery culture with representative characteristics, transform it into film, documentary, animation and other forms in Weibo, WeChat and other micro-communication platform to spread and promote, by telling the story of "Lingnan Chao embroidery", attract users' attention, let the audience feel the inner charm of Lingnan Chao embroidery culture, deepen the understanding and cognition of Lingnan Chao embroidery culture, participate in the process of Lingnan Chao embroidery culture communication, promote the dissemination and inheritance of Lingnan Chao embroidery culture, and further stimulate national pride.

In a word, Lingnan Chao embroidery culture should form a diversified and multi-level communication system based on "big data", "Lingnan masses" as class background, and new media such as Weibo and WeChat as platform. On this basis, we should gradually develop and innovate, actively study the latest communication ideas, contact the most advanced communication platform, integrate and transform them, strengthen communication and communication with traditional media, break down the "barriers" between new media and traditional media, and finally establish the whole media culture communication platform belonging to Lingnan itself.

6   Reflections on Enhancing the Cultural Power of Lingnan Tide Embroidery in the Context of New Media

In the new media era, the bottleneck of Lingnan Chao embroidery culture communication is gradually disintegrating. With the help of new media platforms such as Weibo and

WeChat, Lingnan Chao embroidery culture communication has a broader communication platform. Although various problems emerge in the actual communication process, its important role in the process of Lingnan Chao embroidery culture communication can not be ignored. Lingnan Chao embroidery culture dissemination should be aware of its own advantages and disadvantages, make full use of micro-communication to learn from each other, fully combine the characteristics of micro-communication with the cultural connotation of Lingnan Chao embroidery, with the help of this "Red Sea Market". Lingnan Chao embroidery culture spread to a new height.

(i) **Innovative Ways of Protecting and Inheriting.** In the context of new media, the media is becoming more and more diversified. Digital image technology covers more and more encoded information sequences. Although it can not completely avoid the problem of information loss or ambiguity, it can effectively reduce the misunderstanding caused by subjective images. Because the use of image technology to record objects, compared with the use of written records more objective, the public can repeatedly watch, compare and guess the recorded objects, from which to explore the "meaning" [10]. Through the original holographic recording, the movement track, expression presentation and emotional changes of the object, even the smallest emotional fluctuations will be perfectly captured and accurately recorded, which is the biggest difference between the image recording of the lens and the text written in the text described in the text in the process of communication and the effect of communication. In this way, the innovation of non-posthumous protection and inheritance. For example, the first batch of national intangible cultural heritage list of lake pen production skills are preserved by image recording. The lake pen making process includes 12 big processes, which can not be omitted and reversed. Each big process contains several small processes, with a total of more than 120 large and small processes. By recording the production technology of lake pen in the way of image recording, it can objectively describe and explain the cultural form of lake pen and lake pen itself. It can also promote the public to actively participate in the protection and inheritance of lake pen production skills.

(ii) **Promoting the Deep Exploitation and Utilization of Non-legacy.** The use of new media can objectively present the original appearance of non-heritage, such as digital technology can comprehensively and systematically record all aspects of non-heritage protection and inheritance, form archival data, develop and provide utilization of archival data, form all kinds of cultural products, and promote non-heritage into public view. Make the non-heritage culture of "raised in boudoir unrecognized" knowable, touchable and perceptible, so as to achieve the purpose of wide dissemination and universal education. By means of new media technology and Internet technology, the inheritance, innovation and development of non-posthumous are realized to the maximum extent, and "Internet + non-posthumous" has become a new mode of deep development and utilization of non-posthumous. Such as, Huzhou's "Silk Town" was successfully selected as the first special town in Zhejiang Province in June 2015. Under the development concept of "Internet + Silk", we will actively expand the high-end modern service industries such as "Silk + Leisure", "Silk + Tourism", "Silk + Ecological experience" and so on.

(iii) **Expand the Living Space of Non-posthumous.** By using new media technology such as digital technology to record the original ecological space and field of non-heritage culture, and to generate three-dimensional digital image system by virtual processing, and then to build an online holographic digital museum, The concrete content and artistic essence of non-heritage culture are displayed [11] living mode, so as to inherit non-heritage culture to the greatest extent. Taking Huzhou Shuanglin Aya silk as an example, Shuanglin Aya silk has a history of more than 4700 years and is known as "the flower of silk weaving technology". Aya silk production technology is an important part of the Chinese silk weaving technology culture. If digital technology and three-dimensional animation are used to record and reprocess the silk production technology, a new digital culture form can be generated. Save it in the digital museum, just click the mouse, you can see the virtual silk in a three-dimensional way panoramic display of its technological evolution. In the context of new media, new media technology makes Shuanglin Aya silk radiate new vitality and vitality again, which is of great benefit to cultivate the public's national cultural confidence and enhance the national cultural consciousness. 21 December 2017, The cultural project of Huzhou Chuanjia Town with a total investment of about 15 billion yuan was formally signed at the Municipal Cultural Industry Development Conference. This project takes the non-posthumous protection inheritance as the important grasp, Relying on the natural scenery and cultural heritage of Huzhou, Deep integration and [12] of historical humanities and scientific and technological innovation, traditional culture and modern manufacturing, non-legacy culture and industrial upgrading. Currently, Huzhou actively explores the new mode of cultural tourism development under the new media context, Focus on the construction of non-heritage home area, preaching area and the three core modules, Strive to create a set of skills display, cultural experience, curriculum popularization, product trading in one of the non-posthumous creation space.

7   Conclusion

With the development of global economy and the acceleration of globalization, intangible culture and traditional art are facing a strong multicultural impact. At the same time, young people are not interested in traditional culture now. Make intangible culture and traditional art successor talent missing. Because of the particularity of the industry, the demand for employees is very high, the training time is long, the cost is high, the return is low, the employees are more lacking, and it is difficult to inherit and develop. Another difficulty in the protection of intangible cultural heritage is the lack of funds. Although the government allocates part of the funds to the protection and development of intangible cultural heritage every year, the development and utilization of intangible cultural heritage is a huge system engineering, involving all fields and strata.

In short, there is no effective intangible cultural heritage protection system, its protection system is not perfect, social propaganda organization is not in place. Intangible cultural heritage protection staff themselves lack of understanding of intangible culture, specialization is not strong, leading to the formation of specialized intangible cultural heritage protection team.

Under the new media environment, Lingnan Chaoxiu culture communication has crossed the limitation of time and space, and is gradually eliminating the cultural gap between nationalities.Lingnan region should adhere to the cultural connotation of their own people, strengthen the overall construction of Chaoxiu culture communication, with the help of Weibo, WeChat and other new media development "dividend", the full integration of the two to achieve the Lingnan Chaoxiu culture communication and new media development. It can not be ignored that the micro-communication mode brings convenience to the spread of Lingnan Chao embroidery culture, but also has some drawbacks, especially in the process of "involving national factors" and "social hot events". In order to achieve a special purpose, information is released without proof, which causes negative social emotions, further forms public opinion and has adverse effects on social stability. In order to avoid the adverse social influence caused by the one-sided understanding of Lingnan Chao embroidery culture, every Lingnan Chao embroidery culture practitioner should think and pay attention to it.

References

1. Ya, G.: On the position of minority culture in the construction of national cultural soft power. J. Southwest Univ. Nationalities J. (Humanit. Soc. Sci. Edn.) **07**, 194–197 (2013)
2. Song, Q.: Evolution of minority culture media – taking Guangxi minority region as an example. J. Hubei Univ. Nationalities (Philos. Soc. Sci. Edn.) **2015**(6), 141-146+155 (2015)
3. Hu, B.: Development trajectory and development countermeasures of New China Minority film. Study Nat. Art **2017**(1), 51–58 (2017)
4. Zhang, Z., Zhang, L.: National culture hit the new media, how to break? Press Forum **2015**(01), 68–72 (2015)
5. Zhang, W.: The analysis of cultural communication characteristics in minority areas—taking Qiandongnan autonomous prefecture of Guizhou province as an example. Guizhou Soc. Sci. **2005**(5), 45–47+80 (2005)
6. Gongzhe, T.Z., Kong, R.: Exploration on channels of Miao culture communication in the new media age. Guizhou Ethnic Stud. **2013**(4), 79–82 (2013)
7. Lu, S., Hong, J.: Renmin University of China
8. Some Opinions Strengthening the Reform of the Conservation and Utilization of Cultural Relics. Issued by the State Office of the Guangming Daily 09 Oct 2018
9. Tan, H.: Inheritance of folk literature from the perspective of intangible cultural heritage—with oral literature as the core. J. Guangxi Normal Univ. **2017**(6), 69–78 (2017)
10. Road is quiet. A Brief Study on the Protection of Intangible Cultural Heritage Culture Monthly **2018**(8), 60 (2018)
11. Ling, C.: A study on the digital protection and inheritance of Tibetan intangible cultural heritage in the context of new media. J. Southwest Univ. Nationalities **2010**(11), 39–42 (2010)
12. Qian, H.: Huzhou first large-scale non-legacy space "Huzhou family town" cultural project signed to settle. Zhejiang Daily, 22 December 2017
13. Huang, Y.: Tide Embroidery. Lingnan Art Press (2014)
14. Wang, X.: Understanding of the concept of productive protection of intangible cultural heritage. Art Garden **2011**(2), 97–100 (2011)

CCD in Autonomous Vehicles and Driving

Driver's Perception of A-Pillar Blind Area: Comparison of Two Different Auditory Feedback

Chenxi Cao⑩, Jialing Wei, Xiangyi Wang, and Hao Tan(⊠)

Hunan University, Changsha, Hunan, China

Abstract. Recent years with the development of science and technology, vehicles not only bring convenience to people but also require high quality to ensure the safety of passengers. So, the frame of the car became stronger. In order to improve the safety of the driver, and Consistent with the unity of the body, the car's pillars are also increasingly wide.

As A-pillar widens, their negative impact on the driver's panoramic field of vision increases [1]. At the same time, as the field of view becomes smaller, there is an increased risk that the driver will ignore people or objects outside the vehicle. A-pillar blind area refers to the blind area of vision during driving. Generally, there are three pillars on each side of the car body, and the diagonal pillars on both sides of the windshield are called a pillar. The driver's field of vision will be partially blocked by the A-pillar before the car turns or enters the curve, resulting in a blind area in the field of vision [2]. In order to solve the A-pillar blind area, this project studies the drivers' different perception of sound frequency and rhythm respectively which reflect the distance of obstacles. The results of the study show that both the frequency and rhythm of sound improved drivers' perception of blind areas in their field of vision, and participants had clear preferences in showing distance by sound frequency.

Keywords: Distance · Sound prompt · Safety perception · Traffic · Frequency · Rhythm

1 Introduction

Car manufacturers are trying to keep you safe by building stable chassis around the driver [3]. The survey found that in recent years car manufacturers have sought stronger chassis, as well as wider A-pillars to protect occupants. As a result, the vehicle's panoramic field of view has been decreasing in recent years. Drivers actually have a lot of blind spots while driving, and manufacturers are developing side mirrors to ease the backward blind spots, but people often overlook the forward blind spots caused by the A-pillar (also known as the windshield pillar) [1].

In the actual design process, each automobile factory to minimize the a-pillar field of view blind spot, the purpose is to reduce the resulting traffic accidents. But the problem still exists, but the degree of reduction of manufacturers is different. Studies have shown

© Springer Nature Switzerland AG 2021
P.-L. P. Rau (Ed.): HCII 2021, LNCS 12773, pp. 171–181, 2021.
https://doi.org/10.1007/978-3-030-77080-8_15

that the covering Angle of A-pillar is between 6° and 12°, which is due to the width of A-pillar [4]. Behind a A-pillar with a covering Angle of 6°, an object with a distance of 50 m and a width of 5.2 m can be completely covered [5]. A study about "Body-pillar vision obstruction and lane-change crashes" points out that lane change crashes are increasing with wider A-pillars [6]. In recent years, more and more traffic accidents have been reported in the society because of the forward blind area, and people have gradually realized that the visual blind area caused by A-pillar can have a very bad effect on driving. That's why we start with the A-pillar to solve this problem.

Hearing is the second most important channel of communication to the outside world. There are many indications that sound is suitable for representing higher-dimensional data without overloading the user. Sound displays are ideal for situations where there are many variable parameters or multi-dimensional complex information must be monitored simultaneously. Sound information can be coded in one dimension by using a single feature such as the intensity, frequency and duration of the sound, or in multiple dimensions by combining several features of the sound [2].

The double-ear effect, people through the ears to perceive the external sound. Not only the intensity of the sound, tone, but also can determine the location and distance of the source of sound. Then combine the following two points, first, turn the head to form a multi-point positioning. The most common practice at this time is to subconsciously shake the head. in this way, forming multiple points in space, the brain positioning a sound source in three-dimensional space more than enough. Second, using the human body itself and the surrounding effects on sound. Because people do not rely on two tympanic membrane alone to distinguish sound. sound can be transmitted throughout the human body, especially the skull, so that more than two points can be calculated to be used to locate.

In order to conduct reliable and feasible experiments, we need to choose a suitable element among various elements of sound that have a unified standard of measurement and are not easily influenced by the external environment. Among the commonly used attributes of sound, loudness is easily affected by the external environment, and the volume can be adjusted artificially. The timbre between belong to the nonlinear change, the variety is various, so timbre is out of our consideration. After analysis, we selected the frequency and rhythm that can clearly show the difference and can be standardized according to the numerical value for comparison test [2]. Sound information can use single factors such as loudness, frequency and rhythm of sound as independent variables, and can also combine several factors of sound to transfer information [7].

2 Research Question

A-pillar is an important part of every vehicle and has a great influence on the stability of the vehicle. If there is no A-pillar, when the car has a violent collision, the car is more prone to deformation, seriously endangering the lives of the occupants. The A-pillar not only makes the car stronger, but also plays an important part in the impact. So, it is impossible to remove them without changing the way carmakers make cars today [1]. In this case, in order to minimize its influence on the driver's visual field, the idea proposed in this paper is to help the driver reduce the influence of the blind zone to some extent by setting the obstacle distance warning sound in the car, and possibly avoid some dangers other than obstacles in the blind zone of the visual field. According to the results of the preliminary survey, as a common prompt sound element, sound is easy for subjects to understand and has many forms, thus becoming the subject of our experiment. Based on comprehensive consideration, we proposed the following research question: Which of the different forms of obstacle prompt sound can better enhance the driver's perception of obstacles in the forward blind area? Therefore, in order to carry out preliminary experiments, we need to set up experimental models with strong operability and high similarity to the reality.

3 Related Work

Among the fatalities in 2015, the number of pedestrian fatalities was 5376. It is a 9.5% increase from 4910 pedestrian fatalities in 2014 (National Center for Statistics and Analysis, 2017). The Governors Highway Safety Association predicted an 11% increase in pedestrian fatalities on U.S. roadways [8]. The National Motor Vehicle Crash Causation Survey (NMVCCS), conducted from 2005 to 2007, reported that around 94% of traffic crashes are at least a result of human error [8]. And one of the reasons for these human errors is the A-pillar blind area. In order to better solve the hidden danger of A-pillar blind area, many automobile manufacturers in this problem. A common solution is to open triangular Windows. However, such a design will reduce the visual range of the mirror (see Fig. 1).

Fig. 1. Automobile manufacturer's triangular windows in front window.

There are also some concept cars that better demonstrate the design conjecture to solve the problem of A-pillar and blind area of vision, such as the use of hollow A-pillar. The idea is very bold, but it is still in the conceptual stage, whether the real mass production is still an unknown (see Fig. 2) [9].

Hyundai and Kia have reportedly applied for a patent aimed at solving the A-pillar blind spot problem. The patent is called "pillar display system for blind spot of vehicle. The patent USES a camera and a projector system and is applied to A-pillar, where the camera will take images from inside the vehicle and use the projector to project onto A-pillar to provide the driver with an image of A-pillar's blind area. In the patent

Fig. 2. Automobile A-pillar with cutout design.

application, Hyundai and Kia used a pedestrian test to show that the technology is safe. But when driving, the driver's judgment is the deciding factor. Even a slightly delayed display can have a huge negative impact, increasing the risk of driving (see Fig. 3) [10].

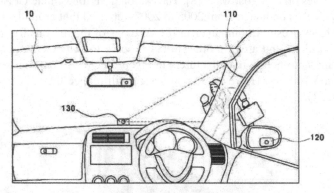

Fig. 3. Application of time projection formed by camera and projector on a-pillar of automobile.

4   Method

To solve the question raised above, we decide to make videos to do a user research. The videos simulate a scene that a pedestrian blocked by an A-pillar approaching a vehicle. During the experiment, the participants are asked to wear headphones to watch all videos in a quiet room, and to fill in tables. Through the sorting and calculation of the final table data, we can get the comparison of two different auditory feedback in some aspects like understandability and etc.

5   Experimental Preparation

We published recruitment information on the Internet to find volunteers to participate in the experiment. Thirty participants were recruited from the age of 18 to 20 who have already had their driver license. They were all students, healthy, with normal vision and hearing level. In addition, each participant was rewarded with a dessert worth 0.83 euros. The confidentiality agreement documents required for the experiment, laptop (Xiaomi Mi Notebook 15.6), headset, two tables, and experiment videos (auditory feedback included) are prepared.

6 Auditory Feedback

We use Adobe Audition CC 2019 (a professional audio editing software) to make our auditory feedback. As mentioned above, we choose to use the rhythm of the sound and the frequency of the sound to show the change of the distance between the obstacle in the blind area of A-pillar and the vehicle. The rhythm range of auditory feedback is from a slow speed about 1,670 ms/note (0.6 note/sec) to 167 ms/note (6.0 notes/sec) for fast, and the frequency range is 299 Hz (low)–5553 Hz (high). The range of rhythm and frequency change is carefully set by us. According to the research and investigation, the rhythm change can be better recognized by people within this range [11], and the frequency range can be better recognized by people under the fixed loudness (60 dB) [10]. The range of the distance (from obstacle to vehicle) is 1–5 m, and each meter corresponds to a piece of audio. Therefore, the sound rhythm change and sound frequency change correspond to five audio segments (see Fig. 4). The first audio is displaying distance by sound rhythm. When the obstacle is within five to four meters of the A-pillar, the sound rhythm is 1667 ms/Note (0.6 note/s) (slow). As obstacles get closer, the rhythm of the sound will become faster and faster. When the obstacle is within one meter of the A-pillar, the rhythm of the sound will reach 167 ms/Note (fast). The second audio is displaying distance by sound frequency. Similarly, as obstacles get closer, the frequency of sound will become higher and higher.

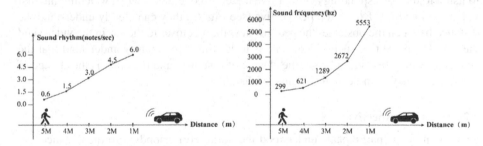

Fig. 4. The sound rhythm change and sound frequency change correspond to five audio segments.

7 Experimental Video

Because the video has both visual and auditory information processing channels, and has a high degree of participation and guidance, we chose to watch the video as an experimental method [11].

We used the director mode in the game Grand Theft Auto V to record the required experimental video which simulate the scene when the pedestrian who was completely blocked by the A-pillar from view of the driver's seat gradually approached the vehicle. In the whole simulation scene, we chose a piece of open space on the beach, the time is 8 o'clock in the morning, and the car is a black SUV. Pedestrian are completely within the range of A-pillar in the driver's view, and walk slowly to the vehicle from a distance (see Fig. 5).

Fig. 5. Pedestrian and car.

7.1 Top Perspective

First, we made a video to let the participants know the audio corresponding to the different distance of obstacles from the A-pillar in advance. Therefore, we ask the participants to listen to the corresponding audio of five meters to one meter when watching the first video. In the video, from a top perspective (see Fig. 6), they can clearly understand the distance between the obstacle (the pedestrian walking closer to the car in the video) and the car. There are two such videos, one is to let the experimenter understand that the change of sound rhythm indicates the change of distance, and the other is the change of sound frequency to indicate the change of distance.

7.2 Driver Perspective

Second, after the participants understood the audio corresponding to the distance, we started to play the second video to allow the experimenter to truly experience the distance of the pedestrian that the audio display is blocked. In this video, the participants in the driver's seat can't see the pedestrians blocked by the A-pillar outside the car (see Fig. 7), but the distance of the pedestrians is actually changing (The order is 5 m, 3 m, 1 m, 4 m, 2 m). And different distances correspond to different audio (see Fig. 4). The video is paused five times for five different distance between the pedestrian and car where the corresponding auditory is also played. So, the participants can take their time to make their distance judge by only hearing the audio because the pedestrian is completely covered by the A pillar and fill out the form that lists the audio in turn.

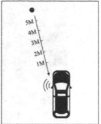

Version 1:Top perspective

Fig. 6. Top perspective.

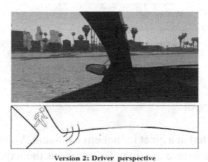

Version 2: Driver perspective

Fig. 7. Guessing the distance by auditory feedback.

7.3 Top-Side Perspective

Third, after the participants make a distance judgment, we need the participants to score the personal impression of displaying distance by the sound which includes Distance Showing, Understandability, Vision Improving, Distraction Degree and User Experience. The score is set to 1–5 points, of which 1 point is not helpful and 5 points is very helpful, except in the Distraction Degree item which 1 point is very distracted and 5 points is non-distracted. Therefore, we will play the third video. From a side-up perspective (see Fig. 8), the participants will see the change in the distance of the blocked pedestrian and the corresponding audio in the second video, so they can find out if their guess is correct. They value the effect of perceiving distance by the frequency change or rhythm change on improving the blind area of A-pillar and then rate this process.

Version 3: Side-up perspective

Fig. 8. Side-up perspective.

8 Experiment Process

The experiment is conducted in a quiet indoor environment. Volunteers need to fill in our confidentiality agreement before starting. To carry out the experiment, it is necessary to wear earphones and watch videos. First, watch the three videos in the rhythm part. When watching the second video, you need to guess the distance and fill in the table. After watching the third video, you need to fill in another table. The process of the frequency part is the same as that of the rhythm part. After the experiment, participants need to submit two completed tables. We use the average data processing and variance data processing methods for data processing.

9 Results

For objective and visual analysis of data, we processed the data by calculating the mean and standard deviation to obtain the average correct rate of the participants' guesses about the distance and their average personal impression of the experiment [1]. The result is as follows.

9.1 Distance Displaying

Average distance estimation of the participants is showed in the Table 1 below. If the gap between the average value and the correct value is small, it means that the participants' judgment on the distance is more accurate, on the contrary, if the gap is large, it means the participants' judgment on the distance is less accurate. The result shows sound frequency changes show greater accuracy in estimating distance.

Table 1. Average distance estimation and standard deviation (SD) in meters.

	Sound rhythm	Sound frequency
Distance	Mean/SD in m	Mean/SD in m
1 m	1.17/0.37	1.17/0.37
2 m	2.33/0.76	2/0.58
3 m	2.67/0.74	2.83/0.37
4 m	4/0.58	4/0.4
5 m	4.83/0.37	5/0.4

9.2 Personal Evaluation

The personal evaluation is showed in the Table 2 below. The higher the score, the higher the participants' satisfaction with this method. Otherwise, the smaller the score, the lower the participants' satisfaction with this method. We know by comparison that the results in the Table 2 also showed that displaying distance by sound frequency is more popular among the participants.

Table 2. Average personal rating between 1 and 5.

	Sound rhythm	Sound frequency
	Mean/SD	Mean/SD
Distance showing	4.33/0.75	4.67/0.47
Understandability	4/0.82	4.75/0.38
Vision improving	3.67/0.75	4.17/0.69
Distraction degree	4.67/0.47	4.74/0.38
User experience	4/0.58	4.33/0.69

(1–5 points, 1 is not helpful while 5 is very helpful).

(1–5 points, 1 point is very distracted, 5 points are not distracted).

10 Discussion

In Table 1, by observing the overall data, we can conclude that participants have a higher accuracy rate in judging the distance of obstacles (pedestrian) blocked by the A-pillars by sound rhythm or frequency. While judging the distance of obstacles (pedestrian) blocked by the A-pillars by sound frequency got a little bit lower accuracy rate by the participants. The error of sound frequency display distance at 4 m and 5 m is very small. This can indicate that both methods of distance display by voices allow participants to judge distances more accurately. It is obvious that participants can finish this experiment easily.

In Table 2, through data analysis, we can find that participants' scores on the effect of distance display and their understandability are high, which indicates that both kinds of displaying distances by sound can be better understood by participants and help participants better know the distance of pedestrian. The user experience score is also high, so we can conclude that participants' overall experience with this experiment is good.

At the same time, from the overall data analysis, it can be seen that the overall score of showing distance by sound frequency is higher than showing distance by sound rhythm. Some participants thought that when the distance changes the sound volume is same, the change of the sound frequency is more obvious than the rhythm change, and it is easy to distinguish. Some participants also think that it may be because high frequencies can give people a stronger auditory stimulus than the rhythm change, so they can have a deeper memory of it.

In general, the experimental results show that both showing distance by sound frequency and the showing distance by sound rhythm can improve the driver's panoramic vision blocked by A-pillar. The distance displayed by the sound is clear and easy to understand. Most participants thought that showing distance by sound frequency is easier to distinguish, which is a more supported solution to the A-pillar blind zone among the participants.

11   Conclusion and Future Work

The result shows that the two kinds of change of sound in the experiment is very good in showing the distance, and the sound frequency gets slightly better impression.

At present, due to safety and condition constraints, the experiment currently allows participants to experience the progress by watching videos. In the future research, we will create a more realistic simulation of a dynamic driving environment and developed an improved solution that is more suitable for balancing internal and external noise in special scenarios.

In the field of autonomous driving, there are also tools for detecting obstacles by external devices, such as detecting other vehicles or bicycles, motorcycles, and pedestrians [12]. In autonomous vehicles, various functions are controlled by software and hardware together, and these functions can work independently without people. Such automation technology can reduce driver stress, improve safety for all users on the road, and reduce fuel consumption [13]. However, in the process of autonomous driving, the driver finds that humans cannot completely rely on autonomous driving, and that automated vehicles have certain limitations [14]. In vehicles, auditory interaction is a tool for communication or warning. It can also keep the driver's attention to a certain extent [15]. Therefore, the on-board auditory feedback device can not only improve the driver's traffic safety during the automatic driving process, but also remind the driver to keep their attention.

Also, some studies have shown that drivers have a good opinion of making 3d prompt sounds in blind spots [12]. Therefore, in future work, we will consider installing auditory feedback on two different A-pillars in the car. When an obstacle is encountered, only the corresponding audible feedback will be played (left or right), which can reduce the driver's distraction in judging the direction of the obstacle.

This experiment reveals that showing the distance by the sound rhythm change and the sound frequency change can improve the panoramic view of the driver. The sound catches the driver's attention so as to rise their awareness of possible obstacle. This study provides an auditory solution to the blind area of the A-pillar, which has proven to be feasible and effective.

Acknowledgement. The paper is supported by Hunan Key Research and Development Project (Grant No. 2020SK2094) and the National Key Technologies R&D Program of China (Grant No. 2015BAH22F01).

References

1. Meschtscherjakov, A., Wanko, L., Batz, F.: LED-a-pillars: displaying distance information on the cars chassis (2015)
2. Baidu Encyclopedia. https://baike.baidu.com/item/theA-pillarblindarea/3459242?fr=aladdin
3. McCarthy, M.G., Walter, L.K., Hutchins, R., Tong, R., Keigan, M.: Comparative analysis of motorcycle accident data from OTS and MAIDS. Published Project Report, vol. 168 (2006). http://www.maids-study.eu/pdf/OTS_MAIDS_comparison.pdf
4. Quigley, C., Cook, S., Tait, R.: Field of vision (Apillar geometry) - a review of the needs of drivers: final report (2001)
5. Beach, R.: Killer Pillars The blinding truth (2004). http://www.safespeed.org.uk/bike005.pdf
6. Sivak, M., Schoettle, B., Reed, M.P., Flannagan, M.J.: Body-pillar vision obstruction and lane-change crashes, Report No. UMTRI-2006–29. http://deepblue.lib.umich.edu/bitstream/handle/2027.42/58719/99778.pdf?sequence=1
7. Neuhoff, J.G., Kramer, G., Wayand, J.: Sonification and the interaction of perceptual dimensions can the data get lost in the map? In: Proceeding of the International Conference on Auditory Display (ICAD). Atlanta, Georgia, USA, pp. 93–98 (2000)
8. Deb, S., Strawderman, L., Carruth, D.W., DuBien, J., Smith, B., Garrison, T.M.: Development and validation of a questionnaire to assess pedestrian receptivity toward fully autonomous vehicles (2017)
9. China Daily. https://baijiahao.baidu.com/s?id=1614532707100708210&wfr=spider&for=pe,2018
10. Li, H., Ge, L., Lu, W.: A study about absolute auditory discrimination at different frequency level (2005)
11. Dowling, W.J., Bartlett, J.C., Halpern, A.R., Andrews, M.W.: Melody recognition at fast and slow tempos: effects of age, experience, and familiartity (2008)
12. Levinson, J., et al.: Towards fully autonomous driving: systems and algorithms. In: Intelligent Vehicles Symposium (IV), pp. 163–168. IEEE (2011)
13. Mersky, A.C., Samaras, C.: Fuel economy testing of autonomous vehicles. Transp. Res. Part C Emerg. Technol. **65**, 31–48 (2016)
14. Vissers, L., Van der Kint, S., Van Schagen, I.N.L.G., Hagenzieker, M.P.: Safe Interaction Between Cyclists, Pedestrians and Autonomous Vehicles. What do We Know and What Do We Need to Know? SWOV Institute for Road Safety Research, The Hague (2016)
15. Ayoub, J., Zhou, F., Bao, S., Yang, X.J.: From Manual Driving to Automated Driving: A Review of 10 Years of AutoUI. General and reference ~ Surveys and overviews

Automated Driving: Acceptance and Chances for Young People

Shiying Cheng, Huimin Dong, Yifei Yue, and Hao Tan[⊠]

Hunan University, Changsha, China
htan@hnu.edu.cn

Abstract. Young people aged 18–24 are the main force of consumption in China, are more inclined to buy self-driving cars and are willing to pay higher prices, which makes them an important potential user of autonomous vehicles. Automated vehicles seem to provide more possibilities in terms of independence and safety. In order to explore the user needs of the young people, we set up a study based on semi-structured interviews and a role play. Our results demonstrated that 90% of the young people (n = 20.) are pleased to drive the automated vehicles, they are looking forward to the aging of automated driving. The remaining 10% of young people were unwilling to try out automated driving systems owing to the nondeterminacy and suspicion (fear of mechanical failures), and they are deeply worried about safety. In the exploration of demand, regardless of the level of acceptance, young people have shown a high demand for safety and security facilities. In addition, we have explored the needs of young people for entertainment and human-computer interaction. This will likely provide arising opportunities for young people.

Keywords: Automated driving · Young people · User needs · User study · Semi-structured interviews · Role play

1 Introduction

In China, young people are a huge potential user for autonomous driving. As of 2017, the youth, aged 18–24 ,accounted for 32.4% of total consumers, ranking first in all age groups, is the main consumer group in the future. At the same time, according to data from the 2019 Automotive Consumer Survey, Chinese consumers are still generally optimistic about the potential advantages of autonomous driving [1]. In addition, Ben [2] has found out that young respondents were willing to pay (WTP) the most, $8,921 or 36% above the initial purchase price, compared to the average WTP of 24% above the purchase price.

On the other hand, in China, due to the low rate of private car ownership in China, a lot of young drivers appeared after passing the driving tests. They were called "carless young drivers", mostly aged between 18–25, already own a driver's license but have little chances to practice driving skills on account of owning no car. So the safety problem related to them has aroused great attention in China. The study showed that young drivers

© Springer Nature Switzerland AG 2021
P.-L. P. Rau (Ed.): HCII 2021, LNCS 12773, pp. 182–194, 2021.
https://doi.org/10.1007/978-3-030-77080-8_16

who don't have their own cars made far more mistakes, attention lapses, and violation behaviors than those who do [3]. Previous researches have shown that the more likely causes of traffic accidents among young driver are a large quantity of decision mistakes (e.g., improper speed), dangerous driving action (e.g., one hand driving), distracted driving (e.g., using mobile phone) and the existence of peers. Although inexperience may have played a part, a plenty of these accident-contributing factors infer poor driving skills. In recent years [4], the road traffic problems caused by young drivers have caused widespread concern in China, and this problem may also became a situation commonly faced by developing countries with similar national conditions and driving environments as China. If this technology is integrated into the lives of young people, it will generally improve the driving standards of young people and greatly reduce the chance of accidents [5]. Therefore, it is necessary to explore and analyze the needs of the main consumer group of young people, while trying to meet their needs. According to a McKinsey survey [6] of about 3,000 consumers in the United States, China and Germany, young people living in big cities more interested in automatic cars. In this work, we analyze exploratively the needs, problems, challenges and the future opportunities of young people in regard to highly automated vehicles (SAE level 4 and 5). By the results, we want to find out how the automotive industry should act for the sake of providing better services to the target group.

2   Related Work

Existing surveys [7] have shown that people are generally positive about automated vehicles. However, these researches ignore that automatic cars have different degrees of automation, and that User Experience (UX) and User Experience Acceptance (UA) with automatic systems diverse in terms of system autonomy. Several studies have shown that not only the apparatus itself, but driver's characteristics such as age, gender and experience have an influence on acceptance of ADAS and automatic vehicles. Joshi [8] et al. find that personality aspects are related to the ADAS's acceptable level. Holtl and Trommer [9] show that drivers who have used navigation equipment have lower acceptance to them than people who have not. Piao [10] et al. find that men and younger drivers are less pleased to spend money on the Intelligent Speed Adaptation System (ISAS) than women and older drivers. In another study, Haboucha, Ishaq, and Shiftan [11] found that the youngsters, students and the people who are more educated will be early adopters of automated vehicles and will spend more time on cars.

In terms of exploring users' needs, automated driving brings different needs of user. An important aspect mentioned by Kun [12], Boll, and Schmidt publishes so-called non-driving related activities or tasks (NDRAs): the "driver" can get a variety of activities and can put their hearts into these activities (for example, putting something down, playing games) when the driving is highly automated or fully autonomous. As Kun et al. specially mentioned, cars can be considered as a place to increase productivity and entertainment. This biases the research focus of needs towards NDRA. Investigations in these areas have been carried out by carinsurance.com, McKinsey and J.D. Power [13] and they have investigated the question of what the driver will "use of the newly released time". Sending message and communicating were the most constantly mentioned activities (26%), and "others" accounted for 21% (including enjoying the scenery along the way), besides, "reading" accounted for 21% followed closely. In addition, "sleeping" accounts for 10%, "watching a movie" accounts for 8%, "playing games" accounts for 7%, and "working" accounts for 7%, these activities were not mentioned often. In the research of Pfleging, Rang, and Broy [14] they made specific research on user's needs in NDRA. Due to technical limitations, analyzing users' needs regarding highly automated driving (HAD) is different from studying traditional AutoUI now. In order to estimate future usage in the absence of users' HAD experience, they used WEB SURVEY, CONTEXTUAL OBSERVATION: SUBWAY, and IN-SITU INTERVIEWS IN SUBURBAN TRAINS to survey users' needs for NDRA. And their final findings show that in addition to highly common activities (chatting with passengers, enjoying music), daydreaming, texting, having something to eat, surfing on the Internet and making phone calls are the most needed activities when people can drive highly automated. And we can know the portable and omnipresent multimedia applications' potential through it.

Although previous studies have shown that people broadly are optimistic about highly automated vehicles, the research does not give a consensus, and the controllability and acceptability of these systems is still a disputed issue. Our research's purpose is exploring the acceptance of young users (ages 18–24) for automated vehicles at L4 and L5 levels. We asked questions by simulating scenes inside autonomous vehicles for the group of people. And we specifically explored the acceptance and demands of young people for highly automated vehicles through the data analysis. In the experiments of Rodel [15] et al. they mainly used online surveys to conduct research, and described disparate standings of autonomy in the shape of scenarios, which can enable participants to visualize disparate standings of autonomy. However, in the experiments by Rodel et al., participants scored their acceptance based on imagination rather than actual experience. While in our study, participants were placed in a small room to simulate the autonomous driving scenarios. These increases the realism of the participants' experience and can further obtain more accurate data. Previous research has shown us the NDRA needs of users in automated driving, but they have not investigated the specific needs of specific group of users, and the setting of the scene is to simulate the autonomous driving scenarios through existing public transport. In terms of exploration users' needs, it is limited to the needs for NDRA and does not study other types of requirements. The research area to explore users' needs is expanding, and it is necessary to study the specific needs of specific group of users. Therefore, we have specifically divided users based on previous research, and try to expand from as many as possible when exploring requirements. Mining the subjective needs of users, and then further tapping the potential needs of users.

3   Method

To investigate the attitude of young people towards automated driving detailed, we conducted semi-structured interviews and organized a role play.

3.1 Interviews

We used a semi-structured interview method and recorded each subject to discover the acceptance and trust of the young people in automated driving, we used a semi-structured interview method and recorded each subject. We interviewed a total of 16 subjects, these subjects were 18 to 24 years old. These results were used when setting up our role-playing scene. At the end of the interview, the five-point scale was used to respondents to rate their overall acceptance of autonomous driving, and used this to classify them into high, medium and low acceptance groups [16].

3.2 Role Play

A total of 8 participants participated in the role-playing, and some of them have participated in previous interviews. We set the role-playing scene in a similar environment to the interior of the car. In role-playing, we let two familiar participants execute exercises together, because our interview experience shows that they will be more willing to express, but a larger team size will reduce the efficiency of the express. Asking about their demographic information and the acceptance of autonomous driving (mainly those who did not participate in the first interview) is the first step. Then, we showed them a video about the development concept of the future autonomous vehicles and a video showing the first perspective of autonomous vehicles. Watching the videos is for the sake of clarity. Afterwards, we explored the subjects' expression in different scenarios, which were developed based on the previous interviews and "eight levels of requirements" [17]. This "eight-level requirement" was once used to explore the requirements in product design. Subjects need to answer several questions witch we set in advance in the "automated car". During the experiment, the interviews of all subjects were recorded throughout the course, and their statements and behaviors were analyzed in the subsequent research process. We use qualitative analysis to analyze our interview data. First, we summarized several theme guides based on the questions we asked, and then formulated questions integrating post-it notes before discussion and during the interview. The combination of these topics and finally formed a thematic framework, which is used to analyze the respondents' quotes. By combining the results of our analysis with the results of the interviews, we finally gained a preliminary understanding of the acceptance and needs of young people for automated driving.

3.3 Question Settings

The questions were developed based on the previous interviews and "eight levels of requirements" (Fig. 1). Firstly, we raised a few general questions in different aspects through "eight levels of requirements", then we made Q & A post-it notes based on the initial answers of the users in the role play. After getting these post-it notes, we based on the initial answers and subjects replies to post for further scene settings and questions to get the specific needs of the user (Table 1).

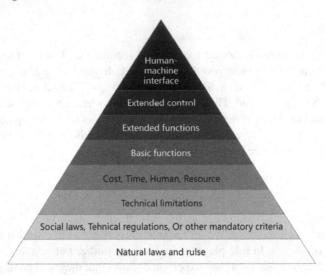

Fig. 1. Eight levels of requirements

Table 1. Question settings

Eight levels of requirements	Questions
Human-machine interface	1. Is it necessary to add a tracking line? (In the videos) 2. Is it necessary to add voice navigation? 3. Wanted control system (input and output) (audio/visual/touch) 4. Is it necessary to connect with another autonomous vehicles?
Exception control	1. Demand in emergency situation (before and after)
Extended functions	1. Special needs (In addition to the basic functions)
Basic functions	1. Fundamental requirements (seats, windows)
Cost, time, human, resource	None
Technical limitation	1. Demand for autonomous driving levels 2. Improvement of specific technology
Social laws, technical regulations or other mandatory criterial	1. Legal needs
Natural laws and rules	None

4 Result

4.1 Acceptance of Young People

Our interviews show that 90% of the participants are willing to drive autonomous vehicles. Highly acceptance accounts for 60%, medium acceptance accounts for 30%, and low acceptance accounts for 10% (Table 2).

Table 2. Acceptance of young people

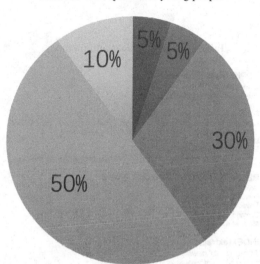

■ Strongly disagree ■ Disagree ■ Not sure ■ Agree ■ Strongly agree

Options	Strongly disagree	Disagree	Not sure	Agree	Strongly agree
Number	1	1	6	10	2

Mostly are enthusiastic of this progress and expect the coming of automated and AI age. Almost half of high acceptance subjects stated they will give priority to cars with self-driving capabilities when buying a car. However, most of them don't trust the technology deeply and worry about security issues recently. The remaining 10% pointed out, that that artificial intelligence is always untrustworthy. They are extremely distrustful of automated vehicles and doubt to the technology. Many of them point out that they will never drive or take automated vehicles.

During the role-playing process, most participants stated that after watching the introductory video, there was an increase in trust and acceptance of autonomous driving. Therefore, we guess that the fear of unknown things is the source of most people's distrust of autonomous vehicles. Once this new thing materializes in their minds, their acceptance of such things will increase to extent. Through the interview, we learned that due to media reports on autonomous vehicle accidents and the lack of popularization of automated cars, making the technology is a major concern for participants, and it also a key issue that affects whether they accept autonomous vehicles.

Meanwhile, the government should strictly control the use of autonomous vehicles, and promulgation relevant laws and regulations, especially the resolution of the problem of liability for accidents in autonomous vehicles, will greatly increase user trust.

4.2   Needs of Young People and Further Chances

Through analyzing qualitative data [18], we sorted out the analysis ideas and methods (Table 3).

Table 3. Analysis ideas and methods

TOPIC GUIDE(extract)
User needs
User needs are reflected in

Social laws : Which aspect ; necessity
Extended functions : What ; why; degree ; interconnected ; data record
Technical limitations: Which level; why; technology needs
Exception control: How; when; what;
Basic functions: Interior layout
Human–machine interface: What; why ; where; degree ; necessity

RESEARCH NOTES AND JOTTINGS
Seat layout
Needs for data record
Privacy needs
Storage function
Medical device

INDEX(extract)
Patterns of needs
1.1 Attribution of legal responsibility
1.2 Traffic laws
1.3 Necessity of law

2.1 Preferred automated driving levels
2.2 Car performance needs
2.3 Other

3.1 Window and seat layout
3.2 Number of windows and seats
3.3 Functions of windows and seats
3 4 Ride experience
3.5 Storage function

4.1 Specific needs for the kitchen
4.2 Specific needs for entertainment
4.3 needs for data record
4.4 Internet needs
4.5 Privacy needs

5.1 Control method
5.2 Control content
5.3 Pre–emergency control
5.4 Control after an emergency

6.1 Necessity and extent of voice control
6.2 Necessity and extent of gesture control
6.3 Car–to–human feedback
6.4 Display identification content and display degree

We first asked the subjects questions and asked them to write their answers on post-it notes (Fig. 2).

Fig. 2. Subjects answers

Then we summarized the subjects' answers on post-it notes, and asked in-depth questions about the parts that could be further explored (Fig. 3).

Fig. 3. Summary

In the end, we summarized the research results into the following table according to the set questions (Table 4).

It is important to solve the problems of young people's daily travel and safety. Under normal circumstances, young people do not have a lot of property at their disposal. The vehicles they mainly come into contact with daily are still public transportation or private cars of others, themselves have little or even no driving experience. In daily travel, young people's travel methods are very diverse. Public transportation, sharing bicycle and their own motorbikes are the most commonly used vehicles. However, even if there are multiple ways to choose from, they said that they still have troubles in weather and impact of insufficient bicycle resources.

Table 4. Questions

Requirements	Results
Social laws, technical regulations or other mandatory criterial	1. Law on attribution
Technical limitation	1. Prefer L5 automated car 2. High-performance cars 3.The identification range and dentification of the car are not affected by the environment 4. New energy
Basic functions	1. Window with single or adjustable light transmittance and large field of view 2. Flexible seats 3. Excellent shock absorption device for office study
Extended functions	1. Information recorder (like the black box of an aircraft) 2. Entertainment device (easy to store) 3. Kitchenette facilities 4. Medical equipment 5. Connected with internet
Exception control	1. Automated alarm (post-accident) 2. More responses (preceding-accident) 3. Timely remind
Human-machine interface	1. Alarms of tracking box 2. Voice and gesture interaction

Automated Cars Level

In terms of the choice of automated cars, they said that they prefer L4 self-driving cars to L5 self-driving cars because they currently have low trust in self-driving cars. They are more willing to go intervene in autonomous driving for the response procedures provided by autonomous vehicles in emergency situations, and thus are more at ease. They also think it is necessary to have an information recorder like the black box on an airplane. Nonetheless, most of them said that they are looking forward to the era of autonomous driving, but also said that they will buy it until the popularity of autonomous vehicles reaches a heavy population.

Technology and Law

For the needs in technology and law, young people hope that use automated cars with security technology and stronger laws. "If there is a car accident, whose responsibility is it? I didn't drive but sitting in the car, whose responsibility is it?" Asked by a subject.

Privacy

Further requirements are having privacy in car. Young people tend to have single light transmission or glass with adjustable light transmission performance windows which can also provide large field of view.

Chair Setting

For the chair settings (Fig. 4), they would like to be rotatable, movable and foldable, and the interior of the car can be flexible to adapt to various scenarios. Some people have suggested that the seat can be laid flat and then spliced to form a bed. Others have suggested that the interior of the car should be able to change color and style. Young people also demand office and entertainment. They pointed out that the car should be equipped with excellent shock absorption devices so that they can read and write along the trip. Subjects also wanted a small table that could be folded and stowed, which some said can be combined with a chair. Young people have a variety of entertainment needs (Fig. 5).

Fig. 4. Subjects placed chairs

Fig. 5. Subjects placed chairs

Entertainment

Many mentioned that they want to install a projector in their car for them to watch movies and television. Entertainment devices such as computers and game consoles in the car which are easy to store also be referred. "It's better to be stored, and the space will look like more comfortable in that case". Said by a subject. In addition, almost all subjects mentioned the need for a small kitchenette, they said that they hope to do same cooking while commuting to save time. It has also been mentioned that it is desirable to provide a space for a picnic device in the car.

Emergency

As almost all the subjects are confused and nervous about how automated vehicles handle emergency situations, it is necessary for vehicles to meet their emergency needs. All subjects said that automated cars need to be equipped with emergency medical boxes, they hope that the car can provide them with more medical assistance in order to respond to emergency situations after the accident.

Before an emergency, quickly analyses of multiple situations are be required. At the same time, they want the car to quickly alert people in the car to prepare. After the emergency, they all want the car to have an automatic alarm function. Some people said that the car should call immediately to save rescue time, and most of the subjects hoped that the car could ask them if they want to call an ambulance before action to avoid unnecessary personnel consumption.

Voice Assistants

In terms of interaction, they prefer intelligent voice assistants to gestures to facilitate their manipulation of the car. "I feel that there(gestures) should be a lot of mistakes. It is the same case with mobile phones, and you gestures may not be fully recognized." Said by a subject. They desire to focus more on voice feedback. They don't need real-time voice of the car to tell them what the car is going to do. They state that it will disturb them to rest. But they desire to be reminded when the car makes more wild movements, such as a sudden braking. And they also want the car to have more intelligent voice navigation, for example, when the car drives to a famous scenic area or representative landmarks in a non-resident city, it can give them a brief introduction and report whether the current scenic area is crowded.

Front Windshield

Most interviewees pointed out that they need self-driving cars to show their identification of the surrounding environment on the front windshield, but there are differences in the level of detail in the selection of objects. The young people with low acceptance of autonomous driving desire to select only small objects that they can't notice, such as roadblocks and distant pedestrians, because they have the strongest needs for driver-assistance. It will not obstruct their view of the front, and facilitate them to drive. Some highly-accepted young people desire to show everything on the road so that they can know the road conditions even in foggy weather or other extreme weather. At the same time, they can also know what self-driving cars have specifically identified. And some highly-accepted young people desire to switch between the two modes.

Association

A part of interviewees believe that it is necessary to conduct road condition analysis by linking with other autonomous vehicles, but they expressed their concerns about privacy leaks at the same time.

5 Conclusion

Most young people are more receptive to autonomous driving for it brings convenience to people and saves people's time. At the same time, autonomous driving systems improve the driving safety of young people and other drivers, which can effectively reduce accidents. Through our research, we found that increasing the expenditure on autonomous driving technology and the publicity of realized technologies can strengthen the support and trust of young people for new technologies to a certain extent. We found in interviews that the distrust of automated cars among young people who do not accept automated cars is mainly related to safety restrictions.

Therefore, meeting the young people's needs for safety is an important opportunity for such people to more accept automated cars. In role-playing, we found that young people like flexible interiors and have high storage requirements for other extended functions. They are looking forward to the convenience that high technology brings to life, but they are skeptical of the technology itself. People who have low acceptance of autonomous driving prefer to use this technology to assist themselves in driving. Therefore, the improvement of autonomous driving technology and interaction technology is necessary. At the same time, most young people hope to buy their own automated cars when autonomous driving has become widespread.

Acknowledgement. The paper is supported by Hunan Key Research and Development Project (Grant No. 2020SK2094) and the National Key Technologies R&D Program of China (Grant No. 2015BAH22F01).

References

1. Deloitte. 2019 Deloitte Global Automotive Consumer Study (2019). https://www2.deloitte. com/cn/zh/pages/consumer-industrial-products/articles/2019-global-auto-consumer-study. htm
2. Ellis, B., Douglas, N., Frost, T.: Willingness to pay for driverless cars. In: Australasian Transport Research Forum (ATRF). Elsevier, Melbourne, Australia (2016)
3. Zhang, Q., Jiang, Z., Zheng, D., Man, D., Xu, X.: Chinese carless young drivers' self-reported driving behavior and simulated driving performance. Traffic Inj. Prev. **14**(8), 853–860 (2013)
4. McDonald, C.C., Curry, A.E., Kandadai, V., Sommers, M.S., Winston, F.K.: Comparison of teen and adult driver crash scenarios in a nationally representative sample of serious crashes. Accid. Anal. Prev. **72**, 302–308 (2014)
5. Poczter, S.L., Jankovic, L.M.: The google car: driving toward a better future? J. Bus. Case Stud. (JBCS) **10**(1), 7–14 (2013)
6. Sovie, D., Curran, J., Schoelwer, M., Björnsjö, A.: 2019 Deloitte Global Automotive Consumer Study (2019)
7. König, M., Neumayr, L.: Users' resistance towards radical innovations: the case of the self-driving car. Transp. Res. Part F Traffic Psychol. Behav. **44**, 42–52 (2017)
8. Joshi, S., Bellet, T., Bodard, V., Amditis, A.: Perceptions of risk and control: understanding acceptance of advanced driver assistance systems. In: Gross, T., Gulliksen, J., Kotzé, P., Oestreicher, L., Palanque, P., Prates, R.O., Winckler, M. (eds.) INTERACT 2009. LNCS, vol. 5726, pp. 524–527. Springer, Heidelberg (2009). https://doi.org/10.1007/978-3-642-03655-2_58

9. Höltl, A., Trommer, S.: Driver assistance systems for transport system efficiency: influencing factors on user acceptance. J. Intell. Transp. Syst. **17**(3), 245–254 (2013)
10. Piao, J., McDonald, M., Henry, A., Vaa, T., Tveit, O.: An assessment of user acceptance of intelligent speed adaptation systems. In: Proceedings 2005 IEEE Intelligent Transportation Systems 2005, pp. 1045–1049. IEEE (2005)
11. Haboucha, C.J., Ishaq, R., Shiftan, Y.: User preferences regarding autonomous vehicles. Transp. Res. Part C Emerg. Technol. **78**, 37–49 (2017)
12. Kun, A.L., Boll, S., Schmidt, A.: Shifting gears: user interfaces in the age of autonomous driving. IEEE Pervasive Comput. **15**(1), 32–38 (2016)
13. Mohr, D., et al.: Competing for the connected customer. Technical report. McKinsey & Company (2015). https://www.mckinsey.de/sites/mck_files/files/competing_for_the_connected_customer.pdf
14. Pfleging, B., Rang, M., Broy, N.: Investigating user needs for non-driving-related activities during automated driving. In: Proceedings of the 15th International Conference on Mobile and Ubiquitous Multimedia, pp. 91–99 (2016)
15. Rödel, C., Stadler, S., Meschtscherjakov, A., Tscheligi, M.: Towards autonomous cars: the effect of autonomy levels on acceptance and user experience. In: Proceedings of the 6th International Conference on Automotive User Interfaces and Interactive Vehicular Applications, pp. 1–8 (2014)
16. Guion, L.A., Diehl, D.C., McDonald, D.: Conducting an in-depth interview. McCarty Hall, FL: University of Florida Cooperative Extension Service, Institute of Food and Agricultural Sciences, EDIS (2001)
17. Chen, Z.Y., Zeng, Y.: Classification of product requirements based on product environment. Concurrent Eng. **14**(3), 219–230 (2006)
18. Ritchie, J., Spencer, L.: Qualitative data analysis for applied policy research. In: Bryman, A., Burgess, B. (Eds.) Analyzing Qualitative Data, p. 173

Analyze the Impact of Human Desire on the Development of Vehicle Navigation Systems

Feng Lan$^{(\boxtimes)}$, Chunman Qiu, Weiheng Qin, Peifang Du, and Hao Tan

The School of Design and Art, Hunan University, Changsha, China

Abstract. The driving environment of automobiles is developing towards diversification and the navigation system in automobiles is also undergoing subversion. But understanding people's desire has become the key expectation of the future development of a navigation system. To better analyze the impact of human desire on the development of vehicle navigation systems, a design method based on human psychological needs is adopted. The participants were divided into two groups: 10 aboriginal people (generation info.) in the information age and 20 professionals in society (professional); The two groups of interviewees have very different growth experiences, and they have different experiences in life scenarios and navigation systems. In-depth interview and analysis of them will help to discover the real internal needs of people today. The five-in-one research framework is used to design a series of problems, according to the experience of vehicle navigation system, which to explore people's internal needs and the potential development of automobile navigation. Through analysis, six themes related to human desire and car navigation were obtained. Two groups had significant differences in their views on a few numbers of themes. Professional group (generation info.) tends to think of it as a simple guidance tool when using a navigation system, and the emerging generation (generation info.) prefers to discuss the extension and additional applications of navigation systems based on the emotional attachment. Finally, we conducted further discussions on the results of the two groups of members, and conceived the future development and extended application of the vehicle navigation system based on needs and emotions.

Keywords: Navigation system · Human desire · Automobile · In depth interview · Topic resolution

1 Introduction

The development and popularization of information technology have transformed the positioning and role of vehicles in modern people's lives from simple vehicles to an important part of daily life, and there has been a demand for the further development of intelligent vehicle-mounted equipment.

Moreover, as mobile digital devices such as smart phones are gradually being expanded in all aspects of life [1], the integration and connection of these smart devices

© Springer Nature Switzerland AG 2021
P.-L. P. Rau (Ed.): HCII 2021, LNCS 12773, pp. 195–211, 2021.
https://doi.org/10.1007/978-3-030-77080-8_17

and vehicles is also the focus of today's drivers [2]. In the foreseeable future, allowing the occupants in the vehicle interior environment using various applications on devices or mobile devices can provide more convenience and comfort for the car to travel [3].

Due to the inherent car, along a particular road transport from the functional properties of a point to another point; in these devices and systems for occupants navigation device is closely related to the car's driving contact is particularly prominent.

Navigation devices have a variety of forms of existence and interaction in vehicles, all of which emphasize practicality and ease of use. However, in the existing studies, quantitative studies, such as technology and ergonomics, have always been dominant [2], and only a few studies have attempted to qualitatively explore the relationship between people's desires and needs and the vehicle's navigation system. However, the traditional quantitative method may not be able to effectively describe human desire and other indicators, so it cannot provide strong support for the design of navigation system.

Therefore, we adopted in-depth interviews and topic analysis methods, citing the User sample selection hierarchy chart, in order to recruit the professional group of interviewees; At the same time, we recruited another group of interviewees in the information age, and formulated a detailed research design plan, and then Conducted multilateral semi-structured interviews. Then, describe the results of the interview and perform data analysis.

We quoted Yao's User sample selection hierarchy chart. In Yao's discussion, this model is based on the industry, lifestyle and demand characteristics of the user group. It is to identify user needs and convert it into the process of design's knowledge. This model helps us narrow down the scope of interviews and conduct accurate in-depth interviews, and also facilitates the analysis of later information.

When setting up groups, the professional interview team, including market managers, communication department heads, education authorities, design researchers and others who are in close contact with vehicle navigation systems and related industries. They have a clearer understanding of the development process of navigation systems. They have their own understanding of the future direction; not only that, they are people with rich social experience, their views largely represent most of society, and have a greater impact on interview analysis. The other group of interviews is the indigenous people of the information age, also known as the information age. They have been exposed to the Internet since their birth and dealt with electronic devices and smart devices. They have unusual effects on these devices, including vehicle navigation systems. Insight, they know more about the advantages and disadvantages of these devices. Not only that, this generation of people has unimaginable visions for future scenes. Their inner thoughts greatly influence the future development of these devices.

The purpose of this study is to explore the relationship between current and imaginary future vehicle navigation systems and human desires. In order to achieve this goal, we use the people-oriented design method to design the problem; with the help of the User sample selection hierarchy, we establish the logical structure of recruiting participants, and develop a detailed research design plan, followed by multilateral semi-structured interviews. Then, describe the results of the interview and analyze the data. Finally, from the perspective of sociological phenomena of human needs and values, six themes of desire are discussed and put forward.

2 Research Design

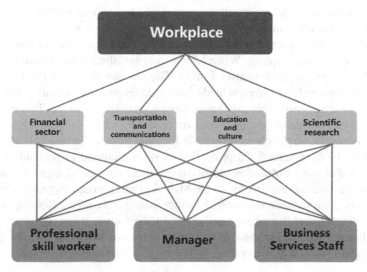

Fig. 1. User sample selection hierarchy chart

2.1 User Sample Selection Hierarchy

Construct a research using user sample selection hierarchy chart [4] to identify the two groups participating in the research, and analyze the interests of different groups according to the research purpose [5]. The most relevant definition of this experiment from Freeman's definition [6] of stakeholder is "includes people, groups, or organizations interested in the performance of the project and the results of planned actions" [5, 7, 8]. In order to recruit the appropriate interviewees, use the benefits user sample selection hierarchy chart. The structure of this model is adapted from the Connected Car of the GSMA [9] (Groupe Speciale Mobile Association) (See Fig. 1).

In order to collect data of different age groups (17–50) and compare and analyze valuable data, the experiment is divided into two stages of the interview. In the first stage, according to the groups, the professional population is composed of "generation X", they are a generation born between 1961 and 1980 [10]. Another group is composed of the "information age" born after 1995 [11, 12]. They are deeply influenced by the Internet, instant messaging, intelligent devices and other technological products [13]. By comparing their needs with those of the first stage population, new interview results are obtained, thus filling the gap of the same group [10–13, 15].

2.2 Non-probability Sampling

Non-probability sampling refers to the method by which investigators draw samples based on their subjective experience or other conditions. Non-probability sampling is mainly applicable to the situation where the sampling frame is incomplete, and the sampling process is not performed according to the random principle. According to statistics, 70% of the items in the market survey are non-probabilistic surveys [15].

In the experiment, the quota sampling method [16] of non probability sampling is adopted, that is, the population elements are classified according to some control indexes or characteristics, and then the sample elements are selected according to random sampling or judgment sampling. When taking samples, the quota in each category is completed according to each control feature. The characteristics of the interviewees selected were: coming from different fields, having different backgrounds, willingness to share their views, and being able to generate new views in the guidance. This method can effectively obtain a wide range of perspectives from interviews [17], so it is selectively used in experiments. So the characteristics of the interviewee can be easily identified.

According to the standard, 20 professionals were recruited to participate through networking and network promotion. Among them, there are 3 car designers, 2 navigator designers and 2 ride-hailing drivers. In the Manufacturing group, there are 2 navigator manufacturers, 2 car manufacturers and 1 factory manager. From the Retail group, there are 2 car salesmen and 2 navigator salesmen. In the fourth group, there are 2 white collars, 2 housewife. They come from different ages, qualifications and backgrounds and have a good balance of views (See Table 1).

Table 1. 20 participants in the professional group.

	Transportation	Manufacturing	Retail	Others
Professinal skill woker	3 Car designers 2 Navigator designers	2 Navigator manufacturers 2 Car manufacturers		2 White collars 2 Housewife
Manager		1 Factory managers		
Business services staff	2 Ride-hailing drivers		2 Car salesmen 2 Navigator salesmen	

In the second phase, 10 members of the information generation were recruited. Unlike the professional group, this group was recruited randomly based on age and gender on a voluntary basis. Payment was offered.

Table 2. Criteria for designing interview question.

To understand the content	To understand the meaning and needs
AEIOU framework Activity Environment Interaction Objects Users	**5Ws and H framework** Who, When, Where, What, Why, How

To understand the action and logic
Event handling framework Users Activity Surroundings Objects Time

2.3 Designing Interview Questions

To set up context related issues, we first used the a.e.i.o.u framework (see Table 2). In conjunction with this framework, we should first consider the needs that people want to be satisfied, how to use the navigation system in a given future situation, how to interact and connect with the navigation system, and their internal expectations for navigation systems [18]. In order to obtain sufficient interview results and various possibilities, and to explain the meaning properly. The designed questions can get more specific and multiple possible results, and can guide the interviewees to think more deeply to get unexpected results. Divide the questions into three categories to judge the truthfulness of the respondents' answers (See Fig. 2). In addition, the semantic difference framework is used to measure meaning [19], and the 5W 1H frameworks are used to balance views [20]. It is worth noting that the questions drawn through the framework are designed to motivate those involved in the interview to envision cutting-edge and emerging technologies that can be applied to navigation systems. In order to obtain unexpected views as much as possible and improve the openness of the interview [21, 22], setting up some guiding questions can make the interviewees better imagine the future. (For example, what is the difference between the future shared car navigation system and the existing form?).

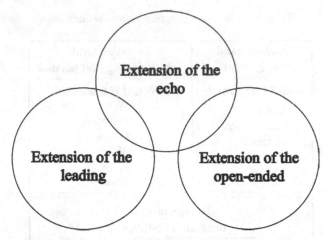

Fig. 2. Problem classification methods for obtaining multiple results

2.4 Question Examples

Extension of the leading:

*Now there is a driverless car. Do you want the car to take you to your destination in one sentence, or guide you through the display? Why?

*In the process of going to the destination, you suddenly want to change the destination, do you want to change the voice, or manually?

*Will you accept the navigation system in shared cars? Personal account login navigation system/privacy.

*The help/impact of the future car interior screen for navigation, compared with AR real navigation.

*When choosing a destination, do you check before departure or in the car?

Extension of the open-ended:

*Do you use AR live navigation on mobile phone map (demo)? Are there any problems during use? Don't you know it or use it and don't want to use it?

* When traveling by car, what should I do if there is a traffic jam in front of me? What are you doing during traffic jam? Will you choose to bypass the road to avoid traffic jam?

*Car music conflicts with navigation voice, what to do, why?

Extension of the echo:

*For example, the Pentium T77 uses holographic projection technology, but the display form is to simulate virtual characters, as a car assistant to provide services to users (or to accompany, reduce loneliness), do you think it will interfere with/help driving, why?

*What are your thoughts on future driving navigation?

2.5 Interview Procedure

All interviewers (n = 24) were interviewed with the same semi-structured interview questions during the meeting. Get the most out of multiple possibilities. The target interview time is 30 to 50 min, which can alleviate the fatigue caused by conversation and cogitation, thereby reducing the bias of interview results. Before the interview, inform the interviewees about the purpose of the experiment and the use of the results to ensure the interviewees' right to know.

2.6 Data Analysis

The purpose of analyzing interview process information is to express people's desire more clearly, to form a connection with the potential development possibility of vehicle navigation, and to get the design theme. Using the method of subject analysis to deal with the interview data [23–25]. In the interview process, the interviewers' questions, to varying degrees, will have an impact on the cognition and answers of the interviewees. Therefore, using the method of subject analysis, the interview questions are divided into three categories (See Fig. 5) to reduce deviation and subjectivity [25, 26]. In order to improve the accuracy and objective multi-dimensionality in keyword extraction and topic resolution, two professionals were invited as reviewers. Among them, they are familiar with psychology and design (23-year-old female) and engaged in interactive research (40-year-old male). People of all ages and backgrounds help to obtain a wide range of perspectives [27].

2.7 Subject Analysis

Transcribe all the recordings of the interview by machine, make compound comparison between the output text and the interview record, read again and adjust manually to ensure the accuracy of the recording.

According to the purpose of the study, words, phrases, proverbs, sentences and paragraphs are analyzed from multiple perspectives, and classified and encoded on the basis of transcoding text. In the process of classification, the above words and sentences are given a highly recognizable description.

Call out the interview text, make a horizontal comparison between the scene information involved in the interview process and the code words, and sort out the scene themes that are consistent with each other; make a secondary sorting of the sorted code words (on a higher semantic system). Examine the transcribed text, search for ignored contextual topics, and identify recurring responses.

For the purpose of objectivity, the two reviewers independently analyzed the transcripts provided, obtained the self-contained codewords and the resolution theme, and compared the resolution theme with the provided theme, and put forward suggestions on some themes. In this regard, in-depth discussions were reworded on the subject of ambiguous definitions.

During the review process, recombining and deleting themes, wording and defining the final theme.

3   Results

The professional group (Professional) and the information generation group (generation info.) came up with six themes, Satisfaction of morality and privacy, Humane factor in an intelligent situation, Integration and connection, Progress with emotion gradation, Accuracy and efficiency of multi-channel input, Efficient output. Among the
professional population group and the information generation group had significant differences in the views of the first two themes. These results can be roughly divided into
two categories in terms of human needs and social values, and the human desires involved
in these results can be used to better explore the possibilities of automobile navigation
systems (Fig. 3).

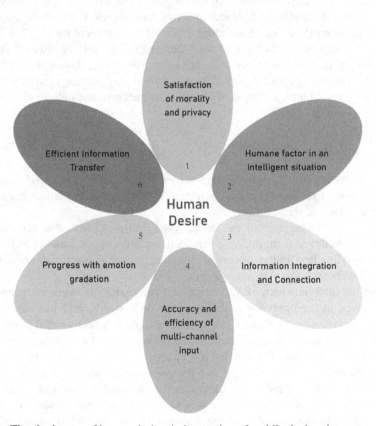

Fig. 3. The six themes of human desires in integration of mobile devices into automobiles

3.1   Satisfaction of Morality and Privacy

In this era of easy access to information, various events caused by frequent disclosure
of personal information on the Internet make people pay more attention to their privacy

rights. In most cases, it is difficult for people to agree with the Internet collection of user information, but in the actual interview process, it is found that the two interview groups have different views on this.

The information age group believes that the Internet age is an open era, and the data collected by the application companies are used for the follow-up development of software. Therefore, they are not too worried about the collection of user information by Internet companies or car navigation companies. Most of them choose to believe in the Internet and car navigation companies.

In the 21st century, we are actually running naked on the Internet. Cloud data is more convenient for our life --- generation info.

The professional team thinks that they are very worried about their personal information being illegally stolen after being collected, but they also say that this is a problem that is difficult to solve. Nowadays, the frequently used addresses (companies, schools, homes) stored in the navigation system have exposed their personal information, but it does improve the user experience.

Information privacy is a false proposition---professional.

People's desire for privacy right in Internet car navigation is a disappearing process (people will gradually recognize it). There is no need to comfort users in the design process. On the basis of privacy, a higher-level problem is raised: the navigation system (such as brain computer interaction) using a new way of interaction is actually risky on the moral level.

When an antisocial person is using the brain computer interface to operate the car navigation system, any of his ideas will be directly and quickly reflected in the vehicle, which is likely to cause major accidents.----Professional.

3.2 Humane Factor in an Intelligent Situation

The "humane factor" in the paper refers to the fact that the operating system can meet the functional and psychological needs of users according to their living habits and operating habits. In the process of using the individual needs, and the system to give users humane care. This part mainly discusses the individuation and humanization in "human care".

Development of users' preference is the most considerable procedure in humanized design, in the light of behavior in daily life, physiological factors, human psychological state, human thinking mode, to optimize the design, so that users can conveniently and comfortably visit and use. As an important auxiliary driving system, vehicle navigation system provides users with pleasure, satisfaction and humanized care. The interviewees have two different views on the humanization of navigation system. One group of people thinks that the navigation system is humanization just as a tool to bring people to the destination. Some people think that the navigation system should be given the function of humanized care so as to provide users with the pleasure and satisfaction of driving or riding.

The navigation system is only responsible for getting me to my destination safely, without guessing if I'm lonely, or if I need any recommendations or chat -- professional

I look forward to it not only as a navigation, but also as a ride companion.-- generation info.

Based on different interests and hobbies of individuals, different behaviors of using navigation, and different moods of using navigation in different situations, the personalized interaction methods and contents of personalized navigation can meet the different needs of different people. Relatively, it can improve the user's degree of freedom, and provide users with endless possibilities in terms of feel. For example, under the premise of ensuring safety and rationality, the control key position can be set individually.

I hope the projected virtual characters will be my favorite cartoon characters -- professional.

I like this control button putting here, it feels very comfortable and convenient to put it in this way----generation info.

The humanized elements of the navigation system are embodied in giving people the pleasure of driving, meeting consumer demand, and highlighting people's life taste as much as possible, bringing people a unique feeling of life. Navigation appearance and interactive graphics can choose a good visual color, to avoid the wrong guidance to people's vision, resulting in fatigue. Route planning should be in line with users' cognition and usage habits to avoid cognitive difficulties.

3.3 Information Integration and Connection

The continuity and integration of information are the key of a vehicle navigation system. Respondent groups use mobile devices such as mobile phones as vehicle navigation systems, and the effectiveness of information transmission will directly affect the user experience. In the future of the popularization of driver-less vehicles, continuity will be particularly important. In order to reduce costs, this kind of vehicle will not be equipped with a navigation system, instead, it is the information bridge between mobile devices and driver-less vehicles.

Before I got in the car, I had searched and confirmed the route on the mobile phone. After I got in the car, the car could read the route information in my phone and take me directly to the destination.----generation info.

Both groups are not satisfied with the real-time and comprehensive information provided by the existing navigation system; the ideas of passengers in the car are changeable, and the information provided by the existing navigation system is relatively isolated, unable to meet the multi-dimensional ideas of passengers. In the big data scenario, the integration of navigation system information will provide a more comprehensive solution for the passengers in the driver-less car and the ordinary car drivers.

In the event of a traffic jam, there is an information bridge between vehicles, which can adjust the route through the vehicle navigation system, and carry out road evacuation or dispatching uniformly.----Professional.

3.4 Progress with Emotion Gradation

On the emotional side, the navigation system can be invoked as a vehicle-mounted intelligent housekeeper, not only managing the recommended route, but also providing additional information along the way. At the same time, it can meet people's emotional needs, provide emotional interactive functions, and reduce loneliness.

During the self-driving tour, the navigation system can remind us to enjoy the scenery along the way when I pass the scenic area----generation info.

Can add more emotional voice communication----professional.

In the process of using public navigation system, as mentioned in Donald Norman's "Emotional Design", users will have an emotional experience progressive process. When seeing the appearance of the navigation system, the user produces instinctive emotions (first level), intuitive beauty will affect the next experience; based on the first level, the series of interactive behaviors of the navigation system will determine the user's ability to understand, that is, behavior Emotion (second level); when completing the two emotional levels of instinct and behavior, users will focus on reflection (third level), compare the previous operating experience with the current one, and find a reasonable operating point. With the gradual progress of the emotive level, the emotional level of the navigation system directly affects the driving experience of the driver or passenger, and even affects the safety during driving.

I just hope the navigation system will make my experience in the car more splendid without increasing the cost of learning----generation info.

3.5 Accuracy and Efficiency of Multi-channel Input Mode

The majority of respondents said they had used more than one type of input in their current usage scenario, often on navigation devices that offered a variety of input options.

It's not suitable for the type input of car navigation and touch screen in the center console. I can use voice when I have a choice----Professional.

When you have to use an input method you don't like, most of the time it's limited for external reasons.

This system enables users to input information in multiple ways. Inevitably, when using different approaches of input, there will be a distinction between accuracy and efficiency. This requires users to choose and choose according to their personal preferences and external conditions.

In addition, the information input by different channels can and should be transformed into general form information as far as possible through technical means for subsequent processing. At the same time, if the user interface can be added with the compensation mechanism of instant selection or confirmation of input content or timely error correction, the user experience will be improved to a great extent through multiple input modes, that is, the so-called multi-channel input [28].

But when only one hand is free, I can just send voice, and then transform voice into texts----generation info.

3.6 Efficient Information Transfer

In previous studies, the visual design of vehicle navigation is to avoid the operation mistakes of driving users, but with the development of the times, driverless vehicles become the future trend, the driving behavior of users will be reduced or even disappear, and the visual design of navigation system will no longer be limited.

This topic mainly relates to how to effectively output information without affecting driving safety. Obviously, it is necessary to optimize the output according to specific driving methods and ergonomics. Secondly, it is necessary to reduce mutual interference with the information of other systems or devices during the output process, which may require the synergy of multiple sensory output methods.

When playing music, I just mute the navigation, watching the graphics of the navigation... When I need to turn, it will be "ding" (vibrate) to remind you----Professional.

Some interviewees mentioned that for image display, simplified or abstracted patterns or models can be displayed in the display interface, which helps to realize the docking of a real scene and graphics and convey information more efficiently.

Like this floor plan, I believe that it is not intuitive enough. It needs a process of transforming the floor plan and the foreground image of the eye---generation info.

You can use clayrender to add to the map, and the guidance for people will be more intuitive---professional.

Six themes were proposed:
1. Satisfaction of morality and privacy 2. Humanized care in intelligent situation 3. Integration and connection 4. Personalized of interaction 5. Accuracy and efficiency of multi-channel input 6. Efficient information transfer. It roughly divides into two categories: human care and practical function. This classification implies requirements similar to the sociological phenomena of human needs and values [29–32].

Themes 1, 2, and 4 express the driver's desire for emotional and care with the vehicle navigation system. Due to the progress of technology, artificial intelligence is increasingly imitating human beings, and people have a tendency to treat people and artificial intelligence in the same way [33]. Therefore, people hope that the navigation system has the function positioning beyond the functional assistance, and they hope that emotions can get special care and consideration in the service. To some extent, this can relieve the loneliness of driving alone [34–36]. In terms of these three points, navigation system needs to meet the needs of maslow's third level, that is, social, emotional and belonging need [37].

Themes 3, 5 and 6 express the desire of drivers and passengers to optimize the practical functions of a vehicle navigation system. It can be seen that due to the development of information technology. The functions of intelligent devices are continuously strengthened, and the possibility of interconnection between various intelligent devices is increasing. Both the driver and the passenger are anxious for the navigation system to better realize its inherent functions and meet their functional desires through the interconnection between multiple devices and a variety of accurate and efficient input/output modes.

4   Discussion

We will explore six topics in terms of needs and emotions.1.Satisfaction of morality and privacy 2. Humane factor in an intelligent situation 3. Integration and connection 4. Progress with emotion gradation 5. Accuracy and efficiency of multi-channel input 6. Efficient information transfer.These six themes are divided into humanistic care and practical functions according to need. This classification implies requirements similar to the sociological phenomena of human needs and values [29–32].Matching and mismatch between people's needs and the functions of the product can lead to a positive emotion, a negative emotion or a neutral emotion. The sentiment formed by people is their evaluation of whether the effectiveness of a product meets the needs [16].After sorting out the six themes from human needs, we try to analyze the emotions after these six desires in order to better understand these needs.

In terms of demand, themes 1, 2, and 4 express the driver's desire for emotion and care with the vehicle navigation system. Due to the progress of technology, artificial intelligence is increasingly imitating human beings, and people have a tendency to treat people and artificial intelligence in the same way [33]. Therefore, people hope that the navigation system has the function positioning beyond the functional assistance, and they hope that emotions can get special care and consideration in the service. To some extent, this can ease the loneliness of driving alone [34–36]. In terms of these three points, navigation system needs to meet the needs of maslow's third level, that is, social, emotional and belonging need [37].

Themes 3, 5 and 6 express the desire of drivers and passengers to optimize the practical functions of a vehicle navigation system. It can be seen that due to the development of information technology, the functions of intelligent devices are continuously strengthened, and the possibility of interconnection between various intelligent devices is increasing. Both the driver and the passenger are anxious for the navigation system to better realize its inherent functions and meet their functional desires through the interconnection between multiple devices and a variety of accurate and efficient input/output modes (Fig. 4).

In terms of emotion, in organization and business environments, people often have distinct roles and workflows, from which functional and non-functional requirements can be extracted. Design psychology believes that each emotion represents a psychological need. Based on the emotional attachment framework, the conclusions are analyzed to provide more valuable analysis for the conclusions.

'Self-expression' is a kind of vision that a person expresses differently from other people's personal identity through emotional thoughts, etc. This emotion is reflected in the personalization of the interaction mode, and the experience of people setting the navigation system through personalization The effect satisfies the need of psychological pleasure, and further highlights the differences with others, indicating the differences between individuals and the uniqueness of the self. In the design of navigation, the designer excavates people's habits when using navigation and classifies the actions and behaviors of different user groups when using the navigation system, so as to provide users with personalized interactive design and appearance design.

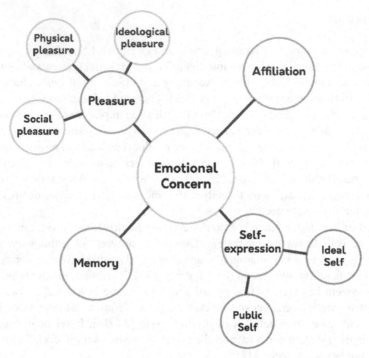

Fig. 4. The emotion attachment model [19]

On the level of "pleasure", it is divided into three directions: social pleasure, physical pleasure and ideological pleasure. Everyone wants all their experiences in the car environment to be comfortable, smooth and private. In other words, users' satisfaction with the privacy in the navigation system is assigned to social pleasure. And multi-channel input is to improve the user information input of mental pleasure. In the future car navigation design, new technologies popular with the user group can be combined, such as the fifth generation mobile communication technology can improve the fluency of the connection between the upcoming smart mobile device and the car navigation system. Thereby enhancing the user's pleasure in using the navigation system.

On the level of "memory", memory connects consumers with products emotionally. Enhances the experience of users' familiarity and brings additional experience effects by contacting with the content and methods familiar to users. This level is mainly reflected in the humanized care in the intelligent situation. With memory, the navigation system can provide users with familiar emotional communication effects, such as providing a familiar voice in the output mode, thus reducing the sense of strangeness.

"Affiliation" is the relationship established between an individual and other things, from which a sense of security and belonging can be achieved. The theme "integration and connection" are related to it. Users want a faster and more efficient connection between smart mobile devices and car navigation to enhance the experience, which increases the connection between people, devices and cars, and obtain an integrated experience. In the future automobile navigation design, designers should fully consider how to bring the sense of belonging to users. For example, this sense of belonging can be reflected in whether the automobile navigation is integrated with the automobile or connected with the automobile as a communicative device so as to bring users a sense of close connection or a sense of portable experience.

During the discussion, it was found that the interviewed groups had significant differences in specific themes. Considering their living habits and backgrounds, this requires designers to devote more attention to the characteristics of these two groups and improve the user experience.

In the theme of "Satisfaction of Ethics and Privacy", the professional group is concerned about the leakage of their personal information, and the information generation is not too worried about the leakage of privacy. For this difference, consider giving appropriate hints of privacy and security, so that professionals could trust the privacy and security of navigation without burdening information generations.

In the theme of "Humane factor in an intelligent situation", the professional group believes that navigation does not need to increase humanization too much, while the information generation believes that humanized care is needed. In view of this difference, in the future design, the integration of humanizing functions should be considered, but the user is given the right to cancel the setting of humanized.

In previous automotive research, it has often focused on improvements in ergonomics and technology [2]. The six themes of the study attempted to outline human desires for car navigation and qualitatively draw conclusions. The six themes derived from this research can eventually be used as design checklists and used as design guidelines for the development of related products and services. In addition, each of the topics can be used as a strategic point to support the generation of certain designs with special uses or purposes, or in-depth research can be conducted around the topic to support the generation of new technical standards or industry specifications.

5   Conclusion and Future Work

In this study, the people-oriented design method is adopted to develop the research plan, the User sample selection hierarchy is established, the in-depth interview method is adopted, and then the interview results and data are collected and summarized to extract, and the requirements of drivers and passengers for the vehicle navigation system are obtained. Finally, six themes that should be taken into account when proposing the vehicle navigation system are summarized. This study identified six topics related to human desire and vehicle navigation system, which may be helpful for the design of future vehicle navigation system, and for each topic, which may be used to generate special design or technical standards and industry specifications. Further investigation, experiment and research are needed to apply the results of this study to practice.

Acknowledgement. The paper is supported by Hunan Key Research and Development Project (Grant No. 2020SK2094) and the National Key Technologies R&D Program of China (Grant No. 2015BAH22F01).

References

1. Berends, L., Johnston, J.: Using multiple coders to enhance qualitative analysis: the case of interviews with consumers of drug treatment. Addict. Res. Theor. **13**(4), 373–381 (2005)
2. Xu, X.Y., Ren, J.: Intelligent man-machine interactive system of car interface design. Design (2015).19th issue 2015
3. Bershidsky, L.: Here Comes Generation Z (2014). https://www.bloombergview.com/articles/2014-06-18/nailing-generation-z. Accessed 20 Mar 2015
4. Yao, X., Cao, X.: Research on User's Lifestyle before Strengthening the Pre-concept Stage of Product Design. Wuhan University of Technology, Wuhan 430070, China (2010)
5. Boyatzis, R.E.: Transforming Qualitative Information: Thematic Analysis and Code Development. Sage (1998)
6. Biederman, K.K.: Moral Responsibility and the Limits of Ignorance. ProQuest LLC. (2008)
7. Bradshaw, J.R.: The taxonomy of social need. In: McLachlan, G. (ed.) Problems and Progress in Medical Care, Oxford University Press (1972)
8. Braun, V., Clarke, V.: Using thematic analysis in psychology. Qual. Res. Psychol. **3**(2), 77–101 (2006)
9. Breazeal, C.: Emotion and sociable humanoid robots. Int. J. Hum. Comput. Stud. **59**, 119–155 (2002)
10. Burnard, P.: A method of analysing interview transcripts in qualitative research. Nurse Educ. Today **11**(6), 461–466 (1991)
11. Capgemini: Cars online 12/13: My car, my way (2012). https://www.capgemini.com/thought-leadership/capgeminicom-cars-online-1213-my-carmy-way. Accessed 28 Feb 2015
12. Wallop, H.: Gen Z, Gen Y, baby boomers a guide to the generations (2014). https://www.telegraph.co.uk/news/features/11002767/Gen-Z-Gen-Y-baby-boomers-a-guide-the-generations.html. Retrieved 15 Mar 2015
13. Wang, B., Wang, Y.: A comparative study of post-80s and post-90s college students' values. Renmin Univ. China Educ. J. **4** (2016). School of Public Management, Renmin University of China, Beijing, China
14. Xiao, H.: Determination of Sample Size in Non-probability Sampling. School of Mathematics and Computer Science, Shanxi Datong University, Datong Shanxi (2018)
15. Wang, B.: Application of quota sampling in letter analysis of accounts receivable. Financ. News **2018**(24) (2018). Tianjin University of Finance and Economics
16. Edward Freeman, R.: Strategic Management: A Stakeholder Approach. Cambridge University Press (2010)
17. Eichler, S., Schroth, C., Eberspächer, J.: Car-to-car communication. In: Proceedings of the VDE-Kongress - Innovations for Europe (VDE Kongress 2006), vol. 6 (2006)
18. Sherkat, M., Mendoza, A., Miller, T., Burrows, R.: Emotional Attachment Framework for People-Oriented Software. School of Computing and Information Systems, The University of Melbourne, VIC, Australia (2018)
19. Giacomin, J.: Human centred design of 21st century automobiles. ATA Ingegneria dell' Autoveicolo **65**(9/10), 32–44 (2012)
20. Han, S., Lee, K., Lee, D., Lee, G.G.: Counseling dialog system with 5W1H extraction(2013)
21. Giacomin, J.: What is human centred design? Des. J. **17**(4), 606–623 (2014)
22. Gkouskos, D., Normark, C.J., Lundgren, S.: What drivers really want: investigating dimensions in automobile user needs. Int. J. Des. **8**(1), 59–71 (2014)
23. Goffman, E.: The Presentation of Self in Everyday Life. Anchor Books (1959)
24. GSMA: Connected car forecast: global connected car market to grow threefold within five years (2013). https://www.gsma.com/connectedliving/wp-content/uploads/2013/06/cl_ma_forecast_06_13.pdf. Accessed 1 Mar 2015

25. GSMA and SBD: 2025 every car connected: forecasting the growth and opportunity (2012). https://www.gsma.com/connectedliving/wp-content/uploads/2012/03/gsma2025everycarconnected.pdf. Accessed 1 Apr 2015

26. Hafner, K.: Coming of age in palo alto: anthropologists find a niche studying consumers for companies in Silicon Valley (1999). https://partners.nytimes.com/library/tech/99/06/circuits/articles/10anth.html. Accessed 30 Mar 2015

27. Henfridsson, O., Lindgren, R.: Multi-contextuality in ubiquitous computing: investigating the car case through action research. Inf. Organ. **15**, 95–124 (2005)

28. Zhang, C., Zhao, J.: State Key Laboratory of Advanced Design and Manufacturing for Vehicle Body, Hunan University, Changsha 410082, China

29. Huber, U.: Mobility of the future. In: Proceedings of the 2013 ACM Workshop on Security, Privacy & Dependability for Cyber Vehicles, pp. 1–2 (2013)

30. James, L.: Data on the Private World of the Driver in Traffic: Affective, Cognitive, and Sensorimotor. Department of Psychology University of Hawaii, Hawaii (1984)

31. Jinho, J.: Future visioning system for designing and developing new product concepts in the consumer electronics industries. Ph.D Dissertation. Brunel University, Uxbridge, UK (2002)

32. Kapsalis, V., Kalogeras, A., Charatsis, K., Papadopoulos, G.: Seamless integration of distributed real time monitoring and control applications utilising emerging technologies. In: Proceedings of the 27th Annual Conference of the IEEE Industrial Electronics Society (IECON 2001), pp. 176–181 (2001)

33. Karwowski, W.: The discipline of ergonomics and human factors, 3rd edn. In: Salvendy, G., (ed.) Handbook of Human Factors and Ergonomics, pp. 3–31. Wiley (2006)

34. Kriglstein, S., Wallner, G.: HOMIE: an artificial companion for elderly people. In Proceedings of the CHI Conference on Human Factors in Computing Systems, pp. 2094–2098 (2005)

35. Lisetti, C., Nasoz, F., LeRouge, C., Ozyer, O., Alvarez, K.: Developing multimodal intelligent affective interfaces for tele-home health care. Int. J. Hum Comput Stud. **59**(1), 245–255 (2003)

36. MacArthur, J.: Stakeholder analysis in project planning: origins, applications and refinements of the method. Proj. Appraisal **12**(4), 251–265 (1997)

37. Mahmassani, H.S., Abdelghany, A.S., Kraan, M.: Providing advanced and real-time travel/traffic information to tourists (1998). https://www.utexas.edu/research/ctr/pdf_reports/1744. Accessed 18 Apr 2015

Acceptance Factors for Younger Passengers in Shared Autonomous Vehicles

Hao Li, Sisi Yu, Jiatai Zheng, Xue Zhao, Peifang Du, and Hao Tan[(✉)]

Hunan University, Changsha 410006, Hunan, China
htan@hnu.edu.cn

Abstract. Once highly automated vehicles become available, the vehicle does not need to be operated by a driver, as a result that the controller of the vehicle will be transformed from drivers to passengers. What can be predicted is, the transition will exert a great influence upon the public transportation, especially in shared vehicle like taxi. With regard to the automated vehicles, younger individuals have been investigated to be more interested in using automated vehicles. However, will the novel type of vehicles be accepted by passengers? In this article, we try to build a model to study the acceptance of autonomous driving shared taxis (hereinafter referred to as SAVS), and accordingly report a survey assessing the acceptance of autonomous driving shared taxi technology among young people (N = 158). The survey results show that more than half of drivers are skeptical of SAVS, but their acceptance may vary considerably. Our paper intends to help shed light on the variance and exhibit relevant acceptance factors for future SAVS usage.

Keywords: Culture and psychology · Design for social market in global markets

1 Introduction

In 2018, SAE classified six levels of self-driving vehicles, ranging from fully manual (L0) to fully automatic (L5), according to which a highly automated vehicle (L4) can be driven autonomously within an operational design domain (ODD). This study is based on level 4 as it is the highest level that can be achieved with existing technologies. Self-driving cars have significant advantages over traditional manual driving in many ways, such as reduced accident rates due to driver error and significantly higher utilization of transportation resources, as studied by Fagnant et al. [9]. Despite the many advantages of self-driving vehicles, the way they are operated is quite different from traditional driving ways, thus the acceptance of self-driving technology by the general public is crucial for both the technology developers and the companies that planning adopt it. This issue has also been studied by a number of scholars in recent years, and as early as 2014, Rödel et al. [20] studied the acceptance and user experience of autonomous levels of autonomous driving technology. Later, Kyriakidis et al. [15] sent a questionnaire to over 5,000 people around the world to survey the public about their perception of automated driving. in 2017, Tennant et al. [22] focused on drivers' attitude toward autonomous vehicles and related factors, based on which a more systematic study was conducted

© Springer Nature Switzerland AG 2021
P.-L. P. Rau (Ed.): HCII 2021, LNCS 12773, pp. 212–224, 2021.
https://doi.org/10.1007/978-3-030-77080-8_18

by Böhm et al. [4] on factors influencing attitudes towards using autonomous vehicles. However, as the technology develops, here comes a new question: will users' acceptance of self-driving technology be affected by the specific scenarios for the application of the technology? A growing number of researchers have focus on this question, and many studies are conducted with severely limited scenarios in which self-driving technology can be used, as an example that Fröhlich et al. [10] applied self-driving technology to the future workplace of automated trucks and studied the acceptance factors of truck drivers. Considering the increased utilization of transportation resources that self-driving technology will bring, Tan et al. [19] vehicle has tendencies to change into transportation systems as well as shared services. On this premise, the technology is more likely to be used as a means of public transportation rather than be used in private vehicles, it is reasonable that using shared autonomous vehicles (SAVs) [15] will become an efficient way to travel in the visible future. Our work focused on the acceptance of self-driving technology by limiting the scenario to SAVs.

As for the acceptance of shared autonomous vehicles, Krueger et al. [14] argued that younger passengers and passengers with multimodal travel are more potential to become SAVs users. We focused our research on the younger 18–27 age group in mainland China. Consequently, a questionnaire study with 158 younger vehicle users in mainland China was conducted as we hope to answer the question:

RQ1: What is the opinion of the younger passengers on the overall acceptance of SAVS? For the time being, little is known about the potential users of SAVs yet, so it is potential and promising to research on the acceptance factors for younger passengers in SAVs. Therefore, the questionnaire also aimed at answering the second question:
RQ2: Which factors are most likely to affect the acceptance of SAVS by young travelers?

Next, we report upon the related work, the process of the research, and the results of the survey. We will next discuss the findings regarding these two research questions. Our contribution is to propose a younger passengers acceptance factor in SAVS, which contributes to the successful design of SAVS.

For the set-up of the questionnaire, we refer to the traditional TAM model as the main framework for subsequent research, but on this basis, we consider the specificity of the application of automated driving technology in shared autonomous vehicles, additionally introducing the concept of trust.

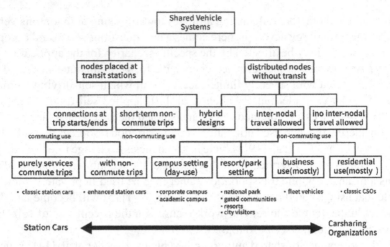

Fig. 1. The SVS model created by Barth et al. [3]

2 Related Work

2.1 Overview

There have been a number of studies focusing on the acceptance of automated cars and automated driving including: a conceptual model explain, predict, and improve user acceptance of driverless pod-like vehicles, and a model of user acceptance for autonomous vehicles (AVAM) [12, 19].

In the domain of taxi, most research about passengers is focused on the taxi passenger demand. A learning model for predicting the temporal and spatial distribution of demand among travelers wanting to use taxis in a short term and recommendations for taxi drivers and those who want a car by using data on passengers' travel patterns and what they know about taxi driver pickup behavior from the taxi's GPS track are good examples (Moreira-Matias et al. 2013; Yuan et al. 2011). However, attributable to the differences in the specific conditions of different cities, it is difficult to establish a unified passenger travel model, but based on the existing data, different ride forecasts can be established based on different city conditions.

2.2 Shared Autonomous Vehicles System

A few results have been investigated in current research on shared vehicle system (SVS), for example, Barth et al. [3] a model (Fig. 1) summarized the use of shared vehicles. SAE Level 4 vehicles can be Regular vehicles with steering wheels and pedals (4R), or pod-like vehicles (4P) with no human driver, driving autonomously under restricted conditions and requiring no driver operation [15]. Based on the related study by Barth et al. [3], the 4P mode was more promising to be used in SAVS. However, the 4P vehicles will be differ from the regular vehicles in controlling, as a result that every individual in 4P vehicles will transform into passengers.

2.3 Cohorts

In recent years, some studies were carried out to research the cohort factors in varied domains. Rogers [22] shows the attitude toward new technology presented a bell-shaped curve on age. To the domain of autonomous vehicles, Men who live in urban areas, have higher incomes and are happy to use new technologies are more willing to pay for new technologies with L4 automation [2], while younger passengers (aged from 18 to 29) and passengers with multimodal travel are more potential to become SAVs users [15]. However, little is known about the younger passengers in SAVs, so it is potential and promising to research on the acceptance factors for younger passengers in SAVs.

2.4 Acceptance

According to TAM [7], perceived usefulness and perceived ease of use are of great significance for the user acceptance to a new technology. With regard of the model's development, the Universal Theory of Usage and Acceptance of Technology (UTAUT) [25] was put forward in early 2000's, which has integrated many of the existing user acceptance models used to explain the user's intent and subsequent behavior in using the system. Moreover, with 1 additional factor, the Car Technology Acceptance Model (CTAM) [19] was designed to fit the acceptance model in car. It is these models that make the future research go on smoothly.

2.5 SAVAM

According to our model, the practicality and ease of use of shared autonomous vehicles will determine the intention of young travellers to use shared autonomous vehicles. On this basis, trust plays a key role in whether users accept autonomous vehicles. In the IT field, the trust of IT technology will affect users' willingness to use IT products [17]. We can infer that the trust in autopilot will affect the attitude of users to use autopilot to a certain extent.

Concerning with the attitude towards SAVS, positive and negative expectations about shared autonomous vehicles (SAVS) were measured through several items which are possible to leave opposite expectations for different individuals, such as safety and driver's behaviors. The complete project can be found in Table 1. Meanwhile, several personal characteristics that were supposed to be potentially related to attitude towards SAVS were included in the questionnaire: age, gender, driving experience, travel modes and technological openness.

In terms of the usefulness of SAVS, one question was presented in relation to few issues which are likely to be improved in SAVS (Especially issues related to improving the cruise flow during travel to the destination). Further, a number of questions considering with trust and ease of use were included.

3 Study and Procedure

3.1 Questionnaire

A questionnaire was surveyed on the acceptance and willingness to use shared driverless taxis among young passengers in mainland China (See Table 1 for details of the questionnaire). Based on the Technology Acceptance Model (TAM) [10], we added technology openness and trust, which constituted our questionnaire. The correlation model can be found in Fig. 2.

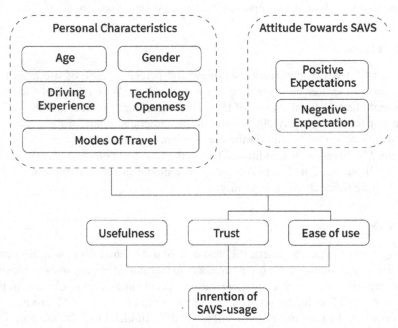

Fig. 2. The model formed by authors

3.2 Preliminary Experiment

A preliminary experiment based on the questionnaire was taken to know whether each question can be understood exactly by study participants or not. Two volunteers (aged 18 and aged 20) were administered in-depth interview on a questionnaire basis.

3.3 Sample

An online questionnaire was distributed via several social network platforms in China Mainland, focusing on college student groups. This questionnaire is written in simplified Chinese only, the reason for this is that this study is mainly for young travellers from mainland China. Of the 177 questionnaires, 19 were considered invalid because they were out of the range of young people (18–29 years old). Table 2 showed the well reliability of the questionnaire.

3.4 Data Analysis

The online SPSS tool was used for data processing. Firstly, the reliability of each question was tested separately, using Cronbach α coefficient, and the results were above good.

Secondly, in sub-topics 5, 6, 7, composition group 1: Technology openness; In sub-topics 8, 9, 10, composition group 2: Indicators of positive expectations; In sub-topics 11, 12, 13, 14, 15, 16, composition group 3: Indicators of negative expectations; In sub-topics 17, 18, 19, 20, composition group 4: Perceived usefulness; In sub-topics 21, 22, 23, 24, composition group 5: Trust; In sub-topics 25, 26, 27, 28, composition group 6: Perceived ease of use; In sub-topics 29, 30, composition group 7: Willingness to use.

New variables were generated from these seven question groups by producing variables-summing; correlation analysis was conducted between question groups 1, 2, 3 and 4, 5, 6 respectively, and the indicators used Pearson correlation coefficients to obtain conclusions; linear regression analysis was conducted between question groups 4, 5, 6, and 7 respectively, to obtain conclusions.

4 Result

In this survey, the effective age of the participants tested was between 18 and 23 years old (male to female ratio was close to 1:1), among which 34.18% (nearly 1/3) participants have driving experience, there are many ways to travel, it suggests that there are several ways that participants choose to travel in their daily life, but it is worth noting that the percentage of choosing a taxi frequently (6.37%) was significantly lower than others. On average, participants reported a moderate or higher degree of technical openness (M = 3.521 SD = 0.803), which is consistent with the researchers' predictions for this group in previous work. Participants tended towards higher positive expectations (M = 3.108), while negative expectation even higher (M = 3.269). Moreover, the standard deviation of negative expectation was less than other dimensions (SD = 0.612). These data show that participants in negative expectations of SAVS showed greater consistency, which is abnormal situation. The details are in Fig. 3.

On average the rates of usefulness, ease of use and trust were fractionally above medium (M = 3.21, M = 3.636, M = 3.177). Through related analysis [1, 17, 24], [30] to the researchers' surprise: the three parameters of usefulness, ease of use, and trust all show positive correlations with the technology openness, Positive expectations and negative expectations (Fig. 4): It's in the author's expectations that usefulness, ease of use, and trust have positive correlations with technology openness and the positive expectations. However, the positive correlations between the negative expectations and the three parameters are out of researchers' expectations, which is interesting. It shows that regardless of the attitude of the participants to SAVS, as long as they have expectations of SAVS, they will trust SAVS and believe it is useful and easy to use.

Regarding Intention of SAVS usage, its average value exceeds the median value (M = 3.437), but less than half of the participants who have a clear willingness to use SAVS (45.57%, which indicates that a considerable part of the population is not negative about SAVS, but do not have a strong will to try SAVS. And through regression analysis [5, 6, 24], [29, 30]. The following results are obtained:

The regression coefficient of usefulness is 0.115 (t = 1.570, p = 0.119 > 0.05), meaning that usefulness does not affect the Intention of SAVS usage.

The regression coefficient of trust is 0.512 (t = 6.861, p = 0.000 < 0.01), meaning that usefulness has a significant positive impact on the Intention of SAVS usage.

The regression coefficient of ease of use is 0.374 (t = 6.295, p = 0.000 < 0.01), which means that ease of use plays an significant active role in the Intention of SAVS usage.

Table 1. Overview about the questionnaire items.

Constructs	Items	Contents
Attitude towards SAVS	Positive expectations	01. I don't have to worry about the unreasonable price of using SAVS 02. Using the SAVS, I can do my own thing on the way without being disturbed 03. I don't have to worry about my personal safety when using SAVS
	Negative expectations	04. I may not be able to get a taxi when there is a problem with my mobile terminal 05. Riding alone in a AV with no one talking along the way makes me feel bored 06. Without a driver, I would feel a little uneasy 07. When SAVS drivers, the Internet of vehicles system may reveal personal information 08. New forms of travel may raise the cost of travel 09. Using SAVSI might have slept too long in the cab because no one reminded me
Usefulness of SAVS	Usefulness	10. Driverless taxis can improve inaccuracy of positioning 11. Driverless taxis can solve the problem of inaccurate road conditions 12. Driverless taxis can solve the problem of poor route planning 13. My communication with the intelligent system can solve the problem of my poor comm-unication with the driver. (such as the driver's dialect is difficult to understand)

(*continued*)

Table 1. (*continued*)

Constructs	Items	Contents
Trust in autonomous vehicle	Trust	14. I think AV works dependably and I don't need to keep an eye on the system 15. If I perceive I'm taking a AV, I would feel at ease 16. I think SAVS will reduce uneasiness factors on highways 17. I believe that the SAVS will work reliably
Ease of use	Ease of use	18. The design of the SAVS room make me feel relaxed 19. The interactive interface of the SAVS room should be easy to understand, even if I first took the ride 20. The interactive approach of the SAVS should be able to provide me with the outside situation in time 21. The SAVS interaction should help me get off the bus
Intention of SAVS-usage	Intention of SAVS-usage	22. I'd like to take a taxi with SAVS 23. I hope that the SAVS can be part of the urban transportation system

It's interesting to find that ease of use and trust plays an significant active role in the Intention of SAVS usage, but usefulness does not affect the Intention of SAVS usage.

Table 2. The Cronbach Alpha of the contents in the questionnaire.

	Cronbach alpha
Technology	0.681
Openess	0.669
Expectations	0.778
Usefulness	0.83
Trust	0.822
Ease of use	0.913
Intention of SAVS usage	0.882

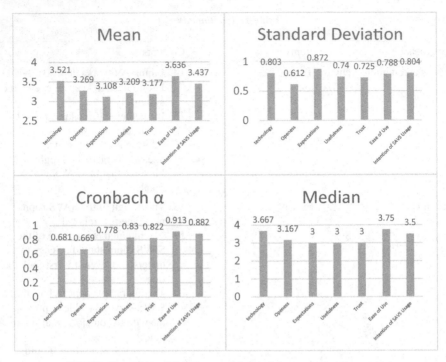

Fig. 3. Overview of the measured technology acceptance aspects

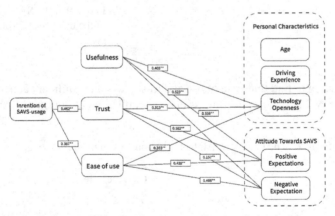

Fig. 4. Pearson correlation coefficients of acceptance factors. Only significant correlations are depicted. Significance levels: *** < 0.01, ** < 0.1, * < 0.5.

5 Discussion

In response to the question on overall acceptance, 45.57% acceptance means that more than half of young commuters currently do not accept the use of SAVS as a new mode of transportation. Explanations for that could be that there is still a strong uncertainty how well SAVS are performing, as well as the special social and culture atmosphere in China Mainland. However, noting that it's no need for us to expect strong opposition to such transportation systems as there are few chances for the individuals to get attach to SAVS prototypes.

According to the results of the multiple linear regression analysis, Ease of use and trust was the main factors affecting the intention of SAVS-usage, while usefulness didn't show a obvious influence on. It suggested that what will SAVS provides was no taken great consideration by the potential users of SAVS, and instead the potential users' intention is strongly related to the trust in the technology as well as ease of use. It is bound to be taken into consideration by business companies in designing future SAVS.

With respect to the result that both positive and negative expectation was positive correlated to trust, usefulness and ease of use. It is striking that even negative expectations like "Riding alone in an AV with no one talking along the way makes me feel bored." Would have a positive influence on trust, usefulness and ease of use. A possible explain might be the younger passengers' curiosity to the unknown SAVS. As our study group of main, younger group is the most curious one and the most possible one to explore their curiosity [22]. Therefore, it is seemly crucial for companies to stimulate the potential users' curiosity and keep SAVS a hot topic constantly to attract more potential users.

However, there are some problems in our research. Although SAVS should be useful, there was not an obvious correlation between usefulness and intention of SAVS-usage. One explanation could be that participants did not have strong feelings about usefulness of SAVS, while another explanation could be that the questions we presented with regard to the usefulness of SAVS might be too specific. Several specific questions based on the problems might be solved in SAVS were included in the questionnaire as the usefulness dimensions of SAVS, however, these questions might be too practical to mislead the participants' understanding of the problems.

In general, our results show that the general attitude of younger passengers towards SAVs' effect will be great in future SAVS. It's not very clear, what factors will enhance younger passengers' trust in SAVS. Meanwhile, how effort should be made to achieve the ease of use, for example, "The design of the SAVS interior space should make me feel relaxed. Further research is also needed.

6 Future Work

What had been found from the study was that the younger passengers' trust in SAVS will greatly affect their intention to use SAVS, and the technical openness may also have an influence on factors like trust in SAVS. Therefore, we might argue that technical openness and trust should be added to the original model of TAM to develop a special technology acceptance model for SAVS. Due to its good reliability and validity, the model was promising to be developed into a mature acceptance model for SAVs (SAVAM) and applied in the future research.

Although this study can demonstrate the influence of the factors presented in this paper on users' intention to use, it still cannot exclude other factors that may have an influence. The SAVAM was required to be tested and consummated in the future works, including enlarging the number of samples, expanding ranges of the cohort's age, regions and so on. In terms of the SAVS, more works were supposed to be done once the SAVS prototype is widely applied.

7 Conclusions

Applying TAM in a specific context now seems to be a commonly used method. We introduce the two parameters of trust and technology openness according to the use scenario of SAVS into the TAM model, and proposed a model that can be used to study the SAVS acceptance factor, named the autonomous driving shared taxi technology acceptance model (SAVAM). Through this model, we have studied the young people in mainland China. We found that the design of SAVS should consider the ease of use in the use process, and the user's trust will become a key factor affecting the acceptance of SAVS.

However, the SAVS user group is not limited to youth groups, groups with low technological acceptance will also become SAVS users. How to make these groups of people accept SAVS may become a problem in the design of SAVS.

Furthermore, due to cultural differences, Chinese younger groups are more tolerant of new technologies than countries such as the United States and Germany. Therefore, the implementation of SAVS in different countries or regions should take into account the differences caused by cultural differences and be targeted and localized.

This experiment was based on the fact that there is no SAVS interactive prototype in our study, so the situation obtained by this conclusion must be different from that when using the SAVS interactive prototype. The author believes that the ability of SAVS interactive prototype design to meet users' expectations for trust and ease of use will greatly affect intention of SAVS usage.

Finally, it is necessary to consider how other stakeholders in the system, such as the operators of SAVS, coordinate between different stakeholders Interest relationship will become one of the topics for future research.

Acknowledgements. The paper is supported by Hunan Key Research and Development Project (Grant No. 2020SK2094) and the National Key Technologies R&D Program of China (Grant No. 2015BAH22F01).

References

1. Arndt, S., Turvey, C., Andreasen, N.C.: Correlating and predicting psychiatric symptom ratings: Spearmans r versus Kendalls tau correlation. J. Psychiatr. Res. **33**(2), 97–104 (1999)
2. Bansal, P., Kockelman, K., Singh, A.: Assessing public opinions of and interest in new vehicle technologies: an Austin perspective. Transp. Res. Part C Emerg. Technol. **67**, 1–14 (2016)
3. Barth, M.J., Shaheen, S.A.: Shared-use vehicle systems: framework for classifying carsharing, station cars, and combined approaches. Transp. Res. Rec. J. Transp. Res. Board **1791**(1), 105–112 (2002)
4. Böhm, P., Kocur, M., Firat, M., Isemann, D.: Which factors influence attitudes towards using autonomous vehicles? In: Proceedings of the 9th International Conference on Automotive User Interfaces and Interactive Vehicular Applications Adjunct, pp. 141–145. Association for Computing Machinery, New York (2017)

5. Cameron, A.C., Trivedi, P.: Microeconometrics: Methods and Applications, 1st edn. Cambridge University Press, New York (2005)
6. Dao-de, S.: Selection of the linear regression model according to the parameter estimation. Wuhan Univ. J. Nat. Sci. 5(4), 400–405 (2000)
7. Dillon, A.: User acceptance of information technology. In: Karwowski, W. (ed.) Encyclopedia of Human Factors and Ergonomics, London (2001)
8. Eisinga, R., Grotenhuis, M., Pelzer, B.: The reliability of a two-item scale: Pearson, Cronbach, or Spearman-Brown? Int. J. Publ. Health 58(4), 637–642 (2013)
9. Fagnant, D.J., Kockelman, K.: Preparing a nation for autonomous vehicles: opportunities, barriers and policy recommendations. Transp. Res. Part A Pol. Pract. 77, 167–181 (2015)
10. Fröhlich, P., Sackl, A., Trösterer, S., Meschtscherjakov, A., Diamond, L., Tscheligi, M.: Acceptance factors for future workplaces in highly automated trucks. In: Proceedings of the 10th International Conference on Automotive User Interfaces and Interactive Vehicular Applications, pp. 129–136 (2018)
11. Green, W.H.: Econometric Analysis, 6th edn. China Social Sciences Press, Beijing (1998). (in Chinese)
12. Hewitt, C., Politis, I., Amanatidis, T., Sarkar, A.: Assessing public perception of self-driving cars: the autonomous vehicle acceptance model. In: Proceedings of the 24 th International Conference on Intelligent User Interfaces, pp. 518–527. Association for Computing Machinery, New York (2019)
13. International SAE. https://www.sae.org/standards/content/j3016_201806/. Accessed 30 Dec 2019
14. Kossowski, T., Hauke, J.: Analysis of the labour market in metropolitan areas: a spatial filtering approach. Quaestiones Geographicae 31(2), 39–48 (2012)
15. Kurueger, R., Rashidi, T.H., Rose, J.M.: Preferences for shared autonomous vehicles. Transp. Res. Part C 69, 343–355 (2016)
16. Kyriakidis, K., Happee, R., Winter, J.D.: Public opinion on automated driving: results of an international questionnaire among 5,000 respondents. Transp. Res. Part F: Traffic Psychol. Behav. 32, 127–140 (2014)
17. Mcknight, H., Carter, M., Thatcher, J., Clay, P.: Trust in a specific technology: an investigation of its components and measures. ACM Trans. Manage. Inf. Syst. 2(2), 12–32 (2011)
18. Moreira-Matias, L., Gama, J., Ferreira, M., Mendes-Moreira, J., Damas, L.: Predicting taxi-passenger demand using streaming data. IEEE Trans. Intell. Transp. Syst. 14(3), 1393–1402 (2013)
19. Nordhoff, S., van Arem, B., Happee, R.: Conceptual model to explain, predict, and improve user acceptance of driverless podlike vehicles. Transp. Res. Rec.: J. Transp. Res. Board 2602(1), 60–67 (2016)
20. Osswald, S., Wurhofer, D., Trösterer, S., Beck, E., Tscheligi, M.: Predicting information technology usage in the car: towards a car technology acceptance model. In: Proceedings of the 4th International Conference on Automotive User Interfaces and Interactive Vehicular Applications, pp. 51–58. Association for Computing Machinery, New York (2012)
21. Rödel, C., Stadler, S., Meschtscherjakov, A., Tscheligi, M.: Towards autonomous cars: the effect of autonomy levels on acceptance and user experience. In: Proceedings of the 6th International Conference on Automotive User Interfaces and Interactive Vehicular Applications, pp. 1–8. Association for Computing Machinery, New York (2014)
22. Rogers, E.: M: Diffusion of Innovations, 4th edn. The Free Press, New York (1995)
23. Tennant, C., Howard, S., Franks, B., Stare, S., Bauer, M.W.: Autonomous Vehicles - Negotiating a Place on the Road: A Study on How Drivers Feel about Interacting with Autonomous Vehicles on the Road. London School of Economics and Political Science, Goodyear Europe, Middle East and Africa (EMEA), London (2016)

24. The SPSSAU project (2019), SPSSAU, (Version 20.0). https://www.spssau.com.
25. Venkatesh, V., Morris, M.G., Davis, G.B., Davis, F.G.: User acceptance of information technology: toward a unified view. MIS Q. **27**(3), 425–478 (2003)
26. Yuan, J., Zheng, Y., Zhang, L.H., Xie, X., Sun, J.Z.: Where to find my next passenger. In: Proceedings of the 1th international conference on Ubiquitous computing, Beijing, pp. 109–118. Association for Computing Machinery, New York (2011)
27. Zhang, H.C., Xu, J.P.: Modern Psychology and Education Statistics, 4th edn. Beijing Normal University Press, Beijing (2009). (in Chinese)
28. Zhou, J.: Questionnaire Data Analysis - Six Kinds of Analysis Ideas to Crack SPSS, 1st edn. Electronic industry press, Beijing (2017). (in Chinese)

Where is the Best Autonomous Vehicle Interactive Display Place When Meeting a Manual Driving Vehicle in Intersection?

Junzhang Li, Haowen Guo, Shuyu Pan, and Hao Tan[✉]

Hunan University, Changsha, People's Republic of China
{Zhanghua7,htan}@hnu.edu.cn

Abstract. Autonomous vehicle technology is becoming more and more practical currently. To drive in urban environment safely, autonomous vehicles need to become able to convey the information to other road users, such as pedestrians or driving drivers, and make them understand their action intent. Although theoretically self-driving cars will let other road users pass through the road firstly, for giving them psychological comfort and sense of safety during road use, certain audiovisual interactions between autonomous car and other road users are helpful and essential. In the current development of autonomous vehicles, there are a lot of Automotive external interactive interfaces designs to show their intention. In this paper, we explore driver's understanding of five interactive display places with three conceptual interactive modalities (text, smile, light strip) in a manual driving car crossing the no traffic signal street situation by video study. Based on different situations and scenes, we compared the error rate of drivers in different display places and modalities, as well as the perception and understanding of drivers on display places and modalities, and we found that door surface is the best place in these places to convey the autonomous car's purpose to the manual driving car's driver.

Keywords: Autonomous vehicle · Video study · Car to car communication · Interactive display place

1 Introduction

With the development of automatic technology, autonomous vehicles are becoming more and more common and universalistic in our lives which based on the urban environment. For the classification of autonomous driving, the NHTSA and the SAE have detailed regulations. It categorized into six levels, the level 0 of SAE is no automation, and the level 5, which is highest, is fully automated [2]. At present, the research as well as the growth of driver assistance systems by car manufacturers on the market has reached the requirements of Level 2 and Level 3, and there are clear plans for Level 4 development, and some have even started Level 5 testing [18].

Functions that help drivers navigate a wide range of traffic conditions, such as Pedestrian Safety and Traffic-aware Cruise Control, can already be found in the market. Also,

© Springer Nature Switzerland AG 2021
P.-L. P. Rau (Ed.): HCII 2021, LNCS 12773, pp. 225–239, 2021.
https://doi.org/10.1007/978-3-030-77080-8_19

the trend is expected to continue [1, 17]. In the complex driving environment, as a manual car driver, safety and control mode, communication mode are very important [19]. But for other road users, communicating with autonomous vehicle is still a puzzle [11, 16]. To ensure that drivers have a pleasant and safe driving experience in a complex environment of autonomous and manual vehicles that has not yet fully transformed into a society filled with autonomous vehicle in the future [10], there have been several methods that could show cars purpose to pedestrians effectively. [3]. So far, the information can be displayed on the exterior surface of the car. There are a lot of design spaces inside and outside the vehicle for transmitting information [4, 20]. The audience for these external displays may be other road users, such as pedestrians, driving drivers or automated systems [15, 20]. A framework for dividing these display places has been initially established and interactive device designs developed for these locations. The research's aim was to make comparison between different modalities and get a better understanding of them from one manual car driver' s perspective and compare the places where these modalities are located and displaying, and find out the display places that are more convenient for MVS drivers to recognize and understand.

A virtual driving situation was set up that two cars meeting with two scene states (day and night), and two driving scenarios (stop and not stop) by creating a 3D-virtual-animations to present different three conceptual interactive modalities (text, smile, light strip) [5–7] on different interactive display places of AVs. We evaluated these display places with modalities by video study via virtual simulation 3D animation video modeling. Participants were asked to watch the video and estimate the purpose of the autonomous vehicle (to stop or not to stop). We recorded the duration of the beginning of the video and the participants' reaction. Then a perception questionnaire was conducted about the five display places and three interactive communication modalities (ICM).

2 Introduction

The research core of this paper is to study the impact of the display place on the self-driving car's understanding of other drivers. In order to reduce the errors caused by different people's different understanding of the same information transmission method, we set up a variety of information interaction modules to be the inter-group variables. Several studies have summarized many alternative research methods for autonomous vehicle behaviors [8], and some of these methods have been used to compare the communicated modalities [3, 11]. In this study, we used an control observational method to explore the communications between the manual vehicle and autonomous vehicle in the video experiment. The video experiment method is easier to immerse the subject in a virtual driving environment than the traditional experimental methods (e, g. Written questionnaires and direct interviews) [8], and it can be more convenient to show and more flexible to compare many kinds of combinations of different display places and different ICM (See Fig. 1).

2.1 Participants

Thirty participants from universities participated in our project for the experiment, including 19 men and 11 women, and the subjects were between 18–36 years old.

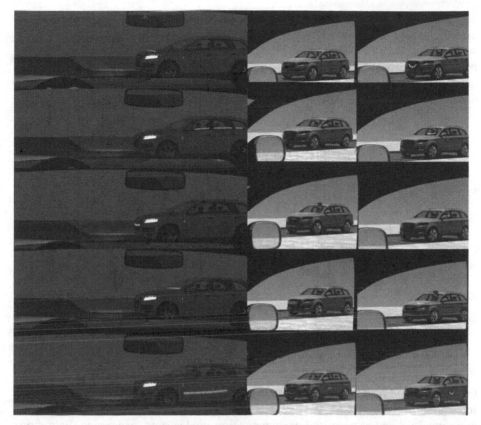

Fig. 1. Five display places with three communicated modalities

All participants have driving qualifications and a certain period of driving experience. For the driving experience of these participants, 13 of them (43%) have been driving for more than one year. Two of them (6%) have been driving for more than 6 years. The other participants (51%) have only about half a year to drive. The Recruitment and evaluation of these participants were conducted from December 17 to December 18, 2019.

2.2 Display Places and Interactive Communicated Modalities

For the division of these display places and the design space of the exterior surface of the vehicle, there have been many researches into it [4, 5]. Some are divided according to the vehicle's external surface [4], according to the interaction method carried on it [5], and some are differentiated according to the perspective of pedestrians [8]. Essentially, cars have used a variety of methods to visually communicate signals to their surroundings. The flashing lights inform other road users of the purpose to turn or overtake, while the brake lights and reversing lights inform others of important information about car movement [4]. However, in a complex urban environment, complex and changeable information also needs new carriers for an autonomous car. For the design of the display places, we build a frame based on the dimension, display area that according to the exterior surface of the vehicle firstly, we choose the parts of this dimension car body surface, bumper grille and windows [4].After this, we divided these three parts into five places (bumper grille, bonnet, front windshield, door surface and side window glass) based on the orthogonal direction of the autonomous car and the perspective of the driver of the manual car in the scene [12] (See Fig. 2).

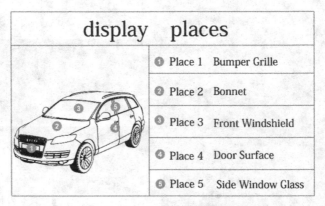

Fig. 2. Design of display places

In order to let participants, judge the intention of autonomous vehicle and reduce the error caused by the same communication modalities to the experiment, we set up inter-group variables- ICM. We designed three sets of ICM according to the existing conceptual vehicle-to-pedestrian communication modes (text, expression, light strip) [5–7, 13]. In addition, different behaviors (or messages) have been designed for each ICM to demonstrate the intent of the car in the following situations: stop and not stop (Whether manual car can cross before the autonomous car) (See Fig. 3).

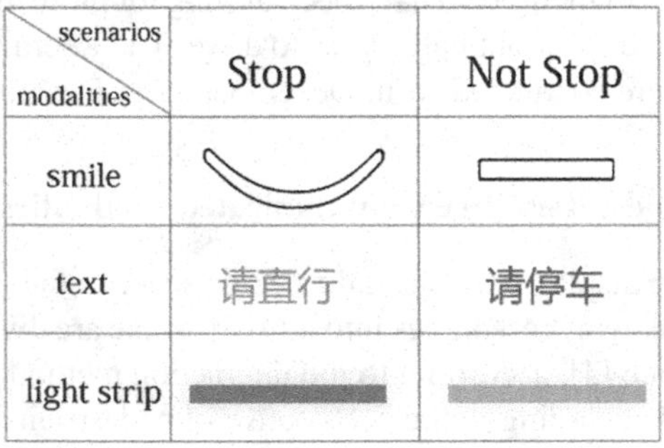

scenarios modalities	Stop	Not Stop
smile	⌣	▭
text	请直行	请停车
light strip	▬▬▬	▬▬▬

Fig. 3. Design of interactive communicated modalities

2.3 Video Setting

Based on the combinations of different display places and different ICM on an autonomous vehicle, we made 60 videos of an autonomous vehicle viewed from the perspective of a manual vehicle's driver to show participants the display places and ICM on an autonomous vehicle which moving on the no traffic signal crossroads. The vehicle model we chosen was the Audi Q7 (5.086 m × 1.983 m × 1.737 m), and its 3D model was obtained from Xue Xi Niu Websites [9]. We also built an environmental scene and some devices (display, projector) which can add to the car model to show different experimental combinations. The videos show a specific traffic situation that an autonomous vehicle approaches an intersection without traffic lights, and a manual vehicle is going to cross the intersection. The approaching car driving at a speed of 12 km/h in a 6-m wide two-way motorway environment. During car driving (video playback), there are three key positions regarding the movement and behavior of the autonomous car (See Fig. 4).

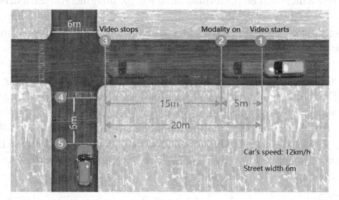

Fig. 4. Top view of virtual environment (The Fig. shows the meeting between the white manual driving vehicle and the yellow autonomous vehicle in the video, the autonomous vehicle is in motion, and the three yellow lines show the distance between the three stages of the driverless vehicle) (Color figure online)

Position 1: When the video starts to play, autonomous vehicle is approaching from the position.
Position 2: Different ICM on different display places start to present different behaviors to convey information (stop or not stop). Currently, the vehicle is 15 m away from the meeting point.
Position 3: When the video ends, the vehicle arrives at the position, and the participant needs to judge the next behavior (stop or not stop).

2.4 Procedure

Set up a laptop with a 15.6-in display for video experiments. Informed consent was obtained from each participant. In the experiment, participants need watch the videos, and after the video ends, judge the purpose of the car (stop or not stop) according to the combination of different display places and different ICM on the vehicle displayed in the video. According to the three interaction modalities (text, expression, light strip), We divided the video experiments into three groups. Each group contains 20 videos, every 4 videos constitute a display place part. Each 4 videos contain 2 videos showing the situations of the approaching autonomous vehicle intending to "stop", while 2 videos show the situations of the autonomous vehicle trying to "do not stop", 2 videos show the scene of autonomous vehicle driving in the daytime while 2 videos show the scene of driving at night. Five part of five different display places constitute to a section of a communicating modalities.

Participants are also divided into three groups and conduct video experiments (watch the video) under the three sections of communicate modalities to reduce the impact of the understanding of the ICM on the display places research. The sequence of videos in every experimental group is random. Before the start of the experiment, the instructions to the participants about the video experiment are as follows: "Hello, you will watch 20 videos in the test which will be performed later. Please imagine you are a driver and you are driving the car in the videos, observing a driverless car that is about to approach to the intersection and meet you. (showing a video screenshot) You will observe it and judge its intent (stop or not stop). Each video can be watched only once and you need to make your judgment on this video. When making a judgment, you don't need to consider the speed of the car. You will observe the following five places (showing five video screenshot). The interaction mode you will observe is this (showing a video screenshot), we will not answer your question until the entire video experiment is over. After the video experiment, Participants are required to make a feeling judgment by filling out the questionnaire about the understanding of various display places and interaction modes. The questionnaire is divided into five levels, "very easy to understand" is level 1, "very difficult to understand" is level 5. (See Fig. 5).

Fig. 5. Screenshot of questionnaire

2.5 Data Measurement

We counted and calculated the error rate of the participants' judgment on the behavior purpose of the autonomous vehicle in the video. The overall error rate was calculated firstly, and then we calculated the error rate of the participant's judgment in two different situations (stop/not stop) and two different scenes (day/night). And we calculated the error rate of the participant's judgment in five different display places (bumper grille, bonnet, front windshield, door surface and side window glass), three different ICM (text, smile, light strip). Also, we recorded the judgement time between each time participants saw the video and responded in five different display places [16].

3 Result and Discussion

For different data, we did ANOVA (analysis of variance), significance tests and descriptive statistics. At different places, these five sets of data on participants' judgement error rate contribute to the consequence that the door surface is better than front windshield, and the door surface will efficiently convey information to other road users by installing ICM which can express autonomous car's purpose to the manual driving car's drivers. And the information on the door surface is more noticeable and understandable than the information on the windshield. This result can be confirmed by the data in the Fig. 6 (the overall error rate was 46.67% in the place 1: bumper grille, 53.33% in the place 2: bonnet, 57.50% in the place 3: front windshield, 38.33% in the place 4: door surface, and 44.17% in the place 5: side window glass.).

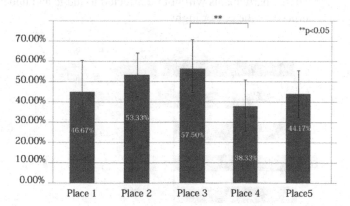

Fig. 6. Error rate: overall

Figure 6 illustrates the error rate when the car intends to stop with bumper grille at 55%, with bonnet at 61.67%, with front windshield at 63.33%, with the door surface at 46.67%, and with side window glass at 48.33%, on the other hand, 38.33% in bumper grille, 45% in bonnet, 51.67% in front windshield, 30% in the door surface, and 40% in side window glass, when the car will not stop. From the information in the chart, the error rate of the first situation (the car will stop) is generally higher than the other situation (the car will not stop), most participants tend to choose not to stop the self-driving vehicle. This experimental result may be caused by the vehicle not slowing down in the video.

Analyzing data shown in Fig. 7 with ANOVA, we found no significant differences in these two situations in all places, from this we can know that the two indication states of the ICM are different for the participants, so the ICM have experimental significance.

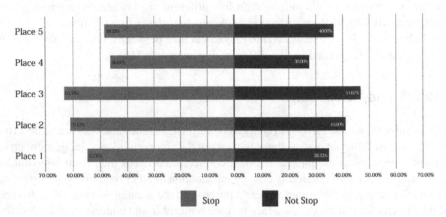

Fig. 7. Error rate: two different situations

Analyzing the results shown in Fig. 8 with ANOVA, we found no significant differences in these two scenes in all places. It also illustrates that in the display places and ICM we designed, the participants will not be affected to judge and understand the autonomous car's purpose by the day or night.

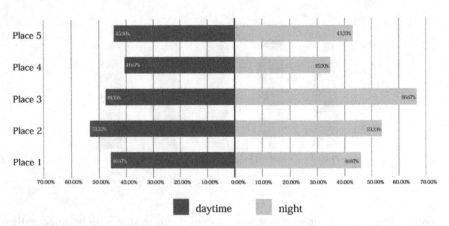

Fig. 8. Error rate: two different scenes

Figure 9 is about the error rate comparison of five display places in the same modalities and the error rate comparison of three ICM in the same places. Among them, the error rate of place 4 is the lowest in the mode of light strip (45%) and the mode of smile

(37.50%), and the mode of text. (32.50%). Figure 10 shows that the error rate of three different ICM was 54.50% in mode "light strip", 47% in mode "smile", 42.50% in mode "text", we found no significant differences in these e three groups of data in these two charts.

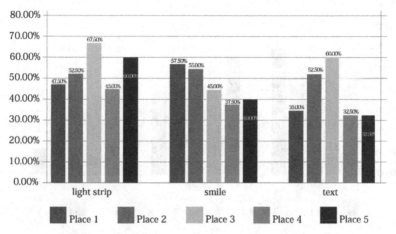

Fig. 9. Error rate: combinations of ICM and display place

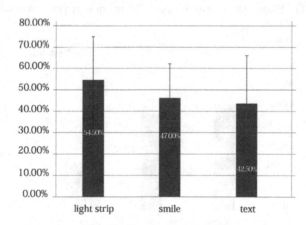

Fig. 10. Error rate: three ICM

From the results of the data analysis, we found no significant differences in different data groups and the data were not statistically significant. Also, for the interactive communicated mode of text, the error rate obtained is too high, which is inconsistent with the pre-experimental results and related paper research [4]. It shows that the text module information is not clear. During the experiment, the set text interaction mode tends to indicate the driver's behavior, and the participants were required to identify the autonomous car's behavior, it will give the participants a certain cognitive illusion, which will cause their cognition to be completely reversed. Based on our main research purpose for display places, we asked the participants and explained the intent of the text mode. The participants asked us to convert the data (100% error rate to 0% error rate). The resulting graph, as shown in the Fig. 11 and Fig. 12.

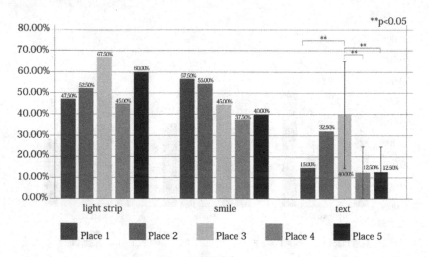

error rate：five different places (Amend)

Fig. 11. Error rate: combinations of ICM and display place (Amend)

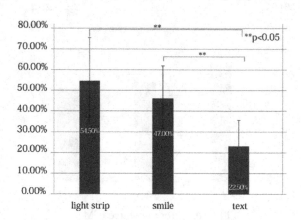

Fig. 12. Error rate: three ICM (Amend)

Figure 11 shows the error rates for places one to five in the text mode are 15%, 32.5%, 40%, 12.5%,12.5%.The error rates for places one to five in the smile mode are 57.5%, 55%, 45%, 37.5%, 40%,and the error rates for places one to five in the light strip mode are 47.5%, 52.5%, 67.5%, 45%, 60%. It illustrates that place 4 and place 5 (Side direction: door surface and side window glass) are better than place 3: windshield to install the mode "text" to communicate with other drivers.

The error rate of mode "text" shown in Fig. 12 was 22.50%. After analyzing the data with ANOVA, the error rate in the mode text show that it is significantly different from the other two error rate. It shows the text will be the best modalities and it could easily convey the information to drivers.

Figure 13 shows mean effect judgement time was 8.45s in place 1: bumper grille, 7.97 s in place 2: bonnet, 7.72 s in place 3: windshield, 6.59 s in place 4: door surface, 8.28 s in place 5: side window glass. Using the ANOVA to analyze the results, it embodies there are significant differences between place 4 and other display places. It embodies that the information in place 4 can be more quickly understood than other display places, with better significance and comprehension. At the beginning of the experiment start, the judgement time was long (13–15 s), as the number of experiments increased, the required time slowly decreased and finally stabilized at 6–7 s, see Fig. 14. It embodies that the participants did not understand what the two modes of the ICM represented at the beginning. After the number of experiments increased (the number of videos watched increased), the participants had his/her own judgment on the interaction mode and mostly judged the autonomous car's purpose based on the display places. Their judgment time is about the end of the video playback.

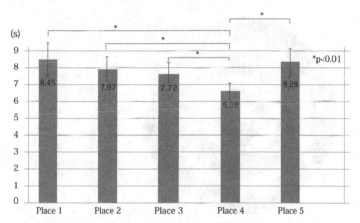

Fig. 13. Judgement time of different places

We collected questionnaires and analyzed the data statistically, the participants' perception of the three ICM and the five display places are respectively illustrated in Fig. 15 and Fig. 16. The former indicates the text is more readable for drivers. The latter indicates that place 1: bumper grille is the most understandable and noticeable display places and place 4: door surface also is a better place than other three.

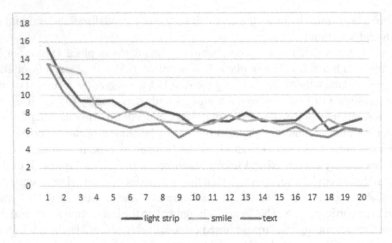

Fig. 14. The trend of judgement time

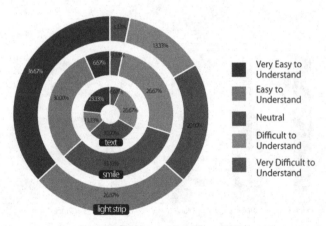

Fig. 15. Understandability of ICM

Our experimental participants are limited, the samples are relatively limited, the number, age span, and driving age span are not large enough. The preparation time for the entire experiment is not enough (but the minimum sample size of 30 degrees of freedom is met). With certain human intervention and errors, the immersion may be improved, and more realistic experimental methods may be used in the future. In the future, there can be more choices and research on the choice of places and ICM.

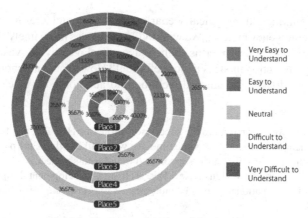

Fig. 16. Understandability of display places

4   Conclusion

This study compares three potential ICM (light strip, smile and text) on five display places (bumper grille, bonnet, front windshield, door surface and side window glass) for autonomous car and a manual driver meeting at a narrow intersection without signal lights. Through show different modalities in different places, vertical and horizontal comparison, our video experiment concluded that place4: door surface is the most understandable and noticeable display places that can efficiently convey information to other road users by installing ICM which can express autonomous car's purpose to the manual driving car's drivers. Figure 6: Place4 has a lower error rate than place3, Fig. 13: Place4 has the shortest judgment time and Fig. 16: Place4 is the most well understood. The results also show that drivers are more likely to believe that the vehicle will not stop if they are not clear about its purposes (Fig. 7). In addition to the display places comparison, we also draw conclusions from the comparison between different interactive communicated modalities. The results show that the text can better express the purpose of the car to the driver than the other four modalities.

5   Future Works

This inquiry is mainly focused on visual interaction between car drivers. There are many types of interactions at this stage. Such as car to car Internet information transfer model. In the future, we will also study a variety of information transmission methods, which will be related to the places explored in this article, and explore the most suitable options. Secondly, other forms of interactions in autonomous cars, such as sound, and specific interactions such as alerts, are worthy of further investigation. According to the different stages of the development of unmanned vehicles, we may establish different models of unmanned vehicles. Its interactive use conditions and functions may change accordingly. Depending on the degree of intelligence, the object-oriented will also be different. For the self-driving cars defined by our experiments today, we mainly explored the perception of location by specific road users in specific scenarios. To further expand our research, we will plan to simulate more typical scenarios, such as simulating parking, parking, overtaking, and emergency braking. We will divide the places more finely. Therefore, we may come up new questions: In terms of the places displayed by the visual interaction device, in which interaction mode is it more efficient? How should autonomous vehicles interact with other drivers, including drivers behind autonomous vehicles? and many more.

Acknowledgments. We are grateful to the supervisor, professor Hao Tan, and his graduate students for their guidance and help. At the same time, we would like to thank the participants of our experiment and Hunan University for the support and help for this project.

The paper is supported by Hunan Key Research and Development Project (Grant No. 2020SK2094) and the National Key Technologies R&D Program of China (Grant No. 2015BAH22F01).

References

1. Lundgren, V.M.: Will There Be New Communication Needs When Introducing Automated Vehicles to the Urban Context? (2017). https://doi.org/10.1007/978-3-319-41682-3_41
2. SAE: SAE On-Road Automated Vehicle.Standards Committee Taxonomy and Definitions for Terms Related to On-Road Motor Vehicle (2016)
3. Chang, C.M., Toda, K., Igarashi, T., Miyata, M., Kobayashi, Y.: A video-based study comparing communication modalities between an autonomous car and a pedestrian. In: Adjunct Proceedings of the 10th International ACM Conference on Automotive User Interfaces and Interactive Vehicular Applications (AutomotiveUI 's18) (2018)
4. Colley, A., Häkkilä, J., Pfleging, B., Alt, F.: A design space for external displays on cars. In: Adjunct Proceedings of the 9th International ACM Conference on Automotive User Interfaces and Interactive Vehicular Applications (AutomotiveUI '17) (2017)
5. Analysing Pedestrian-Vehicle Interaction to Derive Implications for Automated Driving
6. Semcon: The Smiling Car (2016). https://semcon.com/smilingcar/. Accessed 3 Nov 2016
7. Böckle, M.P., Brenden, A.P., Klingegård, M., Habibovic, A., Bout, M.: SAV2P: exploring the impact of an interface for shared automated vehicles on pedestrians' experience. In Proceedings of the 9th International Conference on Automotive User Interfaces and Interactive Vehicular Applications Adjunct, pp. 136–140. ACM, September 2017
8. Rasouli, A., Tsotsos, J.K.: Autonomous vehicles that interact with pedestrians: a survey of theory and practice. IEEE Trans. Intell. Transp. Syst. **21**(3), 900–918 (2019)
9. Xue Xi Niu Website
10. Schroeter, R., Rakotonirainy, A., Foth, M.: The social car: new interactive vehicular applications derived from social media and urban informatics .In: From Proceedings of the 4th International Conference on Automotive User Interfaces and Interactive Vehicular Applications (AutomotiveUI '12) (2012)
11. de Clercq, K., Dietrich, A., Núñez Velasco, J.P., De Winter, J., Happee, R.: External human-machine interfaces on automated vehicles: effects on pedestrian crossing decisions. J. Hum. Factors Ergon. Soc. **61**(8), 1353–1370 (2019)
12. Riegler, A., Riener, A., Holzmann, C.: Towards Dynamic Positioning of Text Content on a Windshield Display for Automated Driving (2019)
13. Clamann, M., Aubert, M., Cummings, M.L.: Evaluation of Vehicle-to-PedestrianCommunication Displays for Autonomous Vehicles, Duke University (2017)

14. Brandli, G., Dick, M.: Alternating current fed power supply. U.S. Patent 4 084 217, 4 November 1978
15. Mahadevan, K., Sanoubari, E., Somanath, S., Young, J.E., Sharlin, E.: AV-Pedestrian Interaction Design Using a Pedestrian Mixed Traffic Simulator (2019)
16. Schwarting, W., Pierson, A., Alonso-Mora, J., Karaman, S., Rus, D.: Social behavior for autonomous vehicles. Proc. Natl. Acad. Sci. **116**(50), 4972–24978 (2019)
17. Camara, F., et al.: Filtration analysis of pedestrian-vehicle interactions for autonomous vehicle control (2018)
18. Jafary, B., Rabiei, E., Diaconeasa, M.A., Masoomi, H., Fiondella, L., Mosleh, A.: A Survey on Autonomous Vehicles Interactions with Human and other Vehicles (2018)
19. Campbell, M., Egerstedt, M., How, J.P., Murray, R.M.: Autonomous driving in urbanenvironments: approaches, lessons and challenges (2018)
20. Habibovic, A., et al.: Communicating Intent of AutomatedVehicles to Pedestrians (2018)

User-Friendliness of Different Pitches of Auditory Cues in Autonomous Vehicle Scenarios

Xinrui Ren[1,2(✉)], Yimeng Luan[1,2], Xue Zhao[1], Peifang Du[1], and Hao Tan[1]

[1] School of Design, Hunan University, Yuelushan, Changsha 410082, Hunan, P.R. China
[2] Department of Industrial Design, University of Science and Technology Beijing, 30 Xueyuan Road, Haidian District, Beijing 100083, P. R. China

Abstract. We conducted a study to evaluate the user-friendliness of different pitches of auditory cues in autonomous vehicle scenarios under the objective and subjective evaluation criteria. First, we designed 6 different pitches of cues to test how intuitive they are at learning and how accurate they are at matching 6 emergency levels. And then we carried out a survey using a questionnaire to collect subjective preference for these pitches. Finally, experimental results demonstrated that the lowest and highest pitches were more intuitive in learning, while those in the middle range required more effort to match and were more likely to cause errors in recognition. The subjective preference showed that cues with relatively low and high pitches received the most popularity from participants. And cues with higher pitches were often more interrupting, and occupied more attention. Combining the results of objective test and subjective preference, we believed that the pitches in the middle range were less user-friendly than the lower and higher ones.

Keywords: Auditory cue · In-vehicle information · Pitch · Informative interruption cues

1 Introduction

Autonomous vehicle has been one of the hottest topics around the globe in 2019. It is commonly predicated that more and more companies will try to develop autonomous vehicles around 2020 [12]. In the dawn of the revolution in autonomous driving, the era is beckoning us to improve the in-vehicle system to create a perfect experience for people.

As cars become more and more autonomous, when autonomous driving technology reaches SAE Level 4 [25], the primary attention of drivers starts to shift from driving the car into some activities that only passengers can do nowadays such as daydreaming, writing text messages and calling [8]. In this way, driving will become a dual-task activity in autonomous vehicle scenarios. So, the diversion of driver's attention must be thoroughly designed to strike a balance between driving-related information and less important information.

P.-L. P. Rau (Ed.): HCII 2021, LNCS 12773, pp. 240–252, 2021.
https://doi.org/10.1007/978-3-030-77080-8_20

In order to provide secondary information for drivers, the concept of in-vehicle information systems (IVIS) are introduced [27]. IVIS include a great variety of information, such as navigation, driving monitoring, road condition updates, hazard warning, and email notifications [13, 14]. However, as the amount of information increases, the information is more likely to influence driver's attention and cause more distractions. Thus, IVIS should be carefully designed to maximize the transmission validity and minimize the disturbance of attention [17].

In previous studies, the effective communication of multimodal warnings (such as auditory, tactile warnings) in the manual driving scenarios have been thoroughly evaluated [2, 13]. However, in autonomous vehicle scenarios, the driving condition and the driver's attention will be quite different than before. In order to match the new criteria and condition, the modes of notifications in IVIS will be reappraised and the alert of different information will be redesigned.

1.1 User-Friendliness

As the basic design guideline, if the in-vehicle information is thoroughly and carefully designed, it should implement its function of delivering information. In the area of IVIS, it is supposed to draw the attention of drivers in the first place (interruption). Then pass the exact content that it carries to the driver, and at the same time allow them to fully and correctly understand (identification). However, even if it is able to deliver the message effectively, it can only be successfully implemented if user's preference can be ensured [26]. Therefore, it should give the driver a better, friendly impression as well as avoid making the driver feel impatient or annoyed (subject preference). In this study, we aimed to evaluate the user-friendliness of different pitches based on various factors, such as interruption, identify difficulty, and subject preference [23, 24].

1.2 Type of Cues

In previous studies, informative interruption cues (IIC) have been proved to be a promising method to achieve the goals of IVIS [1]. The information they convey has two dimensions: the arrival of the coming information and implicit message within the information itself [15]. IIC reduce the information transmission time and thus create a comfortable buffer zone for drivers to manage the diversion of their attentions.

To form this "buffer zone", we need highly efficient IIC. They should meet the criteria of pre-attentive reference, which aims to determine whether signals can be received with minimum attention: 1) The signals are supposed to be received in parallel with ongoing task. 2) It should provide rough messages about what the cue is referring to. 3) Intuitive and natural, it is not supposed to consume unnecessary attention [11]. The auditory, tactile, visual and multimodal notification modality have been evaluated to compare their effectiveness of delivering the message [28].

For driving autonomous cars mostly requires the visual attention, auditory or tactile cues might suit the form of IIC better [9]. Also, according to a large-scale web survey, most people would like to hear or watch something in autonomous cars, such as listening to music, chatting, and looking around [8]. In addition to that, auditory sense triumphs over visual sense in interruption management [10]. Under these evaluation criteria, we picked up a simple alarm sound with a single syllable to carry the message, which was able to avoid the misleading information that melody and voice might bring.

To further design the auditory cues, we should consider which acoustic parameter, such as such as the loudness, pitch, speed, pulse rate and duration, is more significant in autonomous driving scenarios [2, 5, 19, 20]. Those acoustic parameters have been proved to affect the perceived urgency of auditory warnings [29]. However, in different circumstances, people have different perceptions of loudness. What we feel is mostly relative loudness instead of absolute loudness. Besides, in previous studies, the variation of pitch has been proven to be more significant meaning for the change of urgency [2, 5–7], which is the main information for IIC to deliver. Therefore, in this study, we chose auditory cues as IIC and conveyed the emergency levels of messages in pitches rather than the pulse or loudness.

1.3 Design of Cues

We set up 6 priority levels of urgency to match with the common 3 kinds of information in IVIS. Table 1 showed the match between priority levels of urgency and information in IVIS. Level 1 and 2 were matched with non-driving-related information (such as email), representing the lowest priority level of urgency; level 3 and 4 were matched with driving-related unidirectional notification, which is no need for feedback operation (such as weather updates), representing the middle priority level of urgency; and level 5 and 6 were matched with driving-related alert that requires feedback operation, representing the highest priority level of urgency. The whole purpose of design of cues was to seek the most natural and intuitive match between different cues and perception levels of urgency. In this way, people were able to understand the partial information the cue carries with minimum effort.

In order to evaluate how well the IIC matches priority levels of urgency, we chose earcons [18] (such as beep) rather than auditory icons [16]. Due to its abstract sound, it carried absolute pure sound information without any implicit message, and thus eliminated the misunderstanding which other sounds might remind people of [3].

We focused on one sound parameter, the pitch, to explore its extreme effectiveness. First, we chose a short beep from internet as the basic auditory cues. Then we changed its pitch with AU (Adobe Audition) into 6 different versions. The pitch interval between different cues was equal (10 chromatic scales). In the meantime, we guaranteed that the loudness (−15 dB) and length (4 s) of the audio remained the same. Also, the whole pitch range was neither too small to tell the difference, nor too big to make them sounded weird [2]. According to some previous studies, an increase in pitch usually increases the urgency of an auditory cue [2–5]. According to this conclusion, we matched pitches with priority levels of urgency from the lowest to the highest.

Table 1. The match between priority level of urgency and information in IVIS

Level of urgency	Information
1 (Lowest)	Non-driving-related information
2	
3	Driving-related unidirectional notification
4	
5	Driving-related alert
6(Highest)	

1.4 Aim of the Paper

The purpose of our study was twofold: 1) verifying different pitches of auditory cues on the difficulty of learning and perception; and 2) judging the subjective preference caused by the variation of pitches. In this paper, we would like to create several evaluating criteria of objective effectiveness (intuitiveness, interruption) and subjective preference (interference, standard pitch) for user-friendliness of different pitches of auditory cues.

2 Methods

For user-friendliness is mostly influenced by the objective and subjective opinions. The experiment consisted of two parts, objective test (learning and testing) and subjective preference (questionnaire).

2.1 Participants and Procedures

There were twenty participants in the experiment. Nine women, eleven men, aged 18–47, and none of them had hearing problems and obstacles in pitch recognition. After a brief introduction, participants started to learn the auditory cues of different pitches. Then the participants needed to recognize the sound of different pitches that appeared at random time intervals when they were doing the secondary task (playing a game named "happy elimination" on iPad), using this method to verify the learning effect of the first part of the experiment. There were twelve tasks in the test. Six different pitches of audio appeared randomly. And in the end, they filled out a questionnaire about the subjective preference for different pitches of cues.

2.2 Objective Test

A well-designed auditory cue should be received intuitively and with a high rate of accuracy. Therefore, we designed two experiments to test the intuitiveness and recognition accuracy of different cues.

Auditory Learning Tasks. First, the participants were presented with a six-button interface which is shown in Fig. 1. Every button had the same size, and were color-coded.
Six colors changed from green to red (green, yellow-green, yellow-orange, yellow-red,
and red) in sequence correspond to six levels of urgency.

Second, participants were asked to click on six buttons of different colors at will and
then remember how each urgency matches the audio of different pitches. They could
click on the 6 buttons in any order. The number of times for participants to experience
was unlimited. It was entirely up to them. When they were sure that they have learned
how the six different pitches of auditory cues match with the six squares (representing
the six emergency levels), they were supposed to stop and start the test section.

We recorded the number of times they tried on each button to see how much it would
cost to establish the match between the pitches and emergency level.

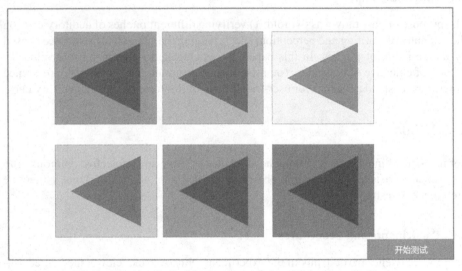

Fig. 1. The interface of the learning session. (Color figure online)

Auditory Cues Recognition Task. Based on what they learned in the previous section,
this section mainly tested how accurately they could match the pitch with the corresponding level of urgency after hearing a cue of a certain pitch. The buttons and the
screen were exactly the same as the learning section.

During the experiment process, Participants were asked to play a game on iPad (as
the secondary task in autonomous vehicles), and in the meantime, click on the button
which corresponds to the pitch when they heard an audio (Fig. 2). The time intervals of
different auditory cues were random (5 to 30 s). There were 12 tests in this section, and
each pitch was tested twice in a random order.

Fig. 2. The identification test on one iPad and the secondary task on the other

2.3 Subjective Preference

Questionnaire. To explore the subjective preference of each pitch, we gave our participants a questionnaire with 4 questions after the previous two tests, using a Rickett method to score from 1 to 10, and a sorting method to rank from 1 to 6.

Condition Preference

Overall Favorite Pitch. Ask the participants to choose one pitch of the cue that they want to use in their future autonomous cars.

Benchmark Pitch for Comparison Judgment. Ask the participants to determine which is the standard pitch when they were identifying.

Comfort Judgment

The Pitch of the Best Recognition. Ask the participants to decide which pitch of the audio is easiest to learn.

The Pitch of the Highest Interference Degree. Ask the participants to recall which pitch of cue sounds most interrupting.

2.4 Measures

To construct the measurement of this study, we used two kind of performance methods and one subjective evaluation method. The first two sections were to test objective factors, learnability and intuition, and the last one was to test subjective factors, such as preferences.

Times of Learning. The number of times participants clicked on each auditory cue proves the cost of learning the pitch. The more times they experienced, the more learning efforts it would cost [21, 22].

Recognition Accuracy. The accuracy of recognition of different pitches in the test proves the results of learning section. If they learnt pretty well in the learning section, the accuracy would be higher.

Subjective Judgment. The interference degree of each audio (makes participants feel offensive, noisy, harsh, dull, etc.) and the simplicity of learning (makes participants feel easy to remember and learn) performs a subjective evaluation. We used a 10-point Likert scale ranging from 1 (strongly disagree) to 10 (strongly agree).

3 Results

3.1 Cue Learning

Times of Learning. The number of times participants experienced of different pitches varies from person to person. In this session, the highest number of times participants played was 14, while the lowest was only 1. According to our observation, we find it was mostly influenced by the character of participants. The more cautious they seemed to be, the more times they might experience. But in general, we can still see how different pitches influenced the number of times of their learning.

Urgency Matching. The less a pitch is likely to be experienced, the less effort participants need to make to match it with the corresponding urgency. On average, the number of times participants experienced in each pitch was 6.8. Figure 3 showed the average amount of times each pitch had been experienced. Comparing the six pitches of auditory cues, ANOVA indicated that the pitches in the middle part (pitch 3, 4) required more effort (8.0 and 7.7 times) for participants to learn (F $(5, 66)$ = 5.36, p < .001). This showed that the lowest and the highest pitch (pitch 1, 6) were more intuitive to match, while the pitches lay in the middle require more effort to match. Especially, when the sound fell into the pitch range which is similar to our daily life (the third or fourth pitch), it became a lot harder for people to identify.

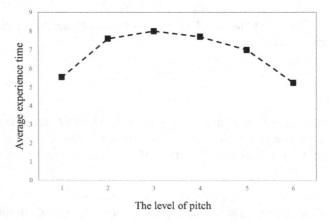

Fig. 3. The average amount of times each pitch had been experienced

3.2 Testing

Recognition Error. The average recognition error in this section was 17.2 times. ANOVA showed that the third and fourth cues were more likely to be misjudged (F $(5,114) = 4.491$, p < .001). The overall recognition errors of each pitch were demonstrated in Fig. 4. As the pitch increases, the number of identification failures started to increase, and reach the top around the third pitch, and then gradually fall back. According to this, it was the highest and lowest pitches (pitch 1, 6) that best fit with participants' perceptions.

Interestingly, even though participants spent the most time on the third or fourth pitch in learning section, they still got confused when it comes to test, which further proved that the medium pitches were less intuitive and more likely to cause identification failures.

In this way, it is better to use the lowest and highest pitches of cues in circumstances where identification accuracy is of greater importance.

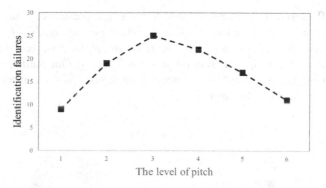

Fig. 4. The total numbers of identification failures of all participants.

3.3 Subjective Preference

Difficulty of Identification. Figure 5 demonstrated the subjective average score of each pitch on identification difficulty (out of 10). ANOVA presented that different pitches had significant impact on identi-fy difficulty (F $(5, 17) = 31.2$, p < 0.01). The result perfectly corresponded to the out-come of objective identification failure in the test section, which added more proof to the point that pitches in the middle range lacked of intuition and required more effort to learn.

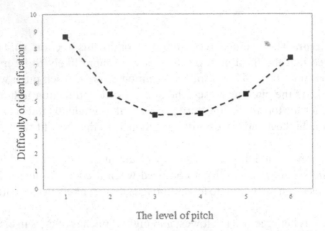

Fig. 5. The subjective evaluation on difficulty of identification

The Overall Preference. Figure 6 showed the average preference score of different pitches. According to the data, the lowest and highest pitches (pitch 1, 6) received the most preference from participants (9.2, 9.5 score). Also, participants relatively preferred higher pitches (pitch 4, 5, 6) than lower pitches (pitch 1, 2, 3). One participant explained that the lower pitches reminded him of noisy sound, such as the car horn, while the higher pitch was clearer and crisper.

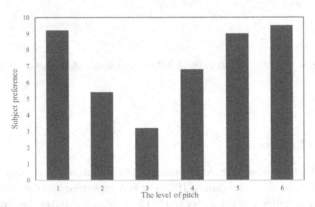

Fig. 6. The subjective preference of six pitches.

Benchmark Pitch. 35% of the participants chose the lowest pitch as their standard when it came to judging. Other than the lowest pitch, it was the highest pitch that they also prefer to choose (Fig. 7). This demonstrated that people tended to choose a pitch with a significant intuitive character for them to establish their standard. It suggests us to match the most intuitive pitch (pitch 1 or 6) with the most important kind of cues, because the more intuitive a standard pitch is, the easier it is for people to establish a judgment system.

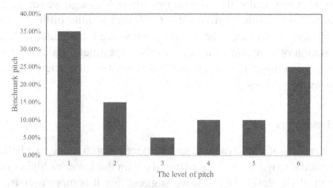

Fig. 7. Benchmark pitch

The Degree of Interference. In terms of the overall trend, as the pitch rose, the auditory cues became more and more interrupting, and consumed more attention (Fig. 8). However, between cues of same priority level of urgency (1 and 2, 3 and 4, 5 and 6), participants believed the cue with a lower pitch is slightly more interrupting than the higher one. The sound of relatively lower pitch was more resonant and occupy more attention. And the sound of higher pitch was more interrupting. So, when it comes to shifting attention, using the highest or the second highest pitch for the auditory cues would be more interrupt-ing and informative.

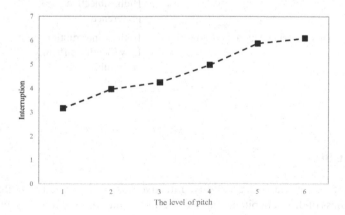

Fig. 8. The degree of interference

4 Discussion

4.1 Design of the Experiment

The design of our experiment still can be improved: 1) Although we set the gap between two pitches as big as possible, participants still found it quite difficult to identify the neighboring pitches, which made them end up guessing rather than judging. 2) Also, due to the limitation of time and conditions in the experiment, the age structure of our participants was rather unsatisfactory, which did not cover all age levels of the potential users of autonomous vehicles.

4.2 User-Friendliness

Combining the results of objective test and subjective preference, we believe that the pitches in the middle range are less user-friendly than the lower or higher ones.

In the aspect of IIC design in IVIS, we suggest that it is more user-friendly to use auditory cues with lower pitches for non-driving-related information (lowest interruption, lowest identify difficulty, and high subjective preference), and higher pitches for driving-related alerts (highest interruption, low identify difficulty, and high subjective preference) in ideal circumstances where there is not any interference of other factors (Table 2).

Table 2. The match between information in IVIS and pitch of cues.

Information in IVIS	Pitch of cues	Criteria
Non-driving-related information	Lowest	Lowest interruption Lowest identify difficulty High subjective preference
Driving-related alert	Highest	Highest interruption Low identify difficulty High subjective preference

5 Conclusions

Aiming to create criteria of learnability, intuitiveness, and subjective preference for user-friendliness of different pitch of auditory cues in autonomous vehicle scenarios. We designed and conducted a study to evaluate the user-friendliness by providing 6 different pitches of cues (of IVIS and IIC) to test how well they matched with 6 levels of priority of urgency. Experimental results demonstrated that the lowest and the highest pitch were more intuitive to match, while the pitches lay in the middle range required more effort to learn. Also, with the increase of pitches, the number of identification failures started to increase, and reached the top around the third pitch, and then gradually fell back. The subjective preference showed that the lower pitches were promising choices to be the standard of identification. Besides, the auditory cues became more and more interrupting as the pitch increased. Generally, cues with low or high pitches received the most preference from participants, and participants relatively preferred higher pitches than lower pitches.

However, the circumstance of autonomous vehicle is usually too complex to measure with few criteria. In different conditions, the importance of urgency, intuition, interruption, and subjective preference might be totally different. Therefore, the user-friendliness of these pitches would vary from situation to situation. In this paper, we provided several criteria on these aspects. Hopefully, in some specific ideal scenarios, we may find the right coefficient for each criterion to unify the factors into one function, and then figure out the specific value of user-friendliness that we are pursuing.

Acknowledgement. The paper is supported by Hunan Key Research and Development Project (Grant No. 2020SK2094) and the National Key Technologies R&D Program of China (Grant No. 2015BAH22F01).

References

1. Cao, Y., Van Der Sluis, F., Theune, M., op den Akker, R., Nijholt, A.: Evaluating informative auditory and tactile cues for in-vehicle information systems. In: Proceedings of the 2nd International Conference on Automotive User Interfaces and Interactive Vehicular Applications (AutomotiveUI 2010), 102–109. ACM, New York (2010)
2. Edworthy, J., Loxley, S., Dennis, I.: Improving auditory warning design: relationship between warning sound parameters and perceived urgency. Hum. Factors **33**(2), 205–231 (1991)
3. Arrabito, G., Mondor, T., Kent, K.: Judging the urgency of non-verbal auditory alarms: a case study. Ergonomics **47**(8), 821–840 (2004)
4. Hellier, E., Edworthy, J.: Quantifying the perceived urgency of auditory warnings. Can. Acoust. [Internet]. **17**(4), 3–11 (1989)
5. Edworthy, J., Hellier, E. and Hards, R.: The semantic associations of acoustic parameters commonly used in the design of auditory information and warning signals. Ergonomics **38**(11), 2341–2361 (1995)
6. Wiese, E., Lee, J.: Auditory alerts for in-vehicle information systems: the effects of temporal conflict and sound parameters on driver attitudes and performance. Ergonomics **47**(9), 965–986 (2004)
7. Haas, E.C., Edworthy, J.: Designing urgency into auditory warnings using pitch, speed and loudness. Comput. Control Eng. J. **7**(4), 193–198 (1996)
8. Pfleging, B., Rang, M., Broy, N.: Investigating user needs for non-driving-related activities during automated driving. In: Proceedings of the 15th International Conference on Mobile and Ubiquitous Multimedia (MUM 2016), pp. 91–99. ACM, New York (2016)
9. Wickens, C.D.: Multiple resources and performance prediction. Theor. Issues Ergon. Sci. **3**(2), 159–177 (2002)
10. Wickens, C.D., Dixon, S.R., Seppelt, B.: Auditory preemption versus multiple resources: who wins in interruption management? In: Proceedings of the Human Factors and Ergonomics Society Annual Meeting, vol. 49, no. 3, pp. 463–466. SAGE Publications, Sage CA: Los Angeles (2005)

11. Woods, D.D.: The alarm problem and directed attention in dynamic fault management. Ergonomics **38**(11), 2371–2393 (1995)
12. Bimbraw, K.: Autonomous cars: past, present and future - a review of the developments in the last century, the present scenario and the expected future of autonomous vehicle technology. In: ICINCO 2015 - 12th International Conference on Informatics in Control, Automation and Robotics, Proceedings, vol. 1, pp. 191–198 (2015)
13. Cao, Y., Mahr, A., Castronovo, S., et al.: Local danger warnings for drivers: the effect of modality and level of assistance on driver reaction. In: Proceedings of the 15th International Conference on Intelligent User Interfaces, pp. 239–248 (2010)
14. Jamson, A., Westerman, S., Hockey, G., Carsten, O.: Speech-based e-mail and driver behavior: effects of an in-vehicle message system interface. Hum. Factors **46**(4), 625 (2004)
15. Ho, C.-Y., Nikolic, M.I., Waters, M.J., Sarter, N.B.: Not now! supporting interruption management by indicating the modality and urgency of pending tasks. Hum. Factors **46**(3), 399–409 (2004). https://doi.org/10.1518/hfes.46.3.399.50397
16. Graham, R.: Use of auditory icons as emergency warnings: evaluation within a vehicle collision avoidance application. Ergonomics **42**(9), 1233–1248 (1999)
17. Peryer, G., Noyes, J., Pleydell-Pearce, K., Lieven, N.: Auditory alert characteristics: a survey of pilot views. Int. J. Aviation Psychol. **15**(3), 233–250 (2005)
18. Blattner, M.M., Sumikawa, D.A., Greenberg, R.M.: Earcons and icons: their structure and common design principles. Hum.-Comput. Interact. **4**(1), 11–44 (1989)
19. Hellier, E.J.: An Investigation into the Perceived Urgency of Auditory Warnings (1991)
20. Hellier, E.J., Edworthy, J., Dennis, I.A.N.: Improving auditory warning design: quantifying and predicting the effects of different warning parameters on perceived urgency. Hum. factors **35**(4), 693–706 (1993)
21. Bacdayan, A.W.: Time-denominated achievement cost curves, learning differences and individualized instruction. Econ. Educ. Rev. **13**(1), 43–53 (1994)
22. Mayley, G.: The evolutionary cost of learning. In: Proceedings of the Fourth International Conference on Simulation of Adaptive Behavior, pp. 458–467 (1996)
23. Marshall, D., Lee, J., Austria, P.: Alerts for in-vehicle information systems: annoyance, urgency, and appropriateness. Hum. Factors **49**(1), 145–157 (2007)
24. Patterson, R.D.: Auditory warning sounds in the work environment. Philos. Trans. Roy. Soc. London B, Biol. Sci. **327**(1241), 485–492 (1990)
25. SAE: SAE Taxonomy and Definitions for Terms Related to On-Road Motor Vehicle Automated Driving Systems (2014). https://saemobilus.sae.org/content/j3016_201609. Accessed August 2018
26. Regan, M.A., Stevens, A., Horberry, T.: Driver Acceptance of New Technology: Theory, Measurement and Optimisation. Human Factors in Road and Rail Transport . Ashgate Publishing Company, Burlington, VT (2014). https://lib.myilibrary.com/detail.asp?id=578350.
27. Oviedo-Trespalacios, O., Nandavar, S., Haworth, N.: How do perceptions of risk and other psychological factors influence the use of in-vehicle information systems (IVIS)? Transp. Res. Part F: Traffic Psychol. Behav. **67,** 113–122 (2019). ISSN 1369-8478. https://doi.org/10.1016/j.trf.2019.10.011
28. Geitner, C., Biondi, F., Skrypchuk, L., Jennings, P., Birrell, S.: The comparison of auditory, tactile, and multimodal warnings for the effective communication of unexpected events during an automated driving scenario. Transp. Res. Part F: Traffic Psychol. Behav. **65**, 23–33 (2019). ISSN 1369-8478. https://www.sciencedirect.com/science/article/pii/S1369847818305473
29. Haas, E.C., Casali, J.G.: Perceived urgency of and response time to multi-tone and frequency-modulated warning signals in broadband noise. Ergonomics **38**(11), 2313–2326 (1995)

Evaluation of Haptic Feedback Cues on Steering Wheel Based on Blind Spot Obstacle Avoidance

Jini Tao[✉], Duannaiyu Wang, and Enyi Zhu

Hunan University, Changsha, China
{Taojini1803,lalupi,SiriusZhu}@hnu.edu.cn

Abstract. Many driving collisions are caused by the existence of blind spot, and traditional warning systems using visual and audio channel produce very little effect. This paper investigates the effect of three novel haptic feedback (thermal feedback, cutaneous push and ultrasonic feedback) on the steering wheel by a series of driving simulator experiments. We evaluated the results by recording the turning recognition accuracy and two subjective scales. Experimental results showed that cutaneous push was the most comfortable feedback, with the least annoyance and complexity. Ultrasonic feedback was best accepted by participants, although it was thought most complex. Thermal feedback was hard to recognize because of the gradual characteristic, however some participants felt comfortable in this condition. Our findings are significant in delivering different type of feedback to drivers according to their preference and the actual condition and maintaining relatively low mental workload at the same time.

Keywords: Haptic feedback · User experience · Steering wheel · Blind spot

1 Introduction

Objects in blind spots are invisible to drivers, and the existence of blind spots in actual driving condition has led to high ratios of risk such as crash and near crash [1]. In China, there are 500,000 traffic accidents every year. 30% of them are caused by blind spots, compared with about 20% in the United States. In 2018, 53.41% of the truck accidents in Shenzhen was caused by blind spots [2], with the death toll accounting for 52.75% of the total.

Blind Spot Warning system (BSW) has a significant impact on road safety. Visual and audial channels are widely used in the existing systems. However, driving is vision-demanding, since 95% of the information is acquired and identified by vision [3]. And audio alarms may be masked by noise (natural or artificial) or may not be understood due to language barrier [4]. Wickens' Multiple Resource Theory proposed that diverting the secondary task to the unoccupied region of brain can enlarge the total processed information amount [5]. Haptic channel is underutilized, and worthy of being researched to reduce driver's workload.

Visual BSW such as flash signal lights on the side mirrors or A-pillars diverts driver's attention from the main driving tasks, so it increases the collision risks to some extent.

© Springer Nature Switzerland AG 2021
P.-L. P. Rau (Ed.): HCII 2021, LNCS 12773, pp. 253–266, 2021.
https://doi.org/10.1007/978-3-030-77080-8_21

Audio feedback, including speech and non-speech [4], can minimize the visual presentation of information, but the results are not always as desired. Haptic feedback is effective to overcome these drawbacks. This paper investigates thermal feedback, cutaneous push and ultrasonic feedback from the recognition accuracy and driver's subjective experience and presented evaluations of haptic warning effect.

2   Related Work

Vibrotactile feedback has been well tested as a warning signal, including BSW when changing lane [6, 7], forward collision warning [8], lane-departure situations warning [6], etc. Typical examples are vibrations applied on the steering wheel or seat pan. However, the location of the exact vibration is hard to identify, and it would be concealed during the drive.

Thermal feedback by heating the steering wheel has been investigated effective for lane changes [9], where 88.57% of the participants changed correctly. Cutaneous push feedback for the palm region can provide distinguishable directional information [10]. It has great significance in blind spot warning.

The development of the tactile network makes it possible for people to physically interact with the real environment through tactile experience [11]. One of the major current trends in the research of tactile display technologies is ultrasonic. Ultrasonic friction reduction is a feeling of air lubrication. The fingertip touches the over-pressurized air layer on the vibrating surface, which feels more lubricative because the squeeze film effect reduces friction [12].

Considering previous studies, we conducted the following study.

3   Study

3.1   Method

The three types of feedback were tested respectively in a driving simulator. We used NASA TLX questionnaire and another 5-point Likert questionnaire to enquire user experience, and then conducted semi-structured interviews to collect participants' comments.

3.2   Experimental Variables

The independent variable is feedback types (thermal, cutaneous push, ultrasonic). The dependent variables were the warning recognition accuracy (including do the task right, wrong and not recognize), perceived mental workload (NASA TLX workload), subjective ratings of the four aspects (comfort and acceptability as positive aspects, annoyance and complexity as negative aspects). Furthermore, we conducted semi-structed interviews to collect users comment on the subjective feelings.

3.3 Hypothesis

Hypothesis 1: Thermal feedback will be more comfortable than cutaneous push.
Hypothesis 2: The acceptability subjective rating for ultrasonic feedback will be higher than the other two type of feedback.
Hypothesis 3: The ultrasonic feedback will be more complex than the cutaneous push.

3.4 Apparatus

The study was conducted in a college Automobile Human–Computer Interaction Laboratory in a driving simulator. There was a Logitech steering wheel equipped with the warning facilities and a computer showing the driving scenario in the city. They drove on two-lane roads and were required to keep a speed of about 50km/h. The default of automobile number is 10%. Participants wore a Sony headphone to mask any external sound and get an immersive experience by presenting real condition noises.

When testing the thermal condition, we affixed a Peltier of 2×2 cm in size [13] and a heating trough to both sides of the steering wheel. When gripping the steering wheel, the Peltier would present warm to the palm region (see Fig. 1).

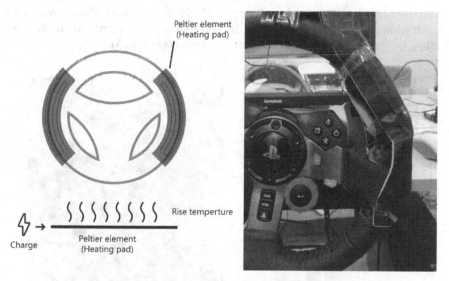

Fig. 1. Thermal haptic equipment with Peltier elements.

The cutaneous push was created by embedding two solenoids into each side of a pre-drilled metal steering wheel according to the experiments of Shakeri et al. [10]. The solenoid pins come out up to 5 mm, exerting a force up to 4.18N (see Fig. 2).

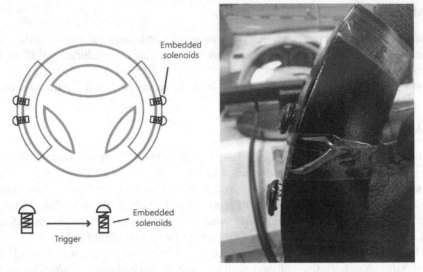

Fig. 2. The steering wheel with two solenoids embedded into each side of it.

The Ultrasonic device is made up by a pair of vibration creators, which create ultrasonic vibration to the display surface [12]. By controlling the frequency of the ultrasonic vibration, we varied the haptic sensation of roughness (see Fig. 3).

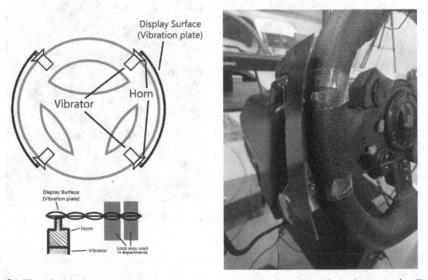

Fig. 3. The ultrasonic was created by the vibrators and transited through the horns to the display surfaces it.

3.5 Experiment Design

Figures No other feedback was given to indicate the obstacle in the blind spot, so the driver should completely depend on the haptic warning. Previous studies give haptic feedback from the side of the turning direction [14]. In our study, we used a questionnaire and open review in advance to collect the participants' preference of whether the turning direction should be consistent with or opposite to the warning direction. 12 participants tended to steer to the opposite direction of the feedback. They said that because the meaning of feedback was a warning of obstacle and danger, it was an instinct reaction to avoid. So in our experiments, the turning direction is the opposite side of the haptic feedback. And at the end of the experiment, 14 participants told researchers that turning in the opposite direction was more appropriate. Two of those who preferred the same side at first came around to the opposite after having a try.

In the thermal feedback condition, the starting temperature of the steering wheel is 32 °C [15], and the participants already adapted to this temperature before the formal experiment. The initial temperature was changed at a rate of 1 °C/sec and resulted in 35 °C for the side of coming obstacles. Participants were instructed to brake on receiving the feedback and turn the steering wheel immediately. Researchers can identify if the cue was recognized correctly by monitoring the drivers' behavior and the computer screen. The temperature would maintain at 35 °C for 3 s and then would drop to the neutral temperature in one second.

When testing the second type of feedback, the solenoid pins from the side of warning pushed the thenar of the palm, keeping constantly for 3 s. Then they would move back to the steering wheel.

The third condition was ultrasonic feedback. The frequency of vibration when ordinary driving was 36.5 kHz, and it rose to 67.5 kHz to create a smoother feeling when presenting the warning [12].

The three types of haptic warning were tested independently. There were twelve obstacle-avoiding tasks in each test, and the time when and the direction from which the obstacle appeared was presented randomly.

When the participant perceived the haptic cue, he/she must turn the steering wheel to the opposite side of the cue and brake until the car fully stopped.

After finishing one task, the participant had enough time to restart the car and get adjusted until the car was running at a constant speed again, except that in each experiment there were 2 tasks presented consecutively to simulate the real road condition.

Since the warning may be presented from either left or right, they were forced to cover the facilities with both hands so that they can perceive to feedback through palm regions.

3.6 Participants

Before the experiment, we asked the participants about their simulated driving experience (including VR, driving games, etc.) and whether they had exposed to tactile feedback (such as mobile phone vibration prompt, etc.). Three of the participants have not experienced virtual driving related equipment and all of them had ever received related tactile cues.

The experiment started by reading and signing a non-disclosure agreement. Also, a questionnaire about demographic data and their preference of turning direction was asked to fill.

The participants drove for about 5 min to adapt to the driving apparatus and the road condition, without being presented with any warning signals. Researchers then introduced all the three feedback and the required actions including pressing the pedal, turning the steering wheel to avoid collision and keeping in the middle lane during the main task. After training, participants wore the headphones to create the driving atmosphere and took part in the first task of thermal feedback (see Fig. 4).

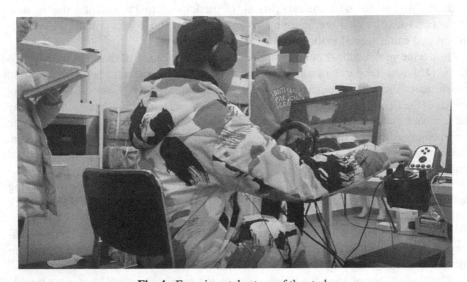

Fig. 4. Experimental set-up of the study.

Drivers usually came to crossroads and curves in order to simulate the real driving condition. Experimenters stood beside to observe and record the performance during the main task for counting the recognition rates and the final interview.

The participants then completed a NASA-TLX questionnaire and another Likert 5-point scale questionnaire rating four subjective feeling aspects of the feedback, including comfort and acceptability as positive ones, and annoyance and complexity as negative ones. Then a semi-structured interview was conducted face-to-face. The questions of the interview were open-ended and covered participants' general experience with the driving simulator, attitudes to the haptic feedback, the obviousness of each type of feedback, and expectations of the improvement of the experiment. Apart from that, to know about their trust in the system, we asked if they would continue to trust the system after reading news about traffic accidents resulting from the driver's dependence on the haptic blind warning system Their comments gave us an insight into users' expectation of a system's feedback. The second and third test were conducted following the same procedures. Finally, the participants were asked to rate the three feedback by the warning effect. They were also asked if turning to the opposite side of the feedback felt appropriate and if the preference of turning side had changed. It took about an hour to finish each experiment, and the participant received 15 yuan as a reward.

4 Results

4.1 Recognition Accuracy

There were 180 overall stimuli presentations in each type of warning. The recognition accuracy was the highest in cutaneous push with 94.4%. Nine tasks were not recognized, and one task was done in a wrong way. 93.9% of the tasks were done correctly in the ultrasonic condition. Ten tasks were ignored and also only one task was done incorrectly. In the thermal condition, 65.6% of the presentations were done correctly, and seven tasks were recognized but done incorrectly, while 58 presentations were completely ignored. The majority of the mistakes occurred in the turning conditions, or in the junctions. When the feedback was presented consecutively, the second one was hard to recognize.

4.2 Subjective Feedback

The 10-point scaled NASA TLX questionnaire collected data of 6 index indices (mental demand, physical demand, temporal demand, performance, effort, frustration) to evaluate participants' subjective workload. Differences in the overall workload were not significant ($Z = 0.710$, $p = 0.478$), and each of the six aspects was also not significant. Because participants generally commented on the feedback "direct and specific" in the interviews, the workload is low. The data were normally distributed and evaluated with LSD (Least Significance Difference Test).

The subjective ratings of acceptability, comfort, complexity and annoyance were normally distributed and evaluated with LSD. Acceptability, comfort and complexity were significantly different for three conditions (acceptability: $Z = 2.409$, $p = 0.016$; comfort: $Z = 2.034$, $p = 0.042$; complexity: $Z = 4.021$, $p < 0.001$). However, there were no significant differences in annoyance ($Z = 1.429$, $p = 0.153$) (see Fig. 5).

For the comfort aspect, cutaneous push gained the highest rating ($M = 3.700$, $SD = 1.4541$), followed by ultrasonic ($M = 3.567$, $SD = 1.995$), and thermal feedback is the lowest ($M = 2.900$, $SD = 1.867$).

The rating of acceptability was similar for cutaneous push ($M = 3.000$, $SD = 1.069$) and ultrasonic feedback ($M = 3.033$, $SD = 1.486$), but for thermal it is much lower ($M = 2.367$, $SD = 1.486$).

The complexity of the three types are significantly different: thermal feedback ($M = 2.833$, $SD = 1.407$), cutaneous push ($M = 2.367$, $SD = 1.387$), ultrasonic feedback ($M = 3.700$, $SD = 1.595$).

Although annoyance have a bad effect on the driving experience in most circumstances, according to P08, in such an emergency, annoyance to a proper extend is necessary. Also, P10 claimed "will choose a relatively annoying type as long as it is obvious and effective".

In the semi-structured interview section, most participants considered the thermal feedback the most difficult to recognize and uncomfortable. P02 said "the heating region

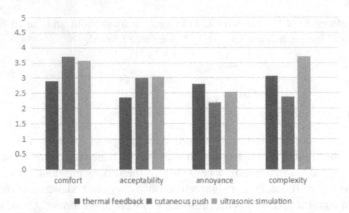

Fig. 5. The ratings for comfort, acceptability, annoyance and complexity of the Likert 5-point questionnaire.

was concentrated in the palm region and it was annoying". Only one person (P13) thought it comfortable. P09 claimed: "The increase in temperature is a gradual process, so it is not appropriate in an emergency which is abrupt and urgent 7 participants mentioned that after several tasks, they got accustomed to the temperature so felt it was imperceptive.

Comfort, annoyance and complexity were significantly better in cutaneous push feedback than in the other types, and its acceptability is similar to ultrasonic feedback. P04 commented on cutaneous push "clearer and more obvious compared with thermal feedback". P14 thought this feedback had "definite directivity". When asked about workload, most participants said that "there is almost no mental or physical need".

Ultrasonic had the highest acceptability although it is considered to be the most complex. Both P02 and P04 thought "it was obvious, but not as comfort as the push warning". Participants trusted in this type most because it was an advanced technology. However, there were three participants who thought it is the worst because of the odd feeling of friction.

Moreover, taking gender into consideration, we found that males and females felt largely different in some aspects of thermal feedback. Comfort ($Z = 2.409$, $p = 0.016$) was statistically significantly worse by males ($M = 2.556$) than females ($M = 3.583$), and the acceptability ($Z = 3.090$, $p = 0.002$) was also worse by males ($M = 1.944$) than females ($M = 3.000$).

5 Discussion

Vibrotactile feedback has been well tested as a warning signal, including BSW when changing lane [6, 7], forward collision warning [8], lane-departure situations warning [6], etc. Typical examples are vibrations applied on.

The recognition accuracy for thermal feedback was the lowest, and it was also the most uncomfortable type because of the heating region and the characteristics of gradual change. While cutaneous push gained the highest rating both in recognition accuracy and comfort rating, as it was obvious with appropriate force. The result was opposite to the claim of Hypothesis 1, so it was not corroborated.

Ultrasonic was the most accepted, and participants showed high trust in it. The result is consistent with the second hypothesis, so Hypothesis 2 was corroborated.

The subjective rating for the cutaneous push in complexity showed statistically significantly better results than ultrasonic feedback. Therefore, Hypothesis 3 was also corroborated.

We also found an interesting phenomenon that for right-handers, the feedback that was not be recognized was usually presented from the right side, and it is the same for left-hander to ignore the feedback from left. We suppose that it may be affected on the cognitive functions of brain but have not drawn a conclusion because the simple size is too small. It is valuable to be further studied in the future.

In this paper, we instructed the participants to keep hands consistently on each side to perceive the feedback. However, in real driving conditions, drivers always hold the steering wheel with one hand. Future work should focus on the haptic feedback presented on one side.

As for the thermal feedback, we suggest that the future tests should be taken in a more realistic outdoors environment and take environmental temperature influence into consideration.

Moreover, it is worth exploring which types is the most effective when the obstacles rush out from both sides,

6 Conclusion

This paper investigates the effect of three haptic feedback providing to drivers through steering wheel in the blind spot warning system. Thermal feedback is hard to recognize and not well-accepted. Ultrasonic gains the best acceptability as a novel type of warning, although it is considered the most complex. Cutaneous push has the highest accuracy of recognition of 94.4%, the highest comfort level, the least annoyance and the least complexity. And its acceptability is almost the same with ultrasonic feedback, only 0.035 lower than it, which means it is pretty effective for blinding spot warning. The results can be a reference to provide appropriate warning feedback to drivers and basis for future studies.

Acknowledgement. The paper is supported by Hunan Key Research and Development Project (Grant No. 2020SK2094) and the National Key Technologies R&D Program of China (Grant No. 2015BAH22F01).

Appendix

Questionnaire 1: NASA-TLX questionnaire [17].
Part 1: Please read the questions carefully and mark the corresponding scale positions of the 6 indicators according to the actual situation of the tasks performed.

NASA Task Load Index

Hart and Staveland's NASA Task Load Index (TLX) method assesses work load on five 7-point scales. Increments of high, medium and low estimates for each point result in 21 gradations on the scales.

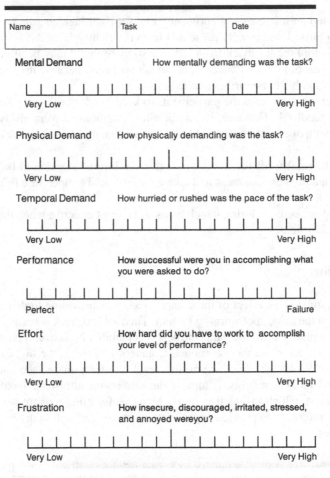

Part 2: Compare and check which of the following indicators is more closely related to the total load.

Effort or Performance	Temporal Demand or Frustration	Temporal Demand or Effort	Physical Demand or Frustration	Performance or Frustration

(*continued*)

(*continued*)

Effort or Performance	Temporal Demand or Frustration	Temporal Demand or Effort	Physical Demand or Frustration	Performance or Frustration
Physical Demand or Temporal Demand	Physical Demand or Performance	Temporal Demand or Mental Demand	Frustration or Effort	Performance or Mental Demand
Performance or Temporal Demand	Mental Demand or Effort	Mental Demand or Physical Demand	Effort or Physical Demand	Frustration or Mental Demand

Questionnaire 2: Likert 5-point questionnaire about user experience (thermal feedback as an example) [18–21].
The questionnaire contained eight statements about subjective experience, all of which were related to the thermal feedback in the experiment. Please check the level of agreement according to your subjective feeling.

1. I felt calm when using this warning system.

○strongly disagree	○disagree	○undecided	○agree	○strongly agree

2. I hesitated to use this feedback because I an afraid to be unable to cope with it.

○strongly disagree	○disagree	○undecided	○agree	○strongly agree

3. Given the opportunity, I would like to buy this warning system.

○strongly disagree	○disagree	○undecided	○agree	○strongly agree

4. I prefer to accept the traditional warning types (visual or auditory) rather than use this feedback.

Ostrongly disagree	Odisagree	Oundecided	Oagree	Ostrongly agree

5. I was in a state of tension when using this feedback.

Ostrongly disagree	Odisagree	Oundecided	Oagree	Ostrongly agree

6. This feedback was easy to perceive.

Ostrongly disagree	Odisagree	Oundecided	Oagree	Ostrongly agree

7. This feedback clearly communicated the warning information.

Ostrongly disagree	Odisagree	Oundecided	Oagree	Ostrongly agree

8. This feedback was annoying.

Ostrongly disagree	Odisagree	Oundecided	Oagree	Ostrongly agree

References

1. Wang, J.S., Knipling, R.R.: Lane change/merge crashes: problem size assessment and statistical description (1994). https://trid.trb.org/view.aspx?id=448125
2. Feng, R., Aqi, D.: Deadly car Blind spots: the first killer on the road, avoiding these areas can save your life! Fenhuang WEEKLY, 29 August 2019. https://mp.weixin.qq.com/s/x_7hn7qNbpQeYh5n53z9tw
3. Shinar, D., Schieber, F.: Visual requirements for safety and mobility of older drivers. Hum. Factors 33(5), 507–519 (1991). https://doi.org/10.1177/001872089103300503

4. Isherwood, S.J., McKeown, D.: Semantic congruency of auditory warnings. Ergonomics **60**(7), 1014–1023 (2016). https://doi.org/10.1080/00140139.2016.1237677
5. Wickens, C.D: Multiple resources and mental workload. Hum. Factors **50**(3), 449–455 (2008). https://doi.org/10.1518/001872008X288394
6. Campbell, J.L., Richard, C.M., Brown, J.L., McCallum, M.: Crash warning system interfaces: human factors insights and lessons learned. National Highway Traffic Safety Administration (2007). https://core.ac.uk/display/65386277
7. Kochhar, D.S., Tijerina, L.: Comprehension of haptic seat displays for integrated driver warning systems. In: Proceedings of the Human Factors and Ergonomics Society Annual Meeting, vol. 50, pp. 2395–2397 (2006). https://doi.org/10.1177/154193120605002210
8. Lee, J.D., Hoffman, J.D., Hayes, E.: Collision warning design to mitigate driver distraction. In: Proceedings of the SIGCHI Conference on Human Factors in Computing Systems (CHI 2004), pp. 65–72 (2004). https://doi.org/10.1145/985692.985701
9. Di Campli San Vito, P., Brewster, S., Pollick, F., White, S., Skrypchuk, L., Mouzakitis, A.: Investigation of thermal stimuli for lane changes. In: Proceedings of the International Conference on Automotive User Interfaces and Interactive Vehicular Applications (AutomotiveU'18), pp. 43–52 (2018). https://doi.org/10.1145/3239060.3239062
10. Shakeri, G., Ng, A., Williamson, J., Brewster, S.A.: Evaluation of haptic patterns on a steering wheel. 2016. In: Proceedings of the 8th International Conference on Automotive User Interfaces and Interactive Vehicular Applications (Automotive' UI 16), pp. 129–136 (2016). https://doi.org/10.1145/3003715.3005417
11. Steinbach, E., Strese, M., Eid, M., Liu, X., et al.: Haptic codecs for the tactile internet. In: Proceedings of the IEEE, vol. 107, no. 2, pp. 447–470, September 2018. https://doi.org/10.1109/JPROC.2018.2867835
12. Watanabe, T., Fukui, S.: A method for controlling tactile sensation of surface roughness using ultrasonic vibration. In: Proceedings of 1995 IEEE International Conference on Robotics and Automation ((ICRA'95) (1995). https://doi.org/10.1109/ROBOT.1995.525433
13. Wettach, R., Behrens, C., Danielsson, A., Ness, T.: A thermal information display for mobile applications. In: Proceedings of the Conference on Human-Computer Interaction with Mobile Devices and Services (Mobile HCI 2007), pp. 182–185 (2007). https://doi.org/10.1145/1377999.1378004
14. San Vito, P.D.C., et al.: Haptic navigation cues on the steering wheel. In: Proceedings of the Conference on Human Factors (CHI 2019), pp. 1–11 (2019).https://doi.org/10.1145/3290605.3300440
15. Jones, L.A., Berris, M.: The psychophysics of temperature perception and thermal-interface design. In: Proceedings of the Conference on Haptic Interfaces for Virtual Environment and Teleoperator Systems (HAPTICS 2002) (2002). https://doi.org/10.1109/HAPTIC.2002.998951
16. Wilson, G., Halvey, M., Brewster, S.A., Hughes, S.A.: Some like it hot? Thermal feedback for mobile devices. In:Proceedings of the SIGCHI Conference on Human Factors in Computing Systems (CHI 2011), pp. 2555–2564 (2011). https://doi.org/10.1145/1978942.1979316.
17. NASA Task Load Index (TLX)
18. https://ntrs.nasa.gov/archive/nasa/casi.ntrs.nasa.gov/20000021488.pdf
19. Spielberger, C.D.: State-Trait Anxiety Inventory (2010). https://doi.org/10.1002/9780470479216.corpsy0943
20. Schlüter, J., Weyer, J.: Car sharing as a means to raise acceptance of electric vehicles: an empirical study on regime change in automobility. Transp. Res. Part F: Traffic Psychol. Behav. **60**, 185201 (2019). https://doi.org/10.1016/j.trf.2018.09.005

21. Edwards, C., Hankey, J., Kiefer, R., Grimm, D., et al.: Understanding driver perceptions of a vehicle to vehicle (V2V) communication system using a test track demonstration. In: Proceedings of the Conference on Human Factors (CHI 2002), pp. 1–8 (2011). https://doi.org/10.4271/2011-01-0577
22. National Highway Traffic Safety Administration: Inappropriate Alarm Rates and Driver Annoyance (1996). https://rosap.ntl.bts.gov/view/dot/13303. Accessed 01 February 1996

The Study of the User Preferences of the Request Channel on Taking Over During Level-3 Automated Vehicles' Driving Process

Qiao Yan$^{(\boxtimes)}$, Yujing Wang, and Jiaru Chen

Huan University, Changsha, Hunan, China

Abstract. When a level-3 automated vehicle fails with an autonomous driving system or encounters an unmanageable traffic situation, the driver needs to control the vehicle to ensure driving safety. The transfer process is called the driving right transfer. This study uses the qualitative research of the Likert scale method. Through experiments and questionnaires, we learn the user preferences of reminder modes in the process of driving right transfer of level-3 automated vehicles. In this research, four different warning modes, including visual takeover warning, auditory takeover warning, tactile takeover warning and multi-mode takeover warning, were tested to conduct a user preference survey and research on the warning mode of driving right takeover of level-3 automated vehicles. Through research, we have concluded that visual- tactile takeover warning is the best warning method.

Keywords: Driving right handover · User preference · Takeover request · User experience

1 Introduction

This research is to study the user preference of warning channel in the process of transferring driving right at the level of level-3 automated vehicle. In 2015, SAE published the "Classification and definition of terms related to road motor vehicle driving automation system", [1] defining six different levels of autonomous driving, ranging from complete non-autonomy at level 0 to complete autonomy at level 5. Our research put a stage at the level of level three, which is automated for limited autonomous driving: Most of the driving time is handed over to the car's self-control, but in some cases, it needs to be converted to manual driving. We intend to find out the preference of the warning channel when the car driving near a toll-gate which means driver needs to drive the car manually. In the research, we used a combination of experiment, questionnaire, and interview. The takeover request should mainly occupy the three channels of hearing, vision, and touch of the subject [2]. So, in the study, we conducted 20 experimental subjects, each of them performed a group of seven experiments: visual warning, auditory warning, tactile warning, visual-auditory warning, visual-tactile warning, auditory-tactile warning, and visual-auditory-tactile warning. After each group of seven experiments, the questionnaire based on Likert scale was used to investigate the opinion of the subject on the

© Springer Nature Switzerland AG 2021
P.-L. P. Rau (Ed.): HCII 2021, LNCS 12773, pp. 267–280, 2021.
https://doi.org/10.1007/978-3-030-77080-8_22

comfort, entertainment and trustworthiness of seven warning channels. After the group of seven experiments, we interviewed the subject to understand his/her preference and opinion on the experiment.

In the sections below, we will detail the methods we used in our experiments and the result of our study.

2   Related Work

The takeover warning information should mainly occupy the auditory, visual, and tactile channels of the subject. The research shows that the three basic warning channels and the multi-mode take-over warning channel can effectively provide drivers with relevant information about driving situations, arouse their attention and help them to takeover successfully [3].

2.1   Visual Takeover Warning Channel

Studies have shown that the biggest advantage of visual warning is that it can provide a mass of visual information at short notice, helping the driver to quickly detect the danger in the environment [4]. Common visual takeover warning channels include the use of LED flashes installed on the steering wheel. The flashing light will bring takeover warning, which not only shortens the driver's takeover response time, but also improves the driver's situational awareness and helps the driver make appropriate decisions [5]. Information about the takeover warning is presented on the central control dashboard, for example, text display, lighting, etc. [6]. The disadvantage of this warning method is that when the warning is issued, the driver has to constantly check the relevant information on the driving scene and dashboard, which leads to the increase of the response time of the driver taking over. The head-up display technology is used to superimpose the warning information directly into the real scene of the driving environment. But this kind of take-over warning occupying the center of the driver's front window can cause some visual impact [7, 8]. However, there are many defects in a single visual warning. Drivers' attention to various warning interfaces (front window, steering wheel, display, etc.) may compete with the execution of driving tasks for visual resources, thus damaging driving performance to varying degrees. In addition, when non-driving related tasks and visual warning occupy the same channel (when drivers are reading newspapers, mobile phones and other immersive content), drivers may also ignore the warning information due to the competition and immersion of visual resources [6].

2.2   Auditory Takeover Warning Channel

The advantage of the auditory takeover warning channel is that it is omnidirectional and does not require the driver's visual resources to be occupied, which can avoid the above problems. Under the auditory warning, the driver can quickly focus on the need to take over. Therefore, audible warning is widely used in driving warning [9]. Common audible warning methods are: common natural sounds such as "beep" and "tick" [8]. This warning method can be effective. It plays a role of reminding the driver. It can also convey different semantics by changing its pitch, frequency, etc. (such as the early warning of forgetting to insert a seat belt, etc.) [6]. The second is the artificially recorded voice information. Compared to pure natural sounds, voices can convey more information and can greatly improve the quality of warnings. There are also methods of using warning signs to warn. The driving signs are generally set to sounds associated with dangerous events such as collisions and tire slippages. Such sounds are easier to understand and cause less driving distraction, which can effectively reduce drivers' response time [11].

However, this method is not universal. It is difficult to find a suitable listening label for warning in the general situation of taking over. The capacity of the auditory channel is limited, and at the same time, the auditory warning information is easily interfered by other sounds in the environment. Secondly, audible warning information and tasks that have no driving requirement in the same channel may also cause competition for resources, causing the driver to miss or fail to receive all the information, which affects the takeover process [2].

2.3 Tactile Warning Channel

Compared with visual takeover warning channel and auditory takeover warning channel, tactile takeover warning channel has more significant advantages [22]. In vehicles' takeover warning, tactile warning is mainly divided into two types: non directional warning and directional warning. Non directional warning has no clear directivity and only contains the warning itself; while directional warning can provide warning to the driver and indicate the position and direction that the driver should pay attention to. For example, when the driver drives the vehicle across the lane, the vehicle provides tactile warning on the left or right side of the steering wheel. The experimenters found that the driver's reaction time to this warning was faster than that of the visual takeover warning channel [22]. If the tactile warning signal directly acts on the human skin, it will basically not be disturbed by the outside world. And it will not increase the driver's hearing and visual load. The common tactile warning methods are: installation on the driver's seat, steering wheel and safety belt [13, 14]. A single tactile warning method has unavoidable shortcomings. The intensity of the vibration stimulus will be affected by the driver's age, gender, individual sensitivity towards the vibration, and the thickness of the driver's clothing. And bumpy roads will increase the possibility of information being ignored [3, 22]. In addition, excessively frequent vibrations will cause the driver to be bored. If frequent vibrations are made over a period of time, the information intelligibility will be reduced [13].

2.4 Multi-model Warning Channel

Multi-modal warning channels can convey more information in unit time [6]. When an external message and the driving task occupy the same channel, it will cause competition in channel resources. So, the advantages of different channels in the multi-channel warning mode are complementary. It can further improve the efficiency and security of takeover requests. Common non-driving tasks (such as reading newspapers, playing mobile phones, etc.) occupy people's vision and hearing, and the tactile warning channel can be used by drivers (such as vibration) to good reminders. And the auxiliary role of the audiovisual and audible alert channels in turn can reduce the risk of haptic information being ignored [2].

3   Methodology

3.1   Selection of Experimental Simulation Methods

Light. During level-3 automated vehicle driving, a visual takeover warning can be expressed by illumining an icon/area on the dashboard [6]. In the light environment with a high color temperature range, people will be more alert. So in actual experiments (see Fig. 1), we use wrap a circle of remote- controllable red led lights on the steering wheel to simulate light warning.

Fig. 1. Experimental process

Sound. Takeover warnings in level-3 automated vehicles can be transmitted with parallel abstract (i.e., non-verbal) warning sounds, such as beeping and ticking sounds [6]. Therefore, we chose to use the phone to play an abstract sound warning that would provide a faster initial response -- the beep.

Vibration. In level-3 automated vehicles, drivers are not likely to take their hands off the steering wheel the whole time. So other items that provide vibration to the driver should be taken into account, for example: seat backs, seat bases, or seat belts. [6] So in the actual experiment, we bought a vibrating massage cushion from the market and had the subject's back rest on the cushion to simulate the vibration of the back seat of the car seat.

3.2 Experiment process

Participants will undergo several experiments, answer questionnaires, and be interviewed at the end of all the experiments. Participants will receive 20 yuan if he/she complete all the steps. Overall, 20 people (11 women and 9 men) aged between 18 and 52 completed the entire survey. All participants had driver's licenses, and 40% of them owned a car or drove regularly for a long time.

- Introduce the experimental situation
- Sign a confidentiality agreement
- Fill in the basic information
- Start the experiment

First, the user plays an immersive game that simulates the user's behavior in level-3 automated vehicles. Then, the experimenter triggers a takeover reminder (a total of seven kinds). And the subject receives the warning and take over the car (touching the steering wheel is as a successful takeover). After each experience, the user is asked to fill out a Likert five-point scale. Experimenter records the whole process of the test, and calculates the time difference between each group of takeover reminder and the user's successful takeover (that is, the hand touching the side plate) in the later stage, so as to objectively analyze the effectiveness of each kind of takeover reminder.

- User-interview

After the seven experiments, in order to further understand the user's feelings about the product based on the scale and deeply explore the reasons for the user's choices in the scale, the user is interviewed. During the interview, all user's verbal information was obtained, so as to summarize the factors influencing user's preference.

- Over the experiment and pay the subject

3.3 Analysis Method

We used SPSS [15] to process and analyze the data collected by Likert scale. We used Cronbach reliability analysis method to judge the reliability quality of scale data [16]. And then basic data description statistics were performed on 20 groups' data [17]. Various Friedman analysis methods and Nemenyi pair-wise comparison methods were used to analyze the correlation between various factors. And Wilcoxon analysis method of paired samples was used to test it [18, 19]. Finally, the corresponding relationship model of all factors was drawn [20], and the gray association method was used to judge [21]. In addition, we also used basic statistical methods to analyze the preferences of users of different genders.

During the interview, we recorded all the words of the users and adopted a sentence-by-sentence analysis method in the subsequent processing of the interview data. And then we extracted keywords from it and drew a statistical table of the frequency of interview coding that affects the preference of the takeover reminder mode.

4 Data

SPSS software was used to analyze the reliability of Likert scale data (Cronbach Alpha coefficient = 0.884, greater than 0.8), indicating that the reliability quality of the research data was high, and further analysis showed that all questions should be retained [16]. Then, descriptive statistical analysis was conducted on the scale data (see Table 1) [17]. We used the central trend analysis method to conduct descriptive statistical analysis on the data of response time and Likert scale [16] and reached a preliminary conclusion. Data showed (see Table 1): the four warning modes that the subjects considered to have tactile warning were very obvious (An = 4.684). Individual tactile warnings and visual-tactile-auditory warnings are most effective (An = 4.737); The subjects' satisfaction

Table 1. Experiment data collection (local)

Influence factors under different modes	AVG	Min	Max	Median [Sample size = 20]
Auditory				
Significant degree	3.632	2	5	4
Validity	3.526	2	5	4
Satisfaction	2.684	3	5	5
Pleasantness	3.053	3	5	4
Interest degree	2.421	4	5	5
Comfort level	3.632	3	5	5
Reliability	2.895	3	5	5
Visual				
Significant degree	3.842	2	5	3
Validity	3.632	2	5	4
Satisfaction	3.474	3	5	5
Pleasantness	3.737	3	5	4
Interest degree	3.000	3	5	5
Comfort level	4.053	3	5	5
Reliability	3.368	3	5	5
Tactile				
Significant degree	4.684	2	4	3
Validity	4.737	2	5	3
Satisfaction	3.316	1	5	4
Pleasantness	2.947	2	4	4
Interest degree	3.632	1	5	3
Comfort level	2.526	2	5	4

(*continued*)

Table 1. (*continued*)

Influence factors under different modes	AVG	Min	Max	Median [Sample size = 20]
Reliability	3.789	2	5	3
Auditory + Visual				
Significant degree	4.211	2	5	3
Validity	4.211	2	5	4
Satisfaction	3.421	1	5	3
Pleasantness	3.526	2	5	4
Interest degree	2.947	1	5	3
Comfort level	3.263	2	5	4
Reliability	3.895	1	5	3
Auditory + Tactile				
Significant degree	4.632	1	4	2
Validity	4.632	1	4	3
Satisfaction	3.000	2	5	4
Pleasantness	3.158	1	4	3
Interest degree	3.211	1	5	4
Comfort level	2.947	2	5	4
Reliability	4.211	2	5	4
Visual + Tactile				
Significant degree	4.579	2	5	4
Validity	4.474	2	5	4
Satisfaction	3.579	1	5	2
Pleasantness	3.421	1	4	3
Interest degree	3.632	1	5	3
Comfort level	3.105	2	5	3
Reliability	4.263	1	5	3
Auditory + Visual + Tactile				
Significant degree	4.684	2	5	3
Validity	4.737	2	5	3
Satisfaction	3.316	2	5	4
Pleasantness	2.947	2	5	4
Interest degree	3.526	3	5	4
Comfort level	3.105	2	5	4
Reliability	4.737	3	5	5

was highest in visual and tactile warning (An = 3.579). What made the subjects most happy and comfortable is the visual warning mode (An = 3.737; The An = 4.053). In addition, under the three warning modes with visual warning, the subjects had higher degree of pleasure and comfort. The subjects considered that tactile warning mode was the most interesting (An = 3.632), and in the three warning modes with tactile warning mode, the subjects felt that the interest level was higher. In addition, the superposition of warning mode can improve the trust degree of the subjects. Taking all factors into consideration, visual-tactile warning mode has the highest score (An = 3.8647).

In order to further verify the warning mode of user preference, we used various Friedman analysis methods to study the differences among factors in the scale. The data showed that: effectiveness, significance, satisfaction, pleasantness, interestingness, comfort and trustworthiness presented significant differences (p = 0.000 < 0.01), indicating statistical differences between the data, thus allowing for further multiple comparisons. We specifically used Nemenyi pair two comparison method to analyze the differences among the factors (see Table 2) [17, 18]. And the results showed that there was no difference between the three factors of pleasure, fun and comfort and satisfaction (p = 0.9), while there was a significant difference between the significant degree, effectiveness and reliability and satisfaction (p = 0.001**). Then, Wilcoxon analysis of paired samples was used to verify the correctness of the above analysis results [19] (see Table 3). Finally, we drew the corresponding relationship model of all factors (see Fig. 1) [20], and judged the influencing factors of user preference more intuitively according to the model.

Table 2. Nemenyi-results of comparison in pairs (local).

(I)Name [(J)Name = Satisfaction]	P
Auditory	
Significant degree	0.001**
Validity	0.001**
Pleasantness	0.9
Interest degree	0.9
Comfort level	0.9
Reliability	0.001**
*p < 0.05 **p < 0.01	

To sum up, we believe that in the actual situation, the six factors of validity, significance, pleasantness, interestingness, comfort and trustworthiness are all correlated with satisfaction, but the three factors of pleasantness, interest and comfort have greater influence on satisfaction than that of significance and validity. By using the grey correlation method, the correlation degrees of pleasure, interest, comfort and satisfaction were calculated to be 0.787, 0.781 and 0.747, respectively (Fig. 2).

Table 3. Wilcoxon analysis results of paired samples.

Name	Paring (median)		Difference (Paring1–2)	Statistical magnitude	P
	Paring1	Paring2			
Satisfaction-Significant degree	3	5	−2	8.227	0.000**
Satisfaction-Validity	3	5	−2	7.850	0.000**
Satisfaction-Pleasantness	3	3	0	0.086	0.931
Satisfaction-Interest degree	3	3	0	0.600	0.544
Satisfaction-Comfort level	3	3	0	0.192	0.848
Satisfaction-Reliability	3	5	−1	5.806	0.000**

*p < 0.05 **p < 0.01

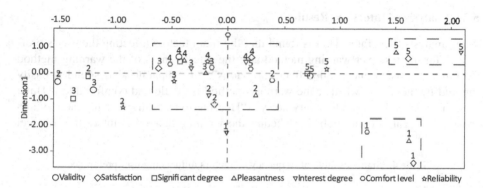

○Validity ◇Satisfaction □Significant degree △Pleasantness ▽Interest degree ○Comfort level ☆Reliability

Fig. 2. Correspondence analysis model of the 7 factors

5 Result and Discussion

5.1 Experimental Data Results

From the perspectives of 20 sets of Likert scale data and reaction time, tactile warning is regarded as an obvious and effective warning mode, while visual warning is regarded as a comfortable and pleasant warning mode. After the description and analysis of all the data, we found that tactile warning and tactile-auditory warning are the two most effective warning channels, while the most satisfying one is visual-tactile warning channel, and the overall optimal is visual-tactile warning channel. After correlation analysis, we believe that the effect of validity and significance on user preference is weaker than that of pleasure, interest and comfort. Users prefer the warning channel that can make them in a better mood, and sees whether the warning is obvious as a supplementary judgment factor, too obvious warning channel will make users feel uncomfortable and even bored. Although the visual-tactile-auditory warning channel is very effective in alerting users, it is too cluttered for users to understand "what's going on". And overmuch prompt forms will make users feel uncomfortable and unpleasant.

In addition, according to our statistical information on human factors, we found that male subjects prefer visual warning channel and single warning channel, the female subjects prefer visual-tactile warning mode and least prefer auditory mode. Combined with the experimental observation, we believe that compared with men, women are more likely to immerse themselves in the entertainment that occupies the auditory channel such as video and music in the process of automatic driving. Besides, according to the data, we find that both men and women prefer the warning channel that they think is pleasant and interesting, and choose the warning channel with moderate comfort level, which means they can receive the reminder without making themselves feel uncomfortable.

To sum up, we believe that when users choose the prompt mode, they only require the mode to have basic "effectiveness" and "obvious degree", but not "the higher the better". However, the factors such as comfort, pleasure and interest are the higher requirements of users for the warning mode. Supported by this view, we believe that visual-tactile warning channel is the most preferred one.

5.2 Analysis of Interview Results

After analysis (See Table 4), it is found that the main factors affecting the user's preference for the takeover warning method are: the effectiveness of the warning method ("I think the visual and tactile effect is obvious", "the separate alert sound is easily ignored by me"), and whether the warning method is gentle and comfortable ("Three combined ways make me feel very noisy", "Hearing-touch will cause my anxiety and tension", "Light is comfortable"), whether the warning method is interesting ("I feel

Table 4. Statistics table of frequency of factors influencing user preference

Interview keywords of different factors	Subject mentions (number/frequency)	Total references (number/frequency)
Validity		
sensation (cue sound) works	1	33/46
(light) can get me enough attention	7/8	
(vibration) plays a major role	2/3	
touch is the fastest and most direct	1/2	
(light + sound) clear enough	2/3	
(light + sound) causes rapid response	1	
(light + vibration) the effect is obvious	2/5	
(cue sound + vibration) effective	1	
(sound + light + vibration) the effect is obvious	5/7	
(vibration) it's a little weak	1	
(light) not obvious	5/8	
(light) limited indication in case of emergency	1	

(continued)

Table 4. (*continued*)

Interview keywords of different factors	Subject mentions (number/frequency)	Total references (number/frequency)
(cue sound) easy to ignore	4/5	
Degree of disturbance/shock		
(cue sound) cause irritation	4/5	29/36
Panics at the cue sound	1	
(light) soft and comfortable	7/9	
(vibration) cause shock	6/7	
(light + sound) comfortable	1	
(light + vibration) makes people feel comfortable	1	
(sound + vibration) frightens people	3/4	
(cue + vibration) causes anxiety and tension	2/3	
(cue + vibration) bored	1	
(sound + light + vibration) give people a sense of oppression	2/3	
(sound + light + vibration) too noisy	1	
Interest		
(vibration) it is interesting	1	3/3
(light + vibration) makes me feel interesting	1	
(sound + vibration) interesting	1	
Personal preferences		
sensitive to sound	1	3/3
don't like the color of the light	1	
like the feeling of the body (vibration)	1	
Information transmission function		
the color of light has the function of transmitting information	1	2/2
(light + vibration) can accurately convey the connection	1	

vibration is very interesting ", "I think tactile-visual is very interesting"), whether the warning method can convey the information ("I think the color of the light can con-vey the takeover information", "the visual-tactile mode can accurately convey the takeover information") and personalized preferences (" I dislike the color of that light ", " I personally like the physical feeling better"). Among them, effectiveness and comfort have become the most important factors affecting the user's preference for taking over the warning mode.

Among the three warning methods of separate lights, sounds and vibration, the response of the light is the best ("light is more comfortable", "light can attract enough attention"), followed by vibration ("vibration is the fastest and most direct "). Of the two warning combinations of the three warning methods, the light-vibration response is the best (" obvious effect ", " This method makes me feel comfortable ") the prompt sound-light is second (" light-prompt sound can cause me fast response"). Of all the seven modes, the light-vibration response is the best. Although the individual light is easy to be ignored, in the combination of the two, the light can play a role of information transmission, allowing users to more clearly understand the takeover situation and optimize the experience of the interaction process. More than half of the partici-pants believe that pairwise combination is better than single and triple combination ("the three will be noisy when they appear at the same time", " I think the combina-tion of the two is enough ") .The single reminder is considered to be insufficiently effective and easy to be ignored. The three combined warning methods are consid-ered to be too disturbing. And it is likely to arouse the user's irritability and frightening mood. On the basis of the above analysis, it can be summarized from the interview results that light-vibration is the most preferred takeover warning mode among users.

6   Conclusions and Future Work

Due to time constraints, we only conducted experiments for the scenario of driving near a toll-gate which means driver needs to drive the car manually. Besides, the experimental samples are small and the age distribution is uneven, which may have some impact on our experimental results. In the future, the experimental samples will be expanded to carry out experiments under three conditions: high emergency and low concentration, low emergency and low concentration, and low emergency and high concentration. We also give some thought to the results of this experiment. The separate light is considered to be mild and comfortable but the prompt is not effective. The combination of the three warning methods is very effective but it is easy to make users feel irritable and frightened. We can design different warning modes according to different situations in the future. For example, users should be reminded in a gentle way at the beginning. If the user does not espond, it will be automatically switched to a stronger warning mode. This humanized design may bring users a better interactive experience (the specific modes and intervals need further experiments). In addition, we can also make specific settings for various modes, such as studying buzzer and voice, ambient light and interactive interface light, backrest vibration and cushion vibration. The factors (effectiveness, comfort, interestingness, information transmission function, etc.) summarized from the experimental and interview data that affect users' preference for take-over warning mode can provide some help and support for the design of automatic driving take-over warning mode in the future.

In addition, according to the conclusion of this experiment, we can further deepen the research. We also do not make a detailed analysis of the situation that more than one person on the vehicle could drive and the driver could not receive a certain warning mode. In addition, for the reminder mode of light + vibration, how to combine the brightness, flicker frequency and vibration frequency of the lantern to bring better experience to users without affecting the reminder effect is also what we need to implement in the future.

References

1. SAE International in United States. Taxonomy and Definitions for Terms Related to Driving Automation Systems for On-Road Motor Vehicles [OL] (2018). https://saemobilus.sae.org/content/J3016_201806/
2. Ma, S., Wei, Z., Shi, J.-L., Zhe, Y.: Human factors in conditional autopilot takeover based on cognitive mechanism. Adv. Psychol. Sci. 1–11 (2019). https://kns.cnki.net/kcms/detail/11.4766.R.20191120.1032.010.html
3. Meng, F., Spence, C.: Tactile warming signals for in-vehicle systems. Accid. Anal. Prevent. **75**, 333–346 (2015)
4. Wege, C., Will, S., Victor, T.: Eye movement and brake reactions to real world brake-capacity forward collision warnings—a naturalistic driving study. Accid. Anal. Prevent. **58**(3), 259–270 (2013)
5. Borojeni, S.S., Chuang, L., Heuten, W., Boll, S.: Assisting drivers with ambient take-over requests in highly automated driving. In: Green, P., Boll, S., Burnett, G., Gabbard, J., Osswald, S. (eds.) Proceedings of the 8th International Conference on Automotive User Interfaces and Interactive Vehicular Applications, pp. 237–244. Assoc Computing Machinery, New York (2016)
6. Bazilinskyy, P., Petermeijer, S.M., Petrovych, V., Dodou, D., de Winter, J.C.F.: Take-over requests in highly automated driving: a crowdsourcing survey on auditory, vibrotactile, and visual displays. Transp. Res. Part F: Traffic Psychol. Behav. **56**, 82–98 (2018)
7. Langlois, S., Soualmi, B.: Augmented reality versus classical HUD to take over from automated driving: an aid to smooth reactions and to anticipate maneuvers. In Proceedings of the 2016 IEEE 19th International Conference on Intelligent Transportation Systems (ITSC), pp. 1571–1578. IEEE, New York (2016)
8. Lorenz, L., Kerschbaum, P., Schumann, J.: Designing take-over scenarios for automated driving: how does augmented reality support the driver to get back into the loop? In: Proceedings of the Human Factors and Ergonomics Society Annual Meeting, pp. 1681–1685. Human Factors and Ergonomics Society, Santa Monica, CA (2014)
9. Bazilinskyy, P., de Winter, J.C.F.: Auditory interfaces in automated driving: an international survey. Peer J. Comput. Sci. **1**, e13 (2015)
10. Politis, I., Brewster, S., Pollick, F.: Language-based multimodal displays for the handover of control in autonomous cars. In: Proceedings of the 7th International Conference on Automotive User Interfaces and Interactive Vehicular Applications, pp. 3–10. ACM, New York (2015)
11. Beattie, D., Baillie, L., Halvey, M.: A comparison of artificial driving sounds for automated vehicles. In: Proceedings of the 2015 ACM International Joint Conference on Pervasive and Ubiquitous Computing, pp. 451–462. Association Computing Machinery, New York (2015)
12. Prewett, M.S., Elliott, L.R., Walvoord, A.G., Coovert, M.D.: A meta-analysis of vibrotactile and visual information displays for improving task performance. IEEE Trans. Syst. Man Cybern. Part C: Appl. and Rev. **42**(1), 123–132 (2012)
13. Petermeijer, S.M., de Winter, J.C.F., Bengler, K.J.: Vibrotactile displays: a survey with a view on highly automated driving. IEEE Trans. Intell. Transp. Syst. **17**(4), 897–907 (2016)
14. Petermeijer, S.M., Doubek, F., de Winter, J.C.F.: Driver response times to auditory, visual, and tactile take-over requests: a simulator study with 101 participants. In: Basu, A., Pedrycz, W., Zabuli, X. (eds.) Proceedings of the IEEE International Conference on Systems, Man, and Cybernetics (SMC), pp.1505–1510. IEEE, Banff, Canada (2017)
15. The SPSSAU project (2019). SPSSAU. (Version 20.0) [Online Application Software]. https://www.spssau.com.xe/[in=epidoc1.in]/?t2000=026564/(100)

16. Eisinga, R., Te Grotenhuis, M., Pelzer, B.: The reliability of a two-item scale: Pearson, Cronbach, or Spearman-Brown? Int. J. Publ. Health **58**(4), 637–642 (2013)
17. Oja, H.: Descriptive statistics for multivariate distributions. Statist. Probabil. Lett. **1**(6), 327–332 (1983)
18. Brown, I., Mues, C.: An experimental comparison of classification algorithms for imbalanced credit scoring data sets. Exp. Syst. Appl. **39**(3), 3446–3453 (2012)
19. Rosner, B., Glynn, R.J., Ting, L.: Incorporation of clustering effects for the Wilcoxon rank sum test: a large-sample approach. Biometrics **59**(4), 1089–1098 (2015)
20. Ziegel, E.R.: Correspondence analysis handbook. Technometrics **35**(1), 103–103 (1993)
21. Qian, W.Y., Dang, Y.G., Xiong, P.P., et al.: Topsis Based on Grey Correlation Method and Its Application, vol. 27, no. 8, pp. 23–25 (2009)
22. Zhang, Z.: Research on tactile warning of highly automatic driving taking over. Sci. Technol. Innov. Appl. **35**, 59–60 (2018)

A Study for Evaluations of Automobile Digital Dashboard Layouts Based on Cognition Electroencephalogram

Hao Yang[1][(✉)], Jitao Zhang[2], and Ruoyu Jia[1]

[1] North China University of Technology, Beijing, China
hao-yang12@ncut.edu.cn
[2] Luxun Academy of Fine Arts, Dalian, China

Abstract. Electroencephalogram (EEG) performed an increasingly important role in user experience research. The study applied P300, one of the main components of event-related potentials (ERPs), into the evaluation of digital dashboard layouts. The amplitude and latency of P300 were used to objectively quantify the cognitive differences of the subjects. By comparing the results with that of a subjective system usability scale, the feasibility of this method was verified. Through statistical analysis, the layout most suitable for the subjects' cognitive characteristics was found. By the amplitude and latency of P300, a support vector machine (SVM) model was established to explore the relationship between P300 and dashboard layout. The results showed a good prediction accuracy of 80%. Analysis of variance (ANOVA) indicated significant differences among the four groups of P300 induced by different layout types ($p < 0.01$). And Type A was the most reasonable one from the perspective of attention allocation. Among the nine schemes of Type A, the one in flat design style with a warm tone presented the highest system usability ($m = 95.833$). The conclusions provided a theoretical basis for the layout evaluation of digital dashboards, and a reference for the design of human-vehicle interface, especially in the era of car-sharing.

Keywords: ERPs · P300 · Digital dashboard · Interface layout · Human-vehicle interaction · SVM

1 Introduction

The display content of digital instruments is different from that of traditional ones. And due to the variety of vehicle models and interior types, it is often necessary to redesign the hardware and appearance when developing instrument panels, which leads to a lot of investment in manpower and capital. In addition, the development of car sharing industry also puts forward new requirements for dashboard design. Because it is difficult for users to quickly understand and master the information architecture of an unfamiliar dashboard when accessing a specific car. The information on the dashboard should be able to attract the user's attention as soon as possible. An off-road glance for more than 2 s will greatly increase driving risks [1].

© Springer Nature Switzerland AG 2021
P.-L. P. Rau (Ed.): HCII 2021, LNCS 12773, pp. 281–295, 2021.
https://doi.org/10.1007/978-3-030-77080-8_23

The overall design idea of digital dashboards is to present the complex information of the car with graphics and animation. Due to the large variety and quantity of information, rather than just numbers, a reasonably designed layout plays an important role in the driver's reading efficiency, which raises new claims for design works. Different application scenarios and various dashboard layouts disperse research resources. In this study, from the perspective of layout, we used electroencephalogram (EEG) as a research means. By the event-related potential (ERP) P300 evoked by the stimulation of the dashboards, the cognitive law of users on digital dashboards was revealed, and the basis and principles of layout design were put forward.

2 Literature Review

In recent years, researches on human-computer interaction interface had become more and more specific and in-depth. In the field of automobile driving, the problem of driver's distraction and inattention caused by the visual cognition of dashboards was becoming a research hotspot. Distraction referred to the fact that the driver was attracted by some activities, objects or people inside or outside the vehicle, which diverted his attention from the basic driving tasks and even delayed his reaction to maintain a safe driving state. Vision was the most important sense for driving. More than 90% of driving information was transmitted to the driver through the visual channel [2].

There were mainly two types of instrument panels: mid-located ones and front facing ones [3], which brought differences to drivers in interactive efficiency. The dashboard would bring about various impacts on drivers' mental load and perception, therefore influencing their information access and driving operations [4]. The visual cognition processing on the shapes (round or linear), pointers and orientation of the instruments would lead to diverse effects. And although the linear ones performed better in vision presentation, people preferred circle gauges subjectively [5].

Using technologies of cognitive neuroscience theory to explore the cognitive mechanism of brain had become a new trend in the study of visual communication design. Although ERPs could not accurately locate the brain, EEG recording instruments with a millisecond time resolution could precisely record the instantaneous change of emotional process from the perspective of time, which was a good tool for emotional research. Handy et al. [6] conducted ERPs experiments on commercial signs with different familiarity from two aspects of emotional preference and visual complexity, and found that N200 and P200 waveforms were induced in the first 200 ms after the stimulus material being presented. Guo et al. [7] used an improved oddball paradigm to carry out a research on the evaluation of user emotion and use intention of different types of mobile phones by taking smart phone interfaces as experimental objects. The results showed that N300 and LPP in the parietal and occipital regions were induced by the pictures of mobile phones which could arouse the intention of using, and the brain activity in the central parietal and occipital regions was more obvious. It further illustrated that the amplitude of ERP components in related brain regions could be used to measure user experience. Hou and Lu [8] studied the cognitive processing mechanism of traffic signs in Chinese national standard by measuring ERPs, and explored semantic and emotional processing. Four ERP components, N170, P200, N300 and N400, were induced in the experiment.

Else et al. [9] classified artistic pictures from both abstract and concrete dimensions, analyzed the cognitive emotional changes of human brain in the process of appreciating pictures by EEG experiments, and found the aesthetic cognitive feedback mechanism of visual art was not only related to individual professional experience, but also related to semantic content.

P300 was an important component of ERPs, which could appear under the oddball paradigm and was related to visual attention, recognition and other cognitive functions [10]. ERP components revealed by researches on working memory and emotion [11] indicated that P300 reflected the activities of neurons in the process of cognition, which was affected by factors such as related tasks, the importance of the stimuli, decision-making, attention, emotion and so on. For human-computer interface (HCI) of industrial products, both concrete modeling parameters in physics and abstract perceptual image in psychology were important factors, which needed to be analyzed synchronously [12]. Perception and attention significantly affected the amplitude of P300, while the physical properties of the stimuli and people's response had little effect on the amplitude. The P300 amplitude, as a measure of information processing capacity, uncovered activation in an event categorization network which was regulated by both attention and working memory. Tommaso et al. [13] found that whether the stimulus was a painting or geometric figure, the one evaluated as beautiful would trigger a larger P300 than a medium or unsightly one, which indicated that people would focus on pleasant visual objects. Additionally, the memory process on the interface layout needed to be further elaborated.

3 Method

3.1 Subjects

18 car owners were invited as subjects, aged between 22 and 28. All of them were postgraduate students or had obtained a master's degree, majoring in industrial design. There were 11 males and 7 females. The subjects had been driving smart cars for 2–5 years and had a deep understanding of the functions and user experience of digital dashboards. All subjects were right-handed and in good physical condition, without a history of neurological and mental illness. Their visual acuity or corrected visual acuity was normal without color blindness or color weakness. All the subjects signed the informed consent, and were promised a certain reward after the end of the experiment as incentive if they contributed effective data.

3.2 Experiment

Dashboard images with different layouts were presented in the experiment. By comparing the latency and amplitude of P300 induced by different dashboard layouts, the relationship between the layout and the driver's excitement and arousal level was judged, which could be taken as a reference for dashboard layout design.

Using electrophysiological methods to conduct psychological experiments needed the researchers to artificially induce different psychological characteristics in a laboratory. Therefore, it was important to draw appropriate layout images of digital dashboards as stimuli. In order to simulate the visual target in road driving, the speed was displayed on the speedometer in the picture. Since 60km/h was the highest speed on the urban roads in China [14], the images with a speed of higher than 60km/h were taken as target stimuli, and the ones with a speed of less than 60km/h were non-target stimuli, so as to test whether the subjects could produce enough response to the dashboard. When the target stimulus images were presented, the subjects were required to respond to the interface by pressing the space bar on a mechanical keyboard to increase their attention and add marks to the recorded EEG data. The reaction time and accuracy could also be collected. The space bar with a larger area was easy for the subjects to click.

According to existing research results [15], automobile dashboard layouts could be summarized into four categories. In order to highlight the layouts, only the tachometer, speedometer, fuel gauge and water temperature gauge were presented in the dashboard prototype images (Fig. 1).

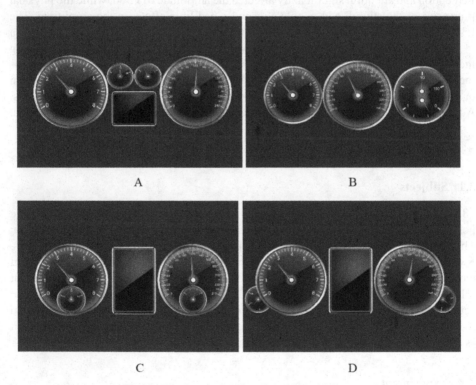

Fig. 1. The four categories of dashboard layouts

For the sake of obtaining reliable data, enough representative layout samples were necessary. We designed nine schemes for each layout. The skeuomorphism style, flat style and minimalism style were adopted. And cold, warm and neutral color tones were given to the design schemes respectively under each style. Therefrom, a total of 4*3*3 = 36 layout schemes were created as the stimuli of the experiment (Fig. 2). The coding and meaning of the stimuli were shown in Table 1.

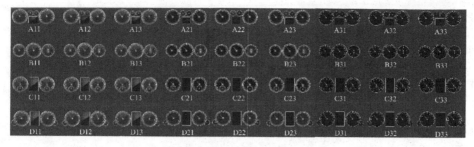

Fig. 2. The 36 layout schemes

Table 1. The coding and meaning of the stimuli of the experiment

Layout type	Design style	Color tone	Layout type	Design style	Color tone
Type A	Skeuomorphism style (A1)	Cold (A11)	Type B	Skeuomorphism style (B1)	Cold (B11)
		Warm (A12)			Warm (B12)
		Neutral (A13)			Neutral (B13)
	Flat design style (A2)	Cold (A21)		Flat design style (B2)	Cold (B21)
		Warm (A22)			Warm (B22)
		Neutral (A23)			Neutral (B23)
	Minimalism style (A3)	Cold (A31)		Minimalism style (B3)	Cold (B31)
		Warm (A32)			Warm (B32)
		Neutral (A33)			Neutral (B33)
Type C	Skeuomorphism style (C1)	Cold (C11)	Type D	Skeuomorphism style (D1)	Cold (D11)
		Warm (C12)			Warm (D12)
		Neutral (C13)			Neutral (D13)
	Flat design style (C2)	Cold (C21)		Flat design style (D2)	Cold (D21)
		Warm (C22)			Warm (D22)
		Neutral (C23)			Neutral (D23)
	Minimalism style (C3)	Cold (C31)		Minimalism style (D3)	Cold (D31)
		Warm (C32)			Warm (D32)
		Neutral (C33)			Neutral (D33)

All the images were in JPEG format with a resolution of 1024*768dpi. We applied the software E-Prime (ver. 2.0) to design the experiment. The stimuli were played on a 21.5 in. monitor, and the distance between the subject's eyes and screen was about 800 mm. All the pictures were presented in the center of the screen according to the oddball paradigm. The probability of target stimulus pictures was slightly less than 20%, and that of non-target stimuli was more than 80%. Considering that the P300 component appeared near the 300th millisecond after the stimulation, in order to ensure the late

P300 component could appear when the stimulation was presented, the presentation time of an image was set to 600 ms, with an interval of 400 ms. The time limit that allowed the subject to react was set to 800 ms.

Each of the 36 dashboard prototype schemes appeared within 30 s repeatedly and presented different speed values. The subjects needed to regularly judge whether they should press the space bar. After that, another scheme would appear in the next 30 s, and so on until the 36 rounds finished. For each subject, the duration of the experiment was about 20 min. The sequence diagram of the experimental material was shown in Fig. 3.

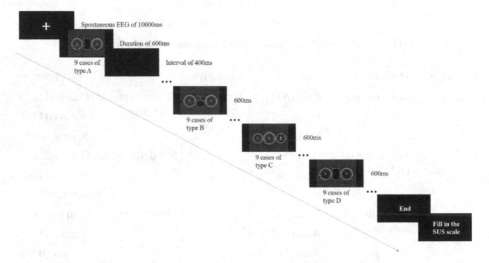

Fig. 3. The sequence diagram of the experimental material

A 64-channel brain electrode cap produced by Neuroscan Co., Ltd. was used in the experiment. The CZ electrode in the center of the head was taken as reference. The sampling frequency was 1000Hz. The acquisition system was the SynAmps2 EEG amplifier with 64 channels.

At the end of the EEG experiment, the subjects were required to evaluate each scheme subjectively with the system usability scale (SUS). System usability is a measure of perceived applicability of a product or system as a whole after users complete a series of tasks. SUS is a standardized scale. The subjects ought to grade each item in the scale by an integer among 1–5 after experiencing a product or interface system [16]. Existing researches showed that SUS was effective in evaluating human-vehicle interface and quantifying user experience reasonably [14, 17].

3.3 Acquisition of P300

In the process of observing and recognizing the dashboard prototype pictures, the visual processing areas of the brain would produce different EEG signals with the changes of psychological emotions. The coordinated, open, beautiful and concise pictures which were positive stimulation would evoke positive EEG signals correspondingly; while the confused, dull, unsightly and complex ones which were negative stimulation would produce negative or inhibitory EEG signals. Analyzing the amplitude and latency of P300 induced by different conditions of stimuli could help to further understand the brain mechanism related to emotion processing.

We used the EEGLAB toolbox (ver. 14.0.0) loaded in MATLAB (ver. R2016a) to analyze the data. Basic FIR filter was applied to filter the raw signals. In order to get a better result in time-frequency analysis, the lower and higher edges of the frequency pass band were sent to 0.5Hz and 35Hz respectively. After filter, we ran the independent component analysis (ICA) to remove the interference signals, including electrooculogram and other obvious artifacts. Then epochs could be extracted according to the marks. For both target and non-target stimuli, the baseline latency range was from 200 ms before the stimulus to the time when the stimulus appeared (-200–0 ms). After preprocessing EEG signals of the 18 subjects, superposition average was made. Therefrom, the ERPs induced by the dashboard images could be obtained. In this means, we drawn the contrast diagram of P300 waveforms, and retrieved the peak and latency data.

3.4 Support Vector Machine (SVM)

In order to judge whether the dashboard layout types were related to the amplitude and latency of P300, SVM model could be useful and supplement the conclusions of analysis on differences between groups. SVM was a kind of generalized linear classifier which classified data according to supervised learning style. Traditional linear processing would result in the loss of part of the information and could not completely duplicate the human processing. However, SVM could approximate any nonlinear function wilfully without falling into the local optimal problem. SVM was able to detect and recognize automobile dashboard characters [18]. And existing researches had used machine learning models such as artificial neural networks and SVM to process P300 data, and achieved good results [19, 20].

The algorithm of SVM could learn from the existing classification data and construct the classification boundary of different data types. After the boundary was established, the algorithm could be used to classify the testing sample dataset. In this study, the LIBSVM toolbox was used to build the recognition and classification model. LIBSVM is an open source SVM algorithm development package, which encapsulates and optimizes the mathematical calculation in SVM algorithm.

4 Results

4.1 Difference of the P300 Components Induced by Four Kinds of Dashboard Layouts

The EEG signals of 18 subjects were superimposed and averaged. And then the signals of the 9 schemes under each dashboard layout were superimposed and averaged to obtain the ERP data of the four groups of schemes. We made analysis of variance (ANOVA) for the P300 amplitude and latency getting from three representative electrodes, and conducted post hoc multiple comparisons to judge whether there were significant differences between the four types of dashboard layouts (Table 2). FZ, CPZ and PZ were chosen as the representative electrodes in frontal, parietal and occipital regions. Generally, the EEG activity in frontal region reflected how people know they were doing something in a certain environment. And the signals from parietal and occipital regions could reflect the location of visual attention and the processing of visual information respectively.

The results of ANOVA showed that there were significant differences in P300 amplitude and latency values induced by the four types of dashboard layouts ($p < 0.01$), indicating the layouts could lead to significant changes in EEG to some extent. However, after pairwise comparisons by the least-significant difference (LSD) test, it could be seen that for any index, the effect between the first group (data of Type A) and each of the other three groups was significant, while the effects among those three groups (data of Type B, C and D) were insignificant ($p > 0.05$). Therefore, in the next section, an SVM model was established to further clarify the relationship between P300 and layout types.

The amplitude of P300 was positively correlated with the amount of psychological resources invested, and its latent period was prolonged with the increase of task difficulty [10]. The higher the average amplitude and the shorter the latency, the more favorable from the perspective of attention allocation. Especially in shared cars where users were not very familiar with the instrument panel interface, if the dashboard could not attract users' attention well, it was easier to cause danger. The average amplitude values of Type A layout from the three electrodes were the highest ($m_{FZ\_A} = 6.986$, $m_{CPZ\_A} = 6.944$, $m_{PZ\_A} = 7.496$), and the latency mean values were relatively short. Among them, the latency from CPZ was the shortest among the four layouts, which was 343.221 ms. Thus, layout A was more reasonable for users' cognitive performance. However, larger P300 amplitude often resulted from higher task loads, which might result in lower accuracy [21]. Therefore, it was still necessary to combine other factors, such as system usability, to judge the rationality of the layouts.

Table 2. The results of ANOVA

		Mean				ANOVA	
		Type A	Type B	Type C	Type D	F	Sig
Amplitude (µV)	FZ	6.986	6.285	6.411	6.215	5.063	0.006
	CPZ	6.944	5.739	6.018	6.223	13.893	0.000
	PZ	7.496	6.563	6.629	6.478	11.772	0.000
Latency period (ms)	FZ	347.104	448.247	342.07	343.631	20.4	0.000
	CPZ	343.221	449.753	359.968	362.212	18.834	0.000
	PZ	405.659	446.816	406.394	397.107	9.363	0.000
SUS score		79.167	64.278	71.944	72.833	2.954	0.047

The statistical results of SUS showed that Type A performed best in system usability, followed by Type D and C. And the mean SUS score of Type B was the lowest. It was approximately consistent with the trend of P300 amplitude and latency. Therefore, we made specific analysis for the SUS scores of the nine layout schemes under Type A, and the results were shown in Table 3.

Table 3. The SUS scores of the nine layout schemes under Type A

Scheme coding	SUS score (n = 18)	
	Mean	Std. deviation
A11	82.083	2.745
A12	67.639	3.147
A13	79.306	2.396
A21	77.083	1.965
A22	95.833	2.1
A23	93.889	2.873
A31	54.444	2.651
A32	77.361	2.775
A33	84.861	2.904
F = 414.599, p < 0.001		

One-way ANOVA indicated that there were significant differences in system usability among the nine schemes ($F = 414.599$, $p < 0.001$). Moreover, the results of LSD test showed that the differences between each pair of schemes were significant ($p < 0.05$) except that between A21 and A32. Among the nine schemes, the SUS score of A22 (flat design style, with a warm color tone) was the highest, reflecting a better system usability for those who read this kind of dashboard. During the 30s when images of A22 were displayed on the monitor, the contrast diagrams of induced P300 waveforms from the FZ, CPZ, PZ electrodes were shown in Fig. 4.

The red curve was the observed EEG of the dashboard images with a value below 60km/h displayed on the speedometer (non-target stimuli), and the blue curve was the EEG of the images on which a speed value above 60km/h was displayed (target stimuli). It could be seen from the figure that the blue curve presented an obvious peak in the range of 340-370 ms after the appearance of the stimuli. The high amplitude reflected the attention and arousal of the subjects.

Brain electrical activity maps (BEAM) also showed that the energy aroused by the target stimuli was mainly concentrated in the parietal and occipital regions. This illustrated that under Type A layout, the design features such as the warm color tone of the luminous band and the flat design style could generate a certain impact on the drivers' cognitive psychology and neurophysiology. Performing design works of dashboards which were based on this scheme would be helpful to prevent driving distraction.

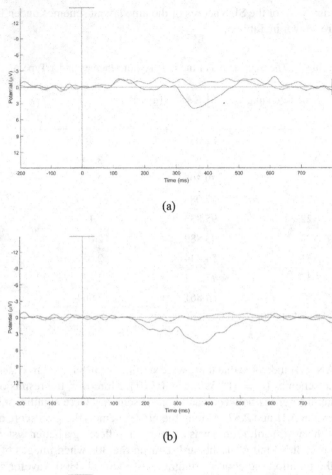

(a)

(b)

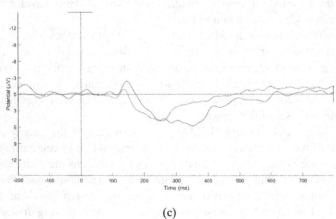

(c)

Fig. 4. P300 waveforms of A22 from the FZ (a), CPZ (b), PZ (c) electrodes

4.2 Classification Prediction Accuracy

According to the energy distribution on BEAM and previous research conclusions, nine electrodes were selected to be analyzed, including FZ in the frontal region, CPZ, CP3, CP4 in the parietal region, and PZ, P3, P4, P7, P8 in the occipital region. The P300 amplitude and latency data induced by the 36 dashboard schemes were taken into the input layer of the SVM model, which was a matrix of 36 rows and 18 columns. The layout category label was used as the output layer data, which was a column vector of 36 rows. By the LIBSVM toolbox in MATLAB (ver. R2016a), the SVM prediction model was built.

The topic studied in this section was a four-classification problem. The conventional SVM model could not be directly applied to multi-classification problems, but some methods could make it effective in this field. The core idea was to transform the multi-classification problems into several binary classification ones [22]. The principal thinking of SVM algorithm was to find a boundary plane between different types of data, so that the two types of samples could fall on both sides of the plane as far away from each other and from the plane as possible. Using radial basis function (RBF) to project the plane into a curved surface could greatly increase the application range of SVM algorithm [23], and make it have a good performance in the classification prediction of this study. Therefore, it was necessary to find the appropriate penalty coefficient C and radial basis kernel parameter g. In this study, we used the cross-validation method. A smaller C made the decision surface smoother, while a larger C aimed to classify all training samples correctly. If the performance of different models was similar, for the sake of cutting down the computing time, a smaller penalty factor C should be the priority. After parameter adjustment, it was determined that C = 6.063, g = 0.66.

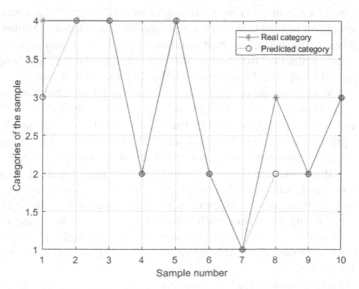

Fig. 5. Prediction accuracy of the SVM model

26 samples were selected randomly as the training set, and the remaining 10 were taken as the testing set. The prediction accuracy was 80%, which was a relatively good result as shown in Fig. 5. It reflected that P300 components induced by observing dashboards could be used to accurately predict the type of layouts to some extent, which indicated a relationship between the layout type of dashboard and P300. This further proved that it was reasonable and scientific to dig out the layout being in line with the characteristics of user attention based on the amplitude and latency of P300.

5 Discussion

This study focused on the application of the combination of EEG signals and subjective test in layout design and user experience evaluation of digital dashboards. For the user experience of different layouts, there was a certain relationship between the subjective evaluation and the P300 component. In interaction design works of dashboards, the design quality of layout could be judged preliminarily according to the subjective evaluation of users.

P300, as a typical endogenous component of ERPs, was often induced by the use and consumption of attention resources. Under the experimental conditions, the average P300 amplitude of Layout A was the highest. The structure of this layout was a left-right symmetrical one, and the positions of the four instruments were independent from each other. In contrast, although Layout D also had this kind of interface structure, the fuel gauge and water temperature gauge were sheltered by the tachometer and speedometer, which interfered the subjects' attention and caused distraction to some extent.

Subjective scale results showed that the design scheme A22 of Layout A which adopted the flat style and a warm color tone had the highest system usability. Michalski and Grobelny mentioned in their research that color had a preprocessing mechanism and played an important role in the visual searching process [24]. Red had a warning effect. Existing research showed that there were significant differences in the cognition of colors during car driving. The drivers' average reaction time of Level 1 alarm color (red) was 2.8s, significantly better than the average decision time 8s of Level 2 and level 3 alarm colors (green and blue) [25]. Compared with the cold and neutral color tones, the red scheme had a greater influence on the driver's attention and decision-making behaviors, which was helpful to their perception. In this study, besides the system usability, the amplitude and latency of P300 induced by the warm tone scheme were also better.

Another problem analyzed in this study was STYLE, including the skeuomorphism style, flat style and minimalism style. Among them, skeuomorphism designs pursued simulating the shape and texture of real objects, and reproducing the real objects by superimposing various effects such as high gloss, texture or shadow on the graphics. Although the cost of cognition and learning was low, more load might be imposed to drivers in fast scanning due to too much color information. On the contrary, flat style design was to abandon the above pursuit of visual effects. And symbolic design elements were used to present an interface. Although minimalism design was also consistent with this point, the control panel with a smaller color area performed worse in operation than that with a larger color area [26], which might be the reason why the dashboard schemes with minimalism style were not good enough. In this kind of dashboards, only the pointer had a single color, and the color area was very small.

In this paper, an SVM model was established and a good prediction accuracy was obtained by learning the amplitude and latency values of P300 components from nine representative electrodes. SVM could detect and recognize the character information of automobile dashboard well [18], and the learning effect of P300 was also good. For example, Ahi et al. studied the calibration time of brain computer interface (BCI) and only four training letters were needed to obtain an average accuracy of 80% [27]. SVM models could automatically achieve the optimal selection of parameters in the process of construction, which could reduce the impact of human intervention. Besides, the generalization ability of SVM was higher than that of back propagation neural networks [28]. It had good robustness, and it did not need to be fine-tuned during operating. According to the subjectivity characteristics of user experience data, we used cross validation method to optimize the selection of the penalty coefficient and kernel function parameters, which improved the classification prediction accuracy. Although the prediction effect might be affected by noises such as physiological and behavioral characteristics and aesthetic preference of the subjects, the processing of SVM in this paper still made the optimized model achieve a high accuracy, which could provide a scientific and effective prediction method for automobile enterprises in product development and management.

6 Limitation

The first limitation came from the less sample size. Especially in the construction of the SVM model, data provided by more subjects would improve the accuracy and robustness of the model. Another limitation was due to restrictions of the experimental conditions. The study adopted an indoor experiment, and the luminous flux was different from that of outdoor driving. This might affect the information transmission effect of design elements such as brightness, tone, line width, font size, etc. In addition, in order to highlight the layout characteristics, the stimulus materials used in this study were simplified images. In next phase of research, it is needed to take the influence of other elements on real dashboards into consideration, such as icons, symbols, color blocks and so on, and to carry out experiments outdoors in real cars. Besides, because the experimental task of this study was relatively simple, there was no significant difference in reading accuracy. However, with the rapid development of intelligent vehicles, there would be more and more instrument panel modeling forms equipped with new interaction modes. In future researches, we will further refine the layout elements to analyze the effect of dashboard layout on reading accuracy under complex cognitive tasks.

7 Conclusion

The study discussed the significance of ERP analysis in the research of automobile digital dashboard design. EEG signals and subjective system usability were collected and analyzed, and the evaluation laws on the dashboard layouts were obtained. The study verified the consistency between subjective assessment and EEG signals. The main conclusions included:

(1) Under the experimental conditions, the P300 amplitude induced by the schemes of Type A layout was the highest. And for the schemes of Type B, the amplitude was the lowest and the latency was the longest, which reflected a poor allocation of attention resources.

(2) Using the P300 amplitude and latency data from the nine representative electrodes collected in the experiment, an SVM model was established and the prediction effect to the layout types was relatively good (accuracy = 80%). By adjusting the parameters, the penalty coefficient C of the model was set to 6.063, and the radial basis kernel parameter g was 0.66.

(3) Under the experimental conditions, the SUS score of Type A layout (m = 79.167) was significantly higher than that of the other three types. And under this layout, the scheme (No. A22) with flat style and a warm color tone performed best in system usability evaluation, which provided a theoretical basis for the layout design of digital dashboards.

Acknowledgments. This research was funded by Scientific Research Program of Beijing Education Commission (grant no. KM202010009003) and Chinese Ergonomics Society & Kingfar Joint Research Fund for Outstanding Young Scholars (grant number CES-Kingfar-2019-001).

References

1. Dingus T., et al.: The 100-Car naturalistic driving study phase II – Results of the 100-Car field experiment. The U.S. Department of Transportation, National Highway Traffic Safety Administration (2006)
2. Underwood, G., et al.: Visual attention while driving: sequences of eye fixations made by experienced and novice drivers. Ergonomics 46(6), 629–646 (2003)
3. Yang, H., Zhao, Y., Wang, Y.: Identifying modeling forms of instrument panel system in intelligent shared cars: a study for perceptual preference and in-vehicle behaviors. Environ. Sci. Pollut. Res. 27(1), 1009–1023 (2019). https://doi.org/10.1007/s11356-019-07001-0
4. Ren, H., Tan, Y.P., Zhang, N.N.: Research on form design of automotive dashboard based on Kansei Engineering. In: 2019 7th International Forum on Industrial Design. IOP Publishing Ltd., Bristol (2019)
5. Francois, M., et al.: Gauges design for a digital instrument cluster: efficiency, visual capture, and satisfaction assessment for truck driving. Int. J. Ind. Ergon. 72, 290–297 (2019)
6. Handy, T.C., et al.: ERP evidence for rapid hedonic evaluation of logos. J. Cogn. Neurosci. 22(1), 124–138 (2010)
7. Guo, F., et al.: Application of evolutionary neural networks on optimization design of mobile phone based on user's emotional needs. Human Factors Ergon Manuf. Serv. Ind. 26(3), 301–315 (2016)
8. Hou, G., Lu, G.: Semantic processing and emotional evaluation in the traffic sign understanding process: evidence from an event-related potential study. Transp. Res. Part F: Traff. Psychol. Behav. 59, 236–243 (2018)
9. Else, J.E., Ellis, J., Orme, E.: Art expertise modulates the emotional response to modern art, especially abstract: an ERP investigation. Front Human Neurosci. 9, 1–18 (2015)
10. Wei, J., Yuejia, L.: Principle and Technique of Event-Related Brain Potentials. Science Press, Beijing (2010)

11. van der Ham, I., van Strien, J., Oleksiak, A., van Wezel, R., Postma, A.: Temporal characteristics of working memory for spatial relations: an ERP study. Int. J. Psychophysiol. **77**(2), 83–94 (2010)
12. Yang, H., Wang, Y., Jia, R.: Dimensional evolution of intelligent cars human-machine interface considering take-over performance and drivers' perception on urban roads. Complexity **2020**, 1–13 (2020)
13. Tommaso, M.D., et al.: Influence of aesthetic perception on visual event-related potentials. Conscious. Cogn. **17**(3), 933–945 (2008)
14. Li, R., et al.: Effects of interface layout on the usability of in-vehicle information systems and driving safety. Displays **49**, 124–132 (2017)
15. Sun, G.L., et al.: Analysis and optimization of information coding for automobile dashboard based on human factors engineering. China Saf. Sci. J. **28**(8), 68–74 (2018)
16. Drew, M., Falcone, B., Baccus, W.: What does the system usability scale (SUS) measure? In: Marcus, A., Wang, W. (eds.) DUXU 2018. LNCS, vol. 10918, pp. 356–366. Springer, Cham (2018). https://doi.org/10.1007/978-3-319-91797-9_25
17. Yang, H., Zhao, Y.: Analysis of effects of interaction modes on IVIS based on sensory information recognition. In: Proceedings of the 2018 2nd International Conference on Big Data and Internet of Things, pp. 198–202. ACM, Beijing (2018)
18. Gao, H.J., et al.: Character segmentation-based coarse-fine approach for automobile dashboard detection. IEEE Trans. Ind. Inform. **15**(10), 5413–5424 (2019)
19. Lee, T., Kim, M., Kim, S-P.: Improvement of P300-based brain–computer interfaces for home appliances control by data balancing techniques. Sensors **20**(19), 5576 (2020)
20. Shukla, P., Chaurasiya, R., Verma, S.: Performance improvement of P300-based home appliances control classification using convolution neural network. Biomed. Sig. Process. Control **63**, 102220 (2021)
21. Niu, Y.: Usability evaluation methods research of digital interface based on brain electrical technology. Southeast University (2015)
22. Li, G., Zhang, F., Yonggang, W.: Influencing factors analysis of multiple vehicle accidents in mountainous expressway based on SVM model. J Wuhan Univ. Technol. Transp. Sci. Eng. **44**(6), 1046–1050 (2020)
23. Najafi, G., et al.: SVM and ANFIS for prediction of performance and exhaust emissions of a SI engine with gasoline-ethanol blended fuels. Appl. Therm. Eng. **95**, 186–203 (2016)
24. Michalski, R., Grobelny, J.: The role of colour preattentive processing in human–computer interaction task efficiency: a preliminary study. Int. J. Ind. Ergon. **38**(3–4), 321–332 (2008)
25. Cui, W., Zhou, R., Yan, Y., Ran, L., Zhang, X.: Effect of warning levels on drivers' decision-making with the self-driving vehicle system. In: Stanton, N.A. (ed.) AHFE 2017. AISC, vol. 597, pp. 720–729. Springer, Cham (2018). https://doi.org/10.1007/978-3-319-60441-1_69
26. Michalski, R.: The influence of color grouping on users' visual search behavior and preferences. Displays **35**(4), 176–195 (2014)
27. Ahi, S., Yoshimura, N., Kambara, H., Koike, Y.: Utilizing fuzzy-SVM and a subject database to reduce the calibration time of P300-based BCI. In: Wong, K.W., Mendis, B.S.U., Bouzerdoum, A. (eds.) ICONIP 2010. LNCS, vol. 6444, pp. 1–8. Springer, Heidelberg (2010). https://doi.org/10.1007/978-3-642-17534-3_1
28. Zhao, Y., Tang, W.: Function fitting about internal stress of ceramic paste based on BP-NN and SVM. In: 2011 International Conference on Information Technology, Computer Engineering and Management Sciences, ICM 2011. IEEE Computer Society (2011)

Explore Acceptable Sound Thresholds for Car Navigation in Different Environments

Yulu Yang[✉], Boxian Qiu, and Xuan Liu

Hunan University, Changsha, China

Abstract. Modern in-vehicle navigation is an important part of car interaction equipment, which is of great significance to drivers. If in a special environment (such as excessive external noise), the inability to deliver the appropriate sound to the driver will become a serious problem. Therefore, investigating the acceptable sound threshold of the car navigation system in different environments plays an important role in determining whether the user is comfortable. This article explores the acceptable sound thresholds of car navigation under low, medium and high external noise. Our results show that the user-acceptable sound threshold is different for these three types of external noise.

Keywords: Car navigation · Driving simulation · Threshold · Noise

1 Introduction

Competition in the car navigation market is increasing. However, in the competition of various performances, the design of interaction is very important, and sound is an important information interaction part of car navigation. In the world, modern vehicle navigation research has a history of more than thirty years. Since the Global Positioning System (GPS) was released by the US Department of Defense in 1944, GPS navigation technology has developed rapidly in the civilian market and has been a hotpot of high-tech company and university research. Existing research shows that auditory perception is also an integral part of car driving. Sound is one of the most familiar information-carrying pathways for human beings. The auditory characteristics of the human ear will extract useful information from many sounds and make decisions through brain judgment. When the vehicle is driving, it is actually an open environment, and external noise will affect the perception of the navigation sound by the human ear. Studying the acceptable sound threshold of car navigation systems in different environments is significant to the future development of car navigation systems.

2 Methods

2.1 Participant

Twenty-two college students without hearing impairment (14 men and 8 women) participated in the study. Ages range from 18 to 22. The distance from each participant to the sounding part of the car navigator is fixed, ranging from 40 to 45 cm.

© Springer Nature Switzerland AG 2021
P.-L. P. Rau (Ed.): HCII 2021, LNCS 12773, pp. 296–303, 2021.
https://doi.org/10.1007/978-3-030-77080-8_24

2.2 Instrument

This research was conducted in a simulated driving environment. Use Biao Kang digital decibel JD-105 (0–188 dB) and iphone7plus built-in navigation (sound range is 40–100 dB). Due to the limitation of equipment, the decibel level of navigation volume cannot be increased gradually, but can only be increased by a fixed numerical gradient.

2.3 Task

After confirming the eligibility and obtaining informed consent, the participants are in a soundproof space. In this space, we place four small speakers on the four corners of a 100 cm * 200 cm rectangle, and the participants are located at the upper left of the rectangle (as shown below).

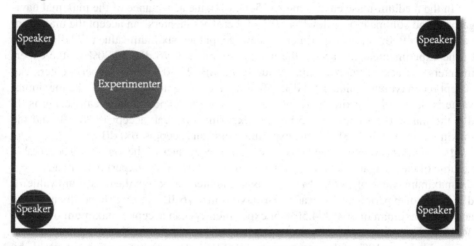

The small speaker emits analog noise with noise decibels of 50–55 dB, 55–60 dB, and 60–65 dB. (The three sections were measured in real time on quiet lanes, normal streets, and roads during peak driving hours). Participants were asked to use the iPhone7plus's built-in navigation. In appropriate condition, participants were instructed to listen to the instructions given by the navigation. The experimenter asked participants to read aloud what they heard. Initially, the decibels of all sentences are 40 dB. After the participant repeated a sentence aloud, the experimenter asked the participant two questions: "Does this sound suit you?" And "Do you think this sound is suitable for use in a car?" If both questions are answered "No", and the experimenter adjusted the sound so that the decibels indicated by the sentence were slightly higher, and repeated the above operation. If the participant answered "yes" to any of the questions, the current decibel level was considered to be the minimum acceptable self-reported sound size. (This method has been used in Explore the comfortable and acceptable text size of a car display, written by Derek Viita, Alexander Muir, which has been proved reliable). Then continue the above operation. If the participant answers "No" to any question, the current decibel level is regarded as the maximum value of the self-reported acceptable sound level. The decibel size of the navigation was adjusted by the experimenter, and each participant was recorded in the cluster display and the acceptable sound threshold.

3   Outcome

3.1   Maximum Value (See Fig. 1)

In a low-noise environment (50–55 dB), the acceptance of the maximal navigable sound volume is as follows: 27.20% of the experimenters can accept the maximum value of 68 dB, 13.70% of the experimenters can accept the maximum value is 72 dB, 22.70% of the experimenters can accept the maximum value of 76 dB, 9.00% of the experimenters can accept the maximum value of 80 dB, 4.50% of the experimenters can accept the maximum value of 84 dB, 0% of the experimenters can accept the maximum value is 88 dB, the maximum value that 13.70% of the experimenters can accept is 92 dB, the maximum value that 0% of the experimenters can accept is 96 dB, and the maximum value that 9.00% of the experimenters can accept is 100 dB.

In the medium-noise environment (55–60 dB), the acceptance of the maximal navigator sound volume is as follows: 4.50% of the experimenters can accept the maximum value of 68 dB, 0% of the experimenters can accept the maximum value is 72 dB, 9.00% of the experimenters can accept the maximum value of 76 dB, 40.90% of the experimenters can accept the maximum value of 80 dB, 22.70% of the experimenters can accept the maximum value of 84 dB, 0% of the experimenters can accept the maximum value is 88 dB, the maximum value that 9.00% of the experimenters can accept is 92 dB, the maximum value that 4.50 of the experimenters can accept is 96 dB, and the maximum value that 9.00% of the experimenters can accept is 100 dB.

In a high-noise environment (60–65 dB), the acceptance of the maximum acceptable volume of the navigator sound volume is as follows: 0% of the experimenter can accept the maximum value of 68 dB, 0% of the experimenter can accept the maximum value 72 dB, 4.50% of experimenters can accept a maximum of 76 dB, 4.50% of experimenters can accept a maximum of 80 dB, 4.50% of experimenters can accept a maximum of 84 dB, 36.40% of experimenters can accept a maximum The value is 88 dB, the maximum value acceptable by 18.10% of the experimenters is 92 dB, the maximum value acceptable by 9.00% of the experimenters is 96 dB, and the maximum value acceptable by 22.70% of the experimenters is 100 dB.

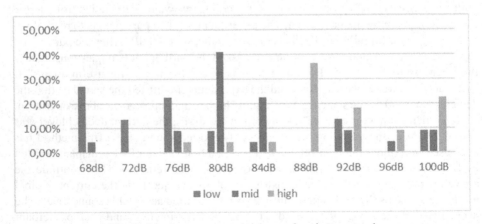

Fig. 1. Maximum accepted volume of navigator sound.

3.2 Minimum Value (See Fig. 2)

In a low-noise environment (50–55 dB), the acceptance of the minimum acceptable volume of the navigator sound volume is as follows: 13.70% of the experimenters have the minimum acceptable value of 48 dB, and 13.70% of the experimenters have the minimum acceptable value of 52 dB, 27.20% of experimenters can accept the minimum value of 56 dB, 27.20% of experimenters can accept the minimum value of 60 dB, 13.70% of experimenters can accept the minimum value of 64 dB, 0% of experimenters can accept the minimum value which is 68 dB, and the minimum value acceptable to 4.50% of the experimenters is 72 dB.

In the medium noise environment (55–60 dB), the acceptance of the minimum acceptable volume of the navigator sound volume is as follows: the minimum acceptable value for 0% of the experimenter is 48 dB, and the minimum acceptable value for the 4.50% of the experimenter is 52 dB, the minimum value that 9.00% of the experimenters can accept is 56 dB, the minimum value that 27.20% of the experimenters can accept is 60 dB, the minimum value that 31.80% of the experimenters can accept is 64 dB, and the minimum value that 22.70% of the experimenters can accept is 68 dB, and the minimum value acceptable to 0% of the experimenters is 72 dB. The minimum acceptable for 4.50% of the experimenters was 76 dB.

In a high noise environment (60–65 dB), the acceptance of the minimum acceptable volume of the navigator sound volume is as follows: 13.70% of the experimenters have the minimum acceptable value of 60 dB, and 4.50% of the experimenters have the minimum acceptable value of 64 dB, the minimum value that 18.10% of experimenters can accept is 68 dB, the minimum value that 31.80% of experimenters can accept is 72 dB, the minimum value that 18.10% of experimenters can accept is 76 dB, and the minimum value that 4.50% of experimenters can accept is 76 dB, and the minimum value acceptable to 9.00% of the experimenters is 80 dB.

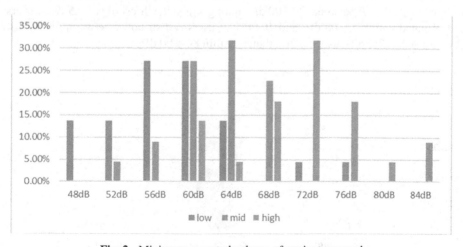

Fig. 2. Minimum accepted volume of navigator sound.

3.3 Threshold Distribution

In a low-noise environment (see Fig. 3), the overall threshold of acceptable sound volume is distributed between 48-100 dB, among which the threshold of 9.00% of users is distributed within 48–60 dB, and the threshold of 40.90% of users is within 60–72 dB. The threshold of 22.70% of users is distributed within 72-84 dB, the threshold of 9.00% of users is distributed within 84–96 dB, and the threshold of 9.00% of users is distributed within 96–100 dB.

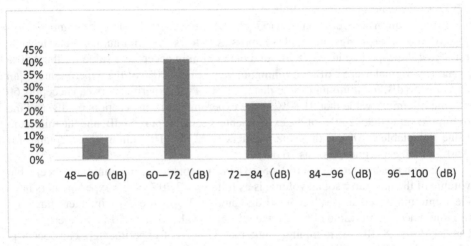

Fig. 3. Threshold value of user acceptable sound in low noise condition.

In a medium noise environment (see Fig. 4), the overall threshold of acceptable sound volume is distributed between 52–100 dB, among which the threshold of 4.50% of users is distributed within 52–68 dB, and 40.90% of users is distributed between 68–84 dB. The threshold of 9.00% users is distributed within 84–100 dB.

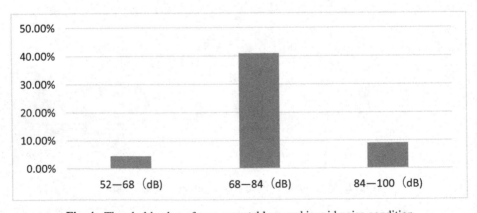

Fig. 4. Threshold value of user acceptable sound in mid noise condition

In high-noise environments (see Fig. 5), the overall threshold of acceptable sound volume is distributed between 60–100 dB, among which the threshold of 13.70% of users is distributed within 60–72 dB, and the threshold of 45.50% of users is distributed within 72–84 dB. The threshold of 31.80% of users is distributed within 84–96 dB, and the threshold of 27.30% of users is distributed within 96–100 dB.

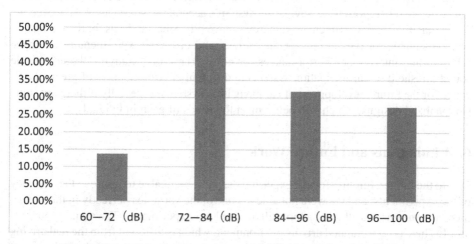

Fig. 5. Threshold value of user acceptable sound in high noise condition

4 Discussion

In three environments (low noise, medium noise, high noise), the minimum value of the sound volume of the navigator increases when the ambient noise increases, and when the environmental noise increases, the maximum value increases at the same time (when the noise increases to a certain level) At decibels, the maximum acceptable volume exceeds the navigation maximum volume). In low-noise environments, the threshold distribution is the widest from 60 to 72 dB. In medium-noise environments, the threshold distribution is the widest from 68 to 84 Db. In high-noise environments, the threshold distribution is the widest at 72–84 Db. Gender wasn't found to be an important factor affecting the acceptable sound level of a self-reported car navigation system.

5 Conclusion

In this article, we explored acceptable sound thresholds for car navigation systems in different environments. The Ipad acts as four speakers that simulate the in-vehicle environment (simulating noise from four directions, that is, noise interference during driving the vehicle), and the experimenter sits in the driver's seat of the simulated cab. Specifically, we conducted an experiment that asked the participants two questions: "Does this sound the right size for you?" And "Do you think this sound is suitable for use in a car?"

to progressively find the maximum and minimum acceptable sound volume. We found that in low-noise environments, the widest range of sound volume thresholds acceptable to experimenters is from 60 to 72 dB, and in medium-noise environments, the widest range of sound volume thresholds acceptable to experimenters is the from 68 to 84 Db. In high- noise environments, the widest range of sound volume acceptable to experimenters is from 72 to 84 Db. Also, gender has not been found to be an important factor affecting the self-reported car navigation system's acceptable sound level. However, we found that different experimenters sometimes have different standards for the navigation volume suitable for the car and for themselves, and different experimenters also have different anti-interference performances for noise. Therefore, studying the acceptable sound threshold of car navigation systems in different environments is of great significance for the future development of car navigation systems, especially if there are some personalized designs, which will better meet the needs of each individual.

6   Limitations and Future Work

Due to the limited funding for our experiments, we were unable to improve the accuracy of the experiments, and we were not able to control the variables perfectly in the experiments. Some uncontrollable factors still affected our experiments. The main limitations are: 1. There is no simulated driving environment, which is different from the real driving environment. 2. The environment of the simulated cab cannot control the influence of noise from the outside. 3. The decibel meter is not precise enough, and the measured data has some errors. It is hoped that in the future, there can be more rigorous experiments to continue research on exploring the acceptable sound threshold of car navigation in different environments.

Acknowledgement. The paper is supported by Hunan Key Research and Development Project (Grant No. 2020SK2094) and the National Key Technologies R&D Program of China (Grant No. 2015BAH22F01).

References

1. Li, L., Jin, M.: Application and prospect of sound perception in autonomous vehicles. Motorcycle Technol. (09) (2018)
2. Chen, X.: Car Navigator—GUI software design and performance optimization
3. Gou, X.: Research on optimal path planning of vehicle navigation system
4. Viita, D., Muir, A.: Explore the comfortable and acceptable text size of a car display
5. Miller, Z.D., et al.: Pennsylvania state forests
6. Hsieh, L., Seaman, S., Young, R.: Effect of emotional speech tone on driving from lab to road: fMRI and ERP studies
7. Gable, T.M., Walker, B.N., Moses, H.R., Chitloor, R.D.: Advanced auditory cues on mobile phones help keep drivers' eyes on the road
8. Burnett, G., Crundall, E., Large, D., Lawson, G., Skrypchuk, L.: A study of unidirectional swipe gestures on in-vehicle touch screens
9. Palinko, O., et al.: Towards augmented reality navigation using affordable technology

10. Kun, A.L., Royer, T., Leone, A.: Using tap sequences to authenticate drivers
11. Riener, A., et al.: Standardization of the in-car gesture interaction space
12. Wilfinger, D., Murer, M., Baumgartner, A., Döttlinger, C., Meschtscherjakov, A., Tscheligi, M.: The car data toolkit: smartphone supported automotive HCI research
13. Macek, T., Kašparová, T., Kleindienst, J., Kunc, L., Labský, M.: Mostly passive information delivery in a car
14. McIlroy, R.C., Stanton, N.A., Harvey, C., Robertson, D.: Sustainability, transport and design: reviewing the prospects for safely encouraging eco-driving

Effects of Multimodal Warning Types on Driver's Task Performance, Physiological Data and User Experience

Yiqiao Zhang[1] and Hao Tan[2]([⊠])

[1] School of Design, Hunan University, Changsha, China
[2] State Key Laboratory of Advanced Design and Manufacturing for Vehicle Body, Changsha, China
htan@hnu.edu.cn

Abstract. Previous studies have compared different multimodal warning types, however, few researchers studied the effects of different multimodal warning types on drivers' task performance, physiological data, and user experience. In our research, we designed a simulated driving experiment to investigate these effects. In the experiment, a small projector, two Bluetooth speakers and two vibration generators were used as signal generators to simulate 6 multimodal warning types, and Mean Deviation, braking reaction time (BRT), normalized GSR, normalized HR, trust, annoyance, satisfaction were used as dependent variables to reflect effects of multimodal warning types on driver's task performance, physiological data and user experience. The main conclusions are drawn in this paper as following: (1) In terms of task performance, there was no significant difference in the effect of different multimodal warning types on driving tasks, but compared with unimodal warning types, both bimodal warning types and trimodal warning type can reduce BRT. (2) In terms of physiological data, normalized GSR and normalized HR were increased by increasing numbers of modalities of warning types. (3) In terms of user experience, trust and satisfaction of multimodal warning types were significantly higher than the other two unimodal warning types. Moreover, the annoyance of warning types which included tactile modality were significantly higher than the warning types which did not included. We speculated that adding tactile signals may increase annoyance of participants, nevertheless, tactile signals still has great potential to increase the trust and satisfaction of multimodal warning types.

Keywords: Multimodal interaction · Warning type · Task performance · Physiological data · User experience

1 Introduction

Vehicle warning systems are designed to direct a driver's attention to an impending danger. Nowadays, with the development of multiple technologies such as natural speech, sensor recognition, and affective computing, warning types based on multi-modality have

© Springer Nature Switzerland AG 2021
P.-L. P. Rau (Ed.): HCII 2021, LNCS 12773, pp. 304–315, 2021.
https://doi.org/10.1007/978-3-030-77080-8_25

been widely used in vehicle warning systems. The warning signals can conveyed in a variety of modalities, such as auditory, visual, and tactile stimuli.

Previous research showed the advantages of auditory, visual, and tactile warning types. The visual and auditory warning types are the most traditional. The auditory warning type is considered to be a safe and effective warning type, because it is gaze-free. However, workloads in perceiving visual warnings and risks to miss visual warnings when driving may be mitigated by a head-up display (HUD). Moreover, visual warning type is straight forward in terms of conveying information [1]. According to the literature [2], tactile warning type is not as widely used as auditory and visual warning types, but it can elicit faster reactions from drivers in response to potential danger than visual or auditory warning types. Moreover, tactile warning type is unlikely to be affected by the level of background noise. Previous research also showed the advantages of the warning types composed of multiple modalities, they can reduce reaction time and lead to higher perceived urgency than unimodal warning types [3, 4].

Although early works have compared different multimodal warning types, few researchers studied the effects of different multimodal warning types on drivers' task performance, physiological data, and user experience. Therefore, in our research, we designed a simulated driving experiment to investigate these effects. The results of our research provided implications for multimodal warning design and reveal how different multimodal warning types can influence drivers' responses during a simulated driving.

2 Method

2.1 Participants

We recruited 30 participants (15 female) by means of online social media and e-mail. The inclusion criteria were to hold a valid driving license; to drive a car at least 5000 km/year, with the two-year trial period passed; to consider oneself perfectly healthy with no adverse medical impact upon car driving; and to have normal vision on both eyes (corrected vision allowed), normal sense of touch, and normal audition. Each participant had a mean age of 32 years (SD = 9.4) and received a voucher with value of about 100 yuan.

2.2 Apparatus

The driving simulation was done with City Car Driving software, the simulated road image (250cm by 180cm) was projected to a wall using a SONY projector. Driving was done on the right lane and the participant shared the road with several cars. The windshield (135 cm by 85 cm), positioned 180cm away form the wall and on top of a table, was made with 0.3cm thick glass, with a 70° tilt [5] (see Fig. 1).

The steering wheel and pedal (Logitech G27) were connected to a personal computer. The iPad served as the center console of the driving simulator. Participants drove with automatic gear shift. Driving simulation software can set the driving environment and the density of road vehicles and pedestrians. The simulator can record the data of the vehicle itself and surrounding vehicles in real time, including operation of the driver and driving state of the vehicle.

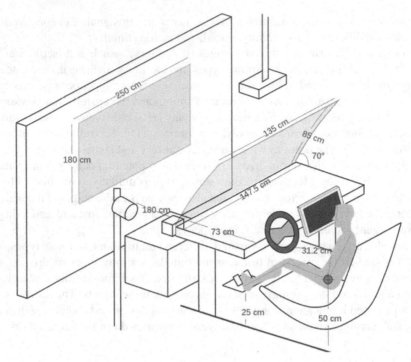

Fig. 1. Experiment apparatus.

2.3   Experiment Design

Multimodal Warning Types. According to the research of Linda-Marie et al.[3], we selected six multimodal warning types of warning system as the independent variables of the driving simulation experiment. six warning types included two unimodal warning types (only visual; only auditory), three bimodal warning types (combined visual-auditory; combined visual-tactile; and combined auditory-tactile) and a trimodal warning type (combined visual-auditory-tactile). Due to the limitations of current technical factors, the multimodal warning types based on smell and taste have not been widely used. In this experiment, they were not selected. In addition, only tactile as a unimodal warning type was difficult to specifically convey relatively complex information content, so it was not selected as an independent variable in this experiment.

In this experiment, based on the research of [3] and [6], the signals of visual warning type, auditory warning type and tactile warning type were sent out by projector, speakers and vibration generators respectively to simulate the head-up display (HUD), voice display and steering wheel vibration in vehicle environment which are currently mature in technology and have broad application prospects in the future.

According to the research of WATSON T et al. [7], we used a small projector as the signal generator of the visual warning type. The projector can project visual signals (warning icon) onto the road in the simulation image (see Fig. 2). According to the research of Liu YC et al. [8], we used two Bluetooth speakers as the signal generator of the auditory warning type. The two speakers were placed on the left and right sides of the participants to send out voice signals (the voice audio "Potential Hazard Ahead!") (see Fig. 2). According to the research of Linda-Marie et al. [3], we encapsulated two vibration generators that can emit tactile vibration signals in a thin plastic material and installed them on the upper left and upper right of the steering wheel to simulate the vibration of the steering wheel (see Fig. 2).

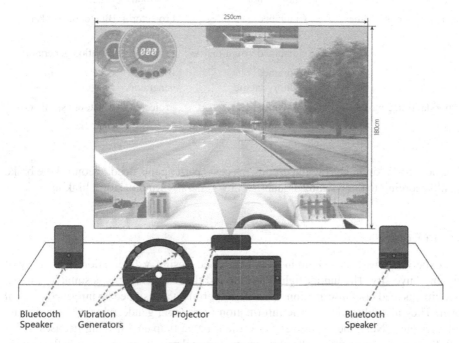

Fig. 2. Signal generators.

We used the signal generators to send out the combined signal to realize the simulation of bimodal and trimodal warning types. The signal was sent out synchronously [3]. For example, for the trimodal warning type (combined visual-auditory-tactile), visual signal generator, auditory signal generators and vibration generators on the steering wheel sent out visual, auditory and tactile signals at the same time. Table 1 shows six multimodal warning types and their signal generators.

In the experiment, each warning type was remotely controlled by the experimenter using the Wizard of OZ method [9], and had no additional impact on the simulated driving process of participants.

Driver Tasks. The simulated driving environment in the experiment was a typical urban road environment, which included roads, buildings, vehicles and pedestrians in the city. In the experiment, in order to simulate the actual scene more realistically, the driving section load factor was set as medium load and the road environment as new urban area in the driving simulation software of City Car Driving. Participant' basic task was to keep the vehicle in the appropriate lane on the right side of the road and drive at a speed

Table 1. Six multimodal warning types and their signal generators.

Modality	Multimodal warning types	Signal generators
Unimodal (one modality)	Only visual	Projector
	Only auditory	Bluetooth speakers
Bimodal (two modalities)	Combined visual-auditory	Projector + Bluetooth speakers
	Combined visual-tactile	Projector + vibration generators
	Combined auditory-tactile	Bluetooth speakers + vibration generators
Trimodal (three modalities)	Combined visual-auditory-tactile	Projector + Bluetooth speakers + vibration generators

of about 35–45 km/h. During the driving process, participants need to control the brake pedal according to the warning signals sent by generators and perform braking task.

2.4 Procedure

The experiment was carried out in the Intelligence Design & Interaction Laboratory of Hunan University. The indoor light was soft and the temperature was suitable. Firstly, participants read brief information about the experiment and signed an informed consent form. They also filled out a basic information form (age, gender, driving-age, driving mileage, etc.). Next, the experimenter verified the participants' driving licenses. After that, the experimenter invited the participants to familiarize themselves with the experimental environment, and introduced the process and tasks involved in the experiment to the participants (the experimenter did not introduce details such as experimental emergencies to the participants to ensure the authenticity of the data). Then, the experimenter introduced the equipment and its operation methods to the participants, and familiarize themselves with the operation (including familiarity with the steering wheel and pedals, familiarity with the experimental road environment, etc.). After that, the participants read the detailed description of the experiment and participated in a 10-min pre-experiment. In the pre-experiment, the experimenter showed each of the six multimodal warning types, and the purpose was to familiarize participants with the overall setup and multimodal warning types in the experiment. After completing the pre-experiment, the participants had a 5-min rest time, then the experiment started.

At the beginning of the formal experiment, firstly, the experimenter pasted sensors and electrodes to the participants, and then played a piece of soft music. When the participants relaxed to a quiet state, the experimenter used BioNex to record the baseline of the participants' physiological signals. Then, the participants started to drive the vehicle and performed the braking task according to the signals of different multimodal warning types. In the experiment, the order of appearance of the six multimodal warning types was arranged according to the Latin square design to eliminate the influence of the different order of appearance on the results. When the participant received a signal of a warning type and completed the corresponding braking task, the participant pulled aside and filled in a user experience questionnaire for the warning type. When all the six multimodal warning types appeared and the participants completed the questionnaires, the experiment ended and the participants received a short interview. During this process, the experimenter continued to record the physiological signals of participants and marked the corresponding time points for later data analysis. The total duration of the experiment was about 30 min.

2.5 Dependent Variables

The dependent variables of interest were as follows: task performance (Mean Deviation and braking reaction time); physiological data (normalized GSR and normalized HR) and user experience (trust, annoyance, satisfaction).

Mean Deviation. Mean Deviation corresponded to the deviation from the actually driven to the optimal driving line [10]. It reflected the driving performance of the driver. Driving performance was negatively correlated with Mean Deviation.

Braking Reaction Time (BRT). Braking Reaction Time (BRT) was defined as the time difference between the onsets of warning and braking. According to the literature [11], the onset of braking was defined as the time at which the brake pedal inclined over 5°.

Physiological Data. According to the literature [12], physiological parameters assessed in this study were galvanic skin response (GSR) and heart rate (HR), which were sensitive to task demand and practical to record. Moreover, GSR and HR proved to be good indicators of stress and arousal [13].

GSR was a measure of the skin conductance. It varied linearly with the overall level of arousal and increases with anxiety and stress and it was considered as a reliable indicator of affective response [14]. HR was a measure of cardiovascular activity. It had been found to increase with stress, workload and difficulty of the task [15]. Since there was a large individual difference in physiological signals, individual baseline had to be taken into account. According to the literature [14, 16], normalized GSR and HR on each task were calculated using the formula (signal-baseline)/baseline for each participant.

User Experience. The user experience of the multimodal warning types was assessed with the questionnaire. Our questionnaire addressed the trust, annoyance and satisfaction:

Trust-"Overall how much do you trust the multimodal warning type?"
Annoyance-"Overall how annoying is the multimodal warning type?"
Satisfaction-"Overall how satisfied are you with the multimodal warning type?"

Each question rated on a five-point Likert scale. The final score of each question was scaled from 1 to 5 (1 = not at all, 5 = very much).

3 Results and Discussion

The mean values of all dependent variables were analyzed using statistical methods. All statistical analysis was performed using SPSS.

3.1 Mean Deviation

Descriptive statistics of the Mean Deviation at different warning types were graphically shown in Fig. 3. A one-way ANOVA was used to analyze the data. The results showed that there was no significant difference among different multimodal warning types, $p > 0.05$.

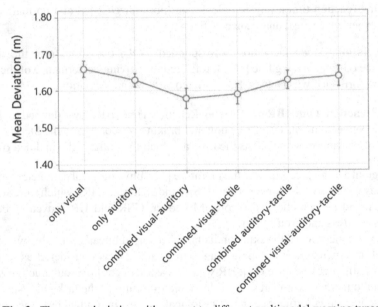

Fig. 3. The mean deviation with respect to different multimodal warning types.

3.2 Braking Reaction Time (BRT)

Descriptive statistics of the braking reaction time (BRT) at different warning types were graphically shown in Fig. 4. The BRT was regarded as the response speed. The BRT of only visual was highest, and the BRT of combined auditory-tactile was lowest. A one-way ANOVA was used to analyze the data. The results showed significant differences among different warning types, $p < 0.05$. Multiple comparisons showed that BRT of only visual (Mean = 1.05) and only auditory (Mean = 1.01) were significantly higher than the other four warning types, $p < 0.05$.

It should be noted that the BRT was significantly decreased from unimodal warning types (only visual and only auditory) to bimodal warning types (combined visual-auditory, combined visual-tactile and combined auditory-tactile), yet was not significantly decreased from bimodal warning types to trimodal warning types. This indicated that compared with unimodal warning types, both bimodal warning types and trimodal warning type can reduce BRT, but there was no significant difference in BRT between bimodal warning types and trimodal warning type.

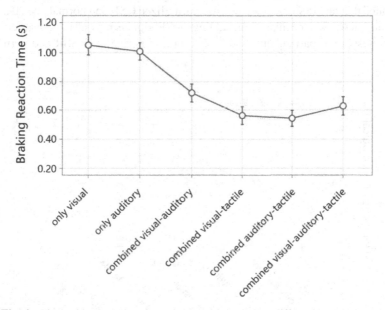

Fig. 4. The braking reaction time (BRT) with respect to different warning types.

3.3 Physiological Data

The physiological data were also analyzed across warning types. Descriptive statistics of the normalized GSR and normalized HR at different warning types were graphically shown in Fig. 5. The tendency that normalized GSR and normalized HR were increased by increasing number of modalities is observed. A one-way ANOVA was used to analyze the data. The results showed significant differences among different warning types, $p < 0.05$. Multiple comparisons showed that normalized GSR of combined visual-auditory-tactile (Mean $= 0.22$), combined auditory-tactile (Mean $= 0.21$), combined visual-tactile (Mean $= 0.19$) and combined visual-auditory (Mean $= 0.14$) were significantly higher than the other two unimodal warning types, $p < 0.05$. The normalized HR was also analyzed in a similar way. The results showed significant differences among different warning types, $p < 0.05$. Multiple comparisons showed that normalized HR of combined visual-auditory-tactile (Mean $= 0.17$), combined auditory-tactile (Mean $= 0.15$), combined visual-tactile (Mean $= 0.15$) and combined visual-auditory (Mean $= 0.09$) were significantly higher than the other two unimodal warning types, $p < 0.05$. These findings indicated that the multimodal warning types (bimodal and trimodal) had a higher normalized GSR and HR than the unimodal warning types. Based on user interviews, we speculated that the result may be due to the fact that multimodal warning types (bimodal and trimodal) may bring more cognitive load and pressure to drivers.

It should be noted that the normalized GSR and normalized HR were significantly increased from only visual to combined visual-tactile, from only auditory to combined auditory-tactile and from combined visual-auditory to combined visual-auditory-tactile. This implied that tactile signals can increase normalized GSR and normalized HR. Based on user interviews, we speculated that adding tactile signals may also increase cognitive load and pressure of participants, and lead to higher normalized GSR and normalized HR.

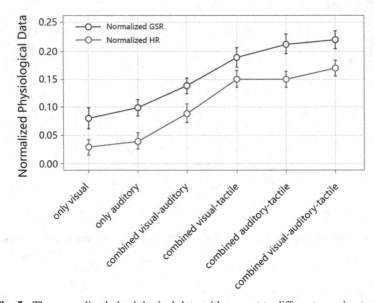

Fig. 5. The normalized physiological data with respect to different warning types.

3.4 User Experience

Descriptive statistics of the trust, annoyance and satisfaction at different warning types were graphically shown in Fig. 6. A one-way ANOVA was used to analyze the data. The results showed significant differences among different multimodal warning types, $p < 0.05$. The trust of combined vvisual-auditory-tactile (Mean = 4.57) is the highest. Multiple comparisons showed that trust of combined visual-auditory (Mean = 4.33), combined visual-tactile (Mean = 3.93), combined auditory-tactile (Mean = 4.23) and combined visual-auditory-tactile (Mean = 4.57) were significantly higher than the other two warning types, $p < 0.05$. This indicated that the multimodal warning types (bimodal and trimodal) had a higher trust than the unimodal warning types. Based on user interviews, we speculated that the result may be due to the fact that multimodal warning types (bimodal and trimodal) were not easy to be ignored by drivers. The annoyance of combined visual-auditory-tactile (Mean = 4.40) is the highest. Multiple comparisons showed that annoyance of combined visual-auditory-tactile (Mean = 4.40), combined visual-tactile (Mean = 3.90) and combined auditory-tactile (Mean = 4.17) were significantly higher than the other three warning types, p < 0.05. This indicated that the multimodal warning types (bimodal and trimodal) which include tactile warning had a higher annoyance. Based on user interviews, we speculated that adding tactile signals may increase annoyance of drivers. The satisfaction of combined visual-auditory (Mean = 4.37) is the highest, multiple comparisons showed that combined visual-auditory-tactile (Mean = 3.83), combined visual-auditory (Mean = 4.37), combined visual-tactile (Mean =

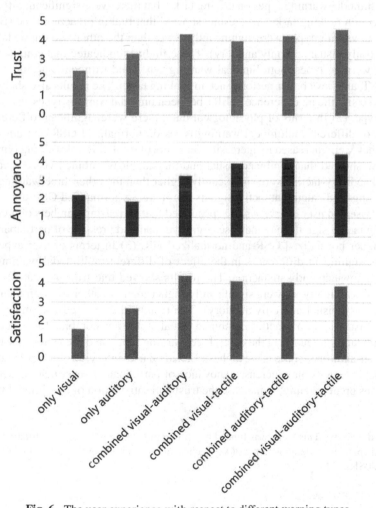

Fig. 6. The user experience with respect to different warning types.

4.13) and combined auditory-tactile (Mean = 4.00) were significantly higher than the other two warning types, $p < 0.05$. This indicated that the multimodal warning types (bimodal and trimodal) had a higher satisfaction than the unimodal warning type.

4 Conclusion

This study performed driving simulator experiment to investigate a common traffic accidental situation, i.e., avoiding danger by braking. Experiment was designed to assess the effects of the different multimodal warning types on the task performance, physiological data, and user experience of drivers.

In accordance to the research, three conclusions are drawn in this paper as following: (1) In terms of task performance, there was no significant difference in the effect of different multimodal warning types on driving tasks, but there were significant differences in the effect on braking tasks. The results showed that braking reaction time (BRT) of multimodal warning types were significantly lower than the other two unimodal warning types (only visual and only auditory). These findings indicated that compared with unimodal warning types, both bimodal warning types and trimodal warning type can reduce BRT, and have better performance in braking tasks. The results also showed that there was no significant difference in BRT between bimodal warning types and trimodal warning type. (2) In terms of physiological data, there were significant differences in the effect of different multimodal warning types on normalized GSR and normalized HR, and they were increased by increasing numbers of modalities of warning types. The results also showed that combined visual-auditory-tactile, combined visual-tactile and combined auditory-tactile were significantly higher than the other three warning types. These findings indicated that tactile signals can increase normalized GSR and normalized HR. Based on user interviews, we speculated that increasing numbers of modalities and adding tactile signals may increase cognitive load and pressure of participants, and lead to higher normalized GSR and normalized HR. (3) In terms of user experience, there were significant differences in the effect of different multimodal warning types on trust, annoyance and satisfaction. The results showed that trust and satisfaction of multimodal warning types were significantly higher than the other two unimodal warning types (only visual and only auditory). The results also showed that annoyance of combined visual-auditory-tactile, combined visual-tactile and combined auditory-tactile were significantly higher than the other three warning types, and the annoyance of combined visual-auditory-tactile is the highest. Based on user interviews, we speculated that adding tactile signals may increase annoyance of participants, nevertheless, tactile signals still has great potential to increase the trust and satisfaction of multimodal warning types.

Acknowledgement. This work was funded by Hunan innovative province construction project: R & D and application demonstration of key technologies of intelligent minitype fire emergency station 2020SK2094.

References

1. Kaizuka, T., Nakano, K.: Effects of urgency of audiovisual collision warnings on response time and accuracy of steering. Int. J. Intell. Transp. Syst. Res. **18**(1), 90–97 (2018). https://doi.org/10.1007/s13177-018-0174-6
2. Meng, F., Spence, C.: Tactile warning signals for in-vehicle systems. Accid. Anal. Prev. **75**, 333–346 (2015)
3. Lundqvist, L.M., Eriksson, L.: Age, cognitive load, and multimodal effects on driver response to directional warning. Appl. Ergon. **76**, 147–154 (2019)
4. Politis, I., Brewster, S., Pollick, F.: Evaluating multimodal driver displays of varying urgency. In: Proceedings of the 5th International Conference on Automotive User Interfaces and Interactive Vehicular Applications, pp. 92–99 (2013)
5. Villalobos-Zúñiga, G., Kujala, T., Oulasvirta, A.: T9+ HUD: physical keypad and HUD can improve driving performance while typing and driving. In: Proceedings of the 8th International Conference on Automotive User Interfaces and Interactive Vehicular Applications, pp. 177–184 (2016)
6. Mcgee-Lennon, M.R., Wolters, M., Mcbryan, T.: Audio reminders in the home environment. In: 13th International Conference on Auditory Display on Proceedings, Montreal, Canada, pp. 437–444 (2007)
7. Watson, T., Cech, L., Eßers, S.: A driving simulator HMI study comparing a steering wheel mounted display to HUD, Instrument Panel and Center Stack Displays for Advanced Driver Assistance Systems and Warnings. SAE Technical Paper (2010)
8. Liu, Y.C.: Comparative Study of the effects of auditory, visual and multimodality displays on drivers' performance in advanced traveller information systems. Ergonomics **44**(4), 425–442 (2001)
9. Green, P., Wei, H.L.: The rapid development of user interfaces: experience with the wizard of OZ method. In: Human Factors & Ergonomics Society Annual Meeting Proceedings, vol. 29, no. 5, pp. 470–474 (1985)
10. Häuslschmid, R., Menrad, B., Butz, A.: Freehand vs. micro gestures in the car: driving performance and user experience. In: 2015 IEEE Symposium on 3D User Interfaces (3DUI), pp. 159–160. IEEE (2015)
11. Merenda, C., Kim, H., Gabbard, J.L., et al.: Did you see me? Assessing perceptual vs. real driving gains across multi-modal pedestrian alert systems. In: Proceedings of the 9th International Conference on Automotive User Interfaces and Interactive Vehicular Applications, pp. 40–49 (2017)
12. Mehler, B., Reimer, B., Coughlin, J.F., et al.: Impact of incremental increases in cognitive workload on physiological arousal and performance in young adult drivers. Transp. Res. Rec. **2138**(1), 6–12 (2009)
13. Forne, M.: Physiology as a tool for UX and usability testing. School of Computer Science and Communication, Master. Royal Institute of Technology, Stockholm (2012)
14. Yao, L., Liu, Y., Li, W., Zhou, L., Ge, Y., Chai, J., Sun, X.: Using physiological measures to evaluate user experience of mobile applications. In: Harris, D. (ed.) EPCE 2014. LNCS (LNAI), vol. 8532, pp. 301–310. Springer, Cham (2014). https://doi.org/10.1007/978-3-319-07515-0_31
15. Lohani, M., Payne, B.R., Strayer, D.L.: A review of psychophysiological measures to assess cognitive states in real-world driving. Front. Human Neurosci. **13**, 57 (2019)
16. Ward, R.D., Marsden, P.H.: Physiological responses to different WEB page designs. Int. J. Human-Comput. Stud. **59**(1–2), 199–212 (2003)

CCD in Virtual Agents, Robots and Intelligent Assistants

Cross-Cultural Design and Evaluation of Robot Prototypes Based on Kawaii (Cute) Attributes

Dave Berque[1]([⊠]), Hiroko Chiba[1], Tipporn Laohakangvalvit[2], Michiko Ohkura[2], Peeraya Sripian[2], Midori Sugaya[2], Kevin Bautista[1], Jordyn Blakey[1], Feng Chen[2], Wenkang Huang[2], Shun Imura[2], Kento Murayama[2], Eric Spehlmann[1], and Cade Wright[1]

[1] DePauw University, Greencastle, USA
{dberque,hchiba,kevinbautista_2021,jordynblakey_2022,
ericspehlmann_2021,cadewright_2021}@depauw.edu
[2] Shibaura Institute of Technology, Tokyo, Japan
{tipporn,peeraya,doly,jnb18109,ma20038,ma20011,
ma18100}@shibaura-it.ac.jp, ohkura@sic.shibaura-it.ac.jp

Abstract. We report on a cross-cultural collaborative project between students and faculty at DePauw University in the United States and Shibaura Institute of Technology in Japan that used cross-cultural teams to design and evaluate robotic gadgets to gain a deeper understanding of the role that kawaii (Japanese cuteness) plays in fostering positive human response to, and acceptance of, these devices across cultures. Two cross-cultural design teams used Unity and C# to design and implement prototypes of virtual robotic gadgets as well as virtual environments for the robots to interact in. One team designed a virtual train station as well as robotic gadgets to operate in the station. The other team designed a virtual university campus as well as robotic gadgets that operated in that environment. Two versions of each robotic gadget were designed, such that the two versions differed with respect to one kawaii attribute (shape, size, etc.) Using these robots, we conducted a formal study that compared perceptions of kawaii robots between American college students and Japanese college students, as well as across genders. The findings revealed that there was not much difference in perception of kawaii across cultures and genders. Furthermore, the study shows that designing a robot to be more kawaii/cute appears to positively influence human preference for being around the robot. This study will inform our long-term goal of designing robots that are appealing across gender and culture.

Keywords: Kawaii · Human-robot interaction · Cross-cultural design

1 Introduction and Motivation

1.1 Kawaii

As robots become increasingly common in daily life, it is critical that roboticists design devices that are accepted broadly, including across cultures and genders. Toward this

© Springer Nature Switzerland AG 2021
P.-L. P. Rau (Ed.): HCII 2021, LNCS 12773, pp. 319–334, 2021.
https://doi.org/10.1007/978-3-030-77080-8_26

end, global collaboration will be pivotal today and in the future. This paper reports on a cross-cultural collaboration between students and faculty at DePauw University in the United States and Shibaura Institute of Technology in Japan. We formed two cross-cultural teams to design and evaluate robotic gadgets to gain a deeper understanding of the role that kawaii (Japanese cuteness) plays in fostering positive human response to, and acceptance of, these devices across cultures.

The word, *kawaii*, is often translated into "cute," "lovely," "adorable", "cool," and sometimes other words depending on the context. There does not seem to be an exact word that can be used as a counterpart in English [1]. That's probably because Japanese people's affection for "kawaii" has been cultivated throughout Japanese history [2]. The sentiment of kawaii seems to have been present in Japanese society since 400 B.C., and the word itself started appearing around the 11th century in literary texts [3, 4]. The meaning of the word has evolved to become a cultural concept or an emotional domain that relates to something or someone lovely, or someone or something that invokes the feeling of "wanting to protect" [4]. In the modern context, the notion of kawaii is embraced as a catalyst to evoke positive feelings [5], as can be seen in designs ranging from Hello Kitty products to road signs to robotic gadgets, to name just a few examples. Kawaii has also been gaining global audiences and customers in the last two decades as well as in Japan [6] through kawaii products. As such, kawaii design principles are incorporated into successful products that are used globally including in robotic gadgets [7, 8].

1.2 Prior Work

Previous studies have examined cross-cultural differences in the acceptance of robots based on various design characteristics. For example, researchers have documented the impact of localizing a robot's greeting style (gestures and language) on acceptance by Japanese versus Egyptian users [9].

Similarly, prior studies have examined perceptions of kawaii including differences in perceptions across cultures and genders. These studies have found gender differences in preferences for various kawaii spoon designs based on shape, color and geometric pattern [10]. A broader study examined the extent to which perceptions of kawaii in 225 photographs differ between male and female Japanese college students [11]. In this study, gender differences were established, depending on the type of object photographed. For example, male subjects found spherical geometric objects to be more kawaii than female subjects [11]. The first two authors extended the original study by presenting 217 of the original 225 images to American college students and gathering data about their perceptions of kawaii-ness in each image. For some types of objects, differences in perceptions of kawaii-ness were found, particularly between Japanese males and American males as well as between Japanese females and all other groups [12].

Prior work that investigates the role of kawaii in user perceptions and user acceptance of robots or robotic gadgets is limited. However, one pair of papers reports on studies of kawaii-ness in the motion of robotic vacuum cleaners [13, 14]. The authors programmed a visually plain version of a Roomba vacuum cleaner to move according to 24 different patterns, including patterns that the authors describe with terms such as: bounce, spiral, attack, spin and dizzy [13, 14]. The studies demonstrated that kawaii-ness can be expressed through motion, even in the absence of more traditional visual kawaii-ness; however, the studies did not consider cultural or gender differences.

2 Cross-Cultural Design of Virtual Environments and Robots

2.1 Overview

Building on the prior research described in the previous section, this paper reports on work, supported by a United States National Science Foundation (NSF) International Research Experiences for Undergraduates (IRES) grant, to gain a deeper understanding of the role that kawaii plays in fostering positive human response to, and acceptance of, robotic gadgets across cultures. More information about the goals of our grant-supported project may be found in [15].

With mentorship from faculty members at Shibaura Institute of Technology and DePauw University, two cross-cultural design teams used Unity and C# to design and implement prototypes of virtual robotic gadgets as well as virtual environments for the robots to interact in.

Each design team was comprised of four students -- two students from Shibaura Institute of Technology in Tokyo and two students from DePauw University in the United States. COVID-19 prevented travel between the United States and Japan. Therefore, all collaboration was conducted virtually, using tools such as Zoom and Slack.

One of our four-person student teams designed a virtual train station and each student on the team designed a pair of robotic gadgets to operate in the station (e.g., robots to clean the floor and to give directions to travelers). The other team designed a virtual college campus and each of its team members designed a pair of robotic gadgets that operated in that environment (e.g., robots to pick up trash and to give campus tours). As shown in the next section, each student designed their pair of robotic gadgets so the two versions differed with respect in one kawaii attribute (shape, size, etc.)

A more detailed description of the design process and tools we used to support our cross-cultural collaboration are provided in [16]. In the remainder of this paper, we present the virtual environments and robotic gadgets that were designed. In addition, we report on a formal cross-cultural study that compared the perceptions of cuteness between American college students and Japanese college students, using the robotic gadgets created by the two teams described above. This study also compared perceptions of the robotic gadgets between genders. This study will inform our long-term goal of designing robots that are appealing across gender and culture.

2.2 Environment Designs

One four-person team designed a university campus as shown in Fig. 1. The video at *tinyurl.com/hcii2021campus* provides a virtual tour of the environment. The university campus served as the context for designing robotic gadgets that served as trash collectors, vending machines, tour guides and stress-reducing human companions as illustrated in Fig. 3 through Fig. 6 in Sect. 2.3.

Fig. 1. University campus environment that provides context for several robotic gadgets

The other four-person team designed a train station as shown in Fig. 2. The video at *tinyurl.com/hcii2021station* provides a virtual tour of the environment.

The train station served as the context for robots that cleaned floors, shared information, assisted with shopping and made announcements. These robots are illustrated in Fig. 7 through Fig. 10 in Sect. 2.3.

Fig. 2. Train station environment that provides context for several robotic gadgets

2.3   Robot Gadget Designs for University Campus

As described previously, each team member designed a pair of robots that operated in the context of one of the virtual environments. These robot pairs are presented, along with contextual information about the robots, in the remainder of this section. The first four pairs were designed to operate on the university campus and the last four pairs were designed to operate in the train station.

Figure 3 shows the first pair of robots that are designed to operate on the university campus. The robots move around the campus and vacuum trash from the ground. The left robot is designed to be more kawaii because it includes animal features. The video at *tinyurl.com/hcii2021vacuum* shows the robot pair in action.

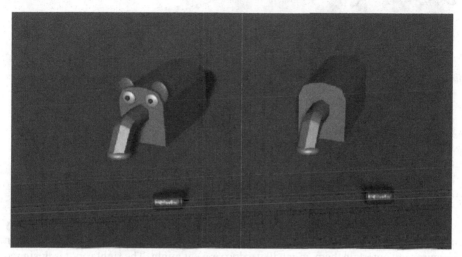

Fig. 3. Vacuum trash removal robots for the university campus

Figure 4 shows a second pair of robots that are designed to operate on the university campus. These vending machine robots move around the campus and ask people if they would like a drink. The right robot is designed to be more kawaii because it is rounder. The video at *tinyurl.com/hcii2021drink* shows the robot pair in action.

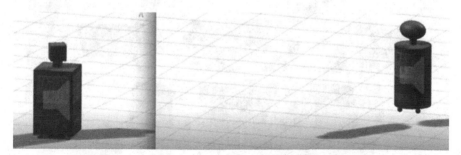

Fig. 4. Vending machine robots for the university campus

Figure 5 shows a third pair of robots that are designed to operate on the university campus. These tour guide robots float through the air and provide verbal information as they guide people around campus. The left robot is designed to be more kawaii because it is round. The video at *tinyurl.com/hcii2021tour* shows the robot pair in action.

Fig. 5. Floating tour guide robots for the university campus

Figure 6 shows the last pair of robots that are designed to operate on the university campus. These companion robots approach nearby people, like the people on the bench in the figure, and entertain them, much like a dog or a cat might. The right robot is designed

Fig. 6. Companion robots for the university campus

to be more kawaii because it is rounder. The video at *tinyurl.com/hcii2021companion* shows the robot pair in action.

2.4 Robot Designs for Train Station

Figure 7 shows the first pair of robots that are designed to operate in the train station. These robots move around the train station and clean the floors automatically. The right robot is designed to be more kawaii because it is rounder. The video at*tinyurl.com/hcii2021floor* shows the robot pair in action.

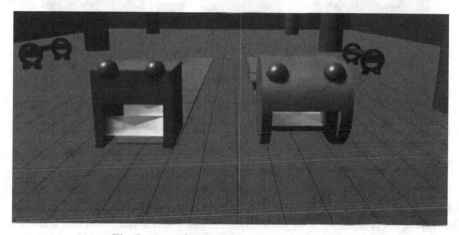

Fig. 7. Floor cleaning robots for the train station

Figure 8 shows the second pair of robots that are designed to operate in the train station. These information kiosk robots provide passengers with information such as train schedules. The right robot is designed to be more kawaii because it is shorter. The video at *tinyurl.com/hcii2021kiosk* shows the robot pair in action.

Fig. 8. Information kiosk robots for train station

Figure 9 shows the third pair of robots that are designed to operate in the train station. The robots store the passenger's shopping packages and bring them to the train. The right robot is designed to be more kawaii because it is softer. Although there is no visual difference between the robots, when the right robot rolls it makes a smoother sound than the left robot. The video at *tinyurl.com/hcii2021shop* demonstrates this sound.

Fig. 9. Shopping assistant robots for train station

Figure 10 shows the last pair of robots that are designed to operate in the train station. These robots move around the train station and make announcements such as letting passengers know that trains are arriving or departing. The right robot is designed to be more kawaii because it is smaller. The video at *tinyurl.com/hcii2021announce* shows the robot pair in action.

Fig. 10. Announcement robots for the train station

3 Evaluation

3.1 Purpose of the Evaluation

We designed and administered an online survey to measure perceptions of the eight robot pairs. The survey, which was administered to Japanese and American males and females, was designed to help us answer three primary questions:

- Determine which robot from each pair the participants judge to be more kawaii/cute so we could see if commonly accepted kawaii design principles extend to our work.
- Determine whether participants prefer to be around robots that they judged to be more kawaii/cute.
- Determine whether there are any gender differences or cultural differences in the survey results.

   The survey also gathered data to let us explore the participant's characterization of the robots with respect to adjectives other than kawaii/cute (e.g., scary, polite, soft). While we will touch on the results of this part of the study in the sections that follow, we will save detailed analysis of these results for future work.

3.2 Participant Demographics

After obtaining Institutional Review Board approval for the study, we recruited student participants from DePauw University (in the United States) as well as from Shibaura Institute of Technology, Chuo University, and the University of Tokyo (all in Japan). Participants had to be at least 18 years old. Because we were studying cultural differences,

DePauw University students could not have lived outside of the United States for more than four months. Similarly, students from Shibaura Institute of Technology and the University of Tokyo could not have lived outside of Japan for more than four months. In effect, this limited participation to students who were fully raised in Japan or who were fully raised in the United States. Participants were also required to have a laptop or tablet to complete the survey. We did not allow participants to complete the survey on a phone because we wanted the embedded videos to be a reasonable size.

In total, 82 eligible students completed the survey. The results from one student are not included in the analysis because the student did not identify primarily as male or as female. Of the remaining 81 participants, there were 40 participants from Japan and 41 participants from the United States. There were 41 females including 20 from Japan and 21 from the United States. There were 40 males including 20 from Japan and 20 from the United States. The participants ranged in age from 18 to 24 with an average age of 21.

3.3 Study Procedure

The survey was originally developed in English and then translated to Japanese by the second author. We used the adjective, cute, for the English counterpart of "kawaii" because it is one of the closest translations, although the interchangeableness has been argued [1]. The English version of the survey was administered to the participants from the United States and the Japanese version was administered to the participants from Japan. Each version was administered as a Google form that embedded the videos shown in Fig. 3. Through Fig. 10.

After accepting the conditions of an online informed consent, and confirming eligibility to participate in the survey, participants provided their age and gender. Participants then watched a sample video and confirmed that they could see the video and hear the associated audio.

Participants from the United States were asked about their familiarity with the Japanese term kawaii. Participants then watched a video (approximately 15 s long) of the first pair of robots (see Fig. 3) and answered twelve multiple choice questions about their perception of differences in the pair of robots. This process was repeated for a video that showed the second pair of robots (see Fig. 4) and then for subsequent videos until all eight videos had been presented.

3.4 Perceptions of Kawaii/Cute for All Participants

After watching each video, participants were asked to compare the left robot and the right robot with regard to each of the adjectives shown in Table 1.

For each adjective, participants selected from one of five choices that compared the robot on the left side of the video to the robot on the right side of the video with respect to the adjective. For example, for the adjective "approachable", participants selected from one of the following choices.

1. The left robot is much more approachable than the right robot.
2. The left robot is somewhat more approachable than the right robot.

Table 1. Adjectives that participants used to compare robot pairs.

Adjectives	
Kawaii	Cool
Cute	Beautiful
Approachable	Polite
Scary	Comfortable
Trustworthy	Soft

3. The left robot and the right robot have about the same level of approachableness.
4. The left robot is somewhat less approachable than the right robot.
5. The left robot is much less approachable than the right robot.

For the purpose of the analysis presented in the remainder of this section, we collapsed responses 1 and 2 into a single category indicating that the participant found the left robot to be more approachable than the right robot. Similarly, we collapsed responses 4 and 5 into a single category indicating that the participant found the left robot to be less approachable then the right robot. In other words, our analysis is based on the following three categories.

1. The left robot is more approachable than the right robot.
2. The left robot and the right robot have about the same level of approachableness.
3. The left robot is less approachable than the right robot.

This study is centrally concerned with the way perceptions of kawaii change when various attributes of a robot are changed. However, only 37% of the participants from the United States indicated that they had a good understanding of the Japanese word kawaii. The English word "cute" is generally considered to be a good proxy for the Japanese word kawaii. Therefore, in what follows, we compare responses from Japanese participants for the adjective kawaii to responses from American participants for the adjective cute.

We performed a series of statistical analyses with a significance level of 0.05. Table 2 shows the results of chi-square tests that compare perceptions of Kawaii/Cute for the left robot versus the right robot in each robot pair for all 81 participants. In the right column of the table, participant responses are summarized in the format x-y-z where x gives the number of participants who judged the left-robot to be more kawaii/cute, y is the number of participants who found no difference in the level of kawaii/cuteness between the two robots, and z is the number of participants who judged the right robot to be more kawaii/cute. For example, the data in the second row of the table indicates that 66 participants found the left robot (with the elephant face) to be more kawaii/cute than the right robot. On the other hand, 9 participants found no difference in the level of kawaii/cuteness while 6 participants found the right robot to be more kawaii/cute than the left robot.

The statistical analyses in Table 2 provide strong evidence that the perceptions of kawaii/cute are not equally distributed. Stronger preferences are indicated by the raw numbers for robots that were more animal-like (Fig. 3), that were rounder (Fig. 4, Fig. 5, Fig. 6, Fig. 7), that were shorter (Fig. 8), and that were smaller (Fig. 10).

The pair of robots in Fig. 9 do not have visual differences. The right robot in this pair makes a smoother sound when it moves, suggesting that the robot is softer which is typically associated with an increased perception of kawaii/cute. Approximately half of the participants, however, indicated that there was no difference in their perception of kawaii/cute between the left robot and the right robot. For those participants who did perceive a difference, more participants thought the left robot was more kawaii/cute than the right robot.

Table 2. Perceptions of Kawaii/cute between left vs. right robot for all participants (N = 81)

Robot pair	Kawaii/Cute
Fig. 3 Vacuum Trash Removal Robots Left: Elephant face Right: No face	Elephant face is more Kawaii/Cute 66-9-6 X^2 (2, N = 81) = 84.7, $p < .001$
Fig. 4 Vending Machine Robots Left: Square Right: Round	Round is more Kawaii/Cute 7-22-52 X^2 (2, N = 81) = 38.9, $p < .001$
Fig. 5 Floating Tour Guide Robots Left: Round Right: Square	Round is more Kawaii/Cute 46-26-9 X^2 (2, N = 81) = 25.4, $p < .001$
Fig. 6 Companion Robots Left: Square Right: Round	Round is more Kawaii/Cute 17-17-47 X^2 (2, N = 81) = 22.2, $p < .001$
Fig. 7 Floor Cleaning Robots Left: Square Right: Round	Round is more Kawaii/Cute 7-11-63 X^2 (2, N = 81) = 72.3, $p < .001$
Fig. 8 Information Kiosk Robots Left: Tall Right: Short	Short is more Kawaii/Cute 13-22-46 X^2 (2, N = 81) = 21.6, $p < .001$
Fig. 9 Shopping Assistant Robots Left: Harsher Sound Right: Smoother Sound	About half selected "no difference" 30-38-13 X^2 (2, N = 81) = 12.1, $p = .002$
Fig. 10 Announcement Robots Left: Big Right: Small	Small is more Kawaii/Cute 28-20-43 X^2 (2, N = 81) = 20.2, $p < .001$

3.5 Perceptions of Kawaii/Cute Across Gender and Culture

To investigate gender differences and the differences that stem from cultural backgrounds, additional statistical tests were employed. A chi-square test of independence was performed to examine the relationship between gender (male vs female) of the participants and their perceptions of kawaii/cute for each of the eight robot pairs. The results show that there was no significant association between gender and how each robot pair was perceived.

Similarly, a chi-square test of independence was performed to investigate the differences in perception between two cultural backgrounds (American vs Japanese). A chi-square test of independence was performed to find the relationship between cultural background and the participants' perceptions of kawaii/cute for each of the eight robot pairs. The result shows that the relationship between these two variables was not significant for any of the pairs.

3.6 Preferences for All Participants

After watching each robot-pair video, participants also answered a question that asked "Which robot would you prefer to be with?" Participants responded to this question by selecting one of the following three choices.

1 The left robot.
2. No preference.
3. The right robot.

Table 3 shows the results of chi-square tests that compares participant's preferences for being around the left robot versus the right robot in each robot pair for all 81 participants. In the right column of the table, participant responses are summarized in the format x-y-z where x gives the number of participants who prefer to be around the left-robot, y is the number of participants who found both robots equally preferable to be around, and z is the number of participants who preferred to be around the right robot. For example, the data in the second row of the table indicates that 45 of the participants preferred to be around the left robot (with the elephant face) as compared to the right robot. On the other hand, 20 participants had no preference for being around one robot as compared to the other while 16 participants preferred to be around the right robot.

As shown in Table 3, for six of the seven robot pairs where there is a visual distinction between the robots, the results show strong evidence that one of the robots in each pair is preferred. In each of these cases, the participants prefer to be around the robots that they judged to be more kawaii/cute. This includes cases when the robots were more animal-like (Fig. 3 and Fig. 7), when they were rounder (Fig. 4, and Fig. 6), when they were shorter (Fig. 8), and when they were smaller (Fig. 10). For the robot pair shown in Fig. 5, while the statistical results did not show differences in preference for the rounder and more kawaii/cute robot, the tendency is in the expected direction.

As noted previously, the pair of robots in Fig. 9 do not differ visually. The right robot in this pair makes a smoother sound when it moves. Participants indicated preference for being around the robot that sounds smoother.

3.7 Preferences Across Genders and Cultures

Further analyses were conducted to investigate the relationships between gender (male vs female) and preferences. A chi-square test of independence was performed to examine

Table 3. Preference for being around left versus right robot for all participants (N = 81)

Robot Pair	Preference to be around Left vs. Right
Fig. 3 Vacuum Trash Removal Robots Left: Elephant face Right: No face	Elephant face is preferred 45-20-16 $X^2 (2, N = 81) = 18.5, p < .001$
Fig. 4 Vending Machine Robots Left: Square Right: Round	Round is preferred 7-24-50 $X^2 (2, N = 81) = 35.1, p < .001$
Fig. 5 Floating Tour Guide Robots Left: Round Right: Square	No significant difference 35-23-23 $X^2 (2, N = 81) = 3.6, p = .165$
Fig. 6 Companion Robots Left: Square Right: Round	Round is preferred 20-14-47 $X^2 (2, N = 81) = 23.1, p < .001$
Fig. 7 Floor Cleaning Robots Left: Square Right: Round	Round is preferred 11-20-50 $X^2 (2, N = 81) = 31.2, p < .001$
Fig. 8 Information Kiosk Robots Left: Tall Right: Short	Short is preferred 16-15-50 $X^2 (2, N = 81) = 29.7, p < .001$
Fig. 9 Shopping Assistant Robots Left: Harsher Sound Right: Smoother Sound	Smoother sound is preferred 29-12-40 $X^2 (2, N = 81) = 14.9, p < .001$
Fig. 10 Announcement Robots Left: Big Right: Small	Small is preferred 16-12-53 $X^2 (2, N = 81) = 38.2, p < .001$

the relationship between gender (male vs female) of the participants and their preferences for each of the eight robot pairs. The results show that there was no significant relationship between gender and how each robot was perceived for 7 of the 8 robot pairs. The exception, as described below, is for the robot pair shown in Fig. 10.

A two-variable chi-square test demonstrates a significant difference between male participants and female participants with respect to preferences for the pair of announcement robots shown in Fig. 10. While males and females both prefer the smaller robot on the right side of the figure, this preference is more significant for females $X^2 (2, N = 41) = 8.1, p = .017$.

Similarly, a chi-square test of independence was performed to investigate the differences in perception between cultural backgrounds (American vs Japanese). A chi-square test of independence was performed to find the relationship between cultural background and the participants' preferred robots. The result shows that there is not significant association between cultural backgrounds and preferences for any of the eight robot pairs.

4 Discussion

This study demonstrates that designing a robot to be more animal-like, rounder, shorter, and smaller increases participant's perceptions that the robot is kawaii/cute. More importantly, the study shows that designing a robot to be more kawaii/cute appears to positively influence human preference for being around the robot. These findings hold across Japanese and American culture and across males and females.

While both males and females prefer smaller robots to larger ones, this preference appears to be more significant for females. However, no other differences were found between genders or between cultures.

5 Future Work

In this paper, we have only reported the results of initial chi-square tests, but for future work, we will conduct post hoc tests to more clearly explore relationships among robot pairs.

In this paper we have confined our formal analysis to the adjectives kawaii and cute. The data suggests that the adjectives "approachable," "beautiful," "comfortable" and "soft" may be correlated with the adjectives "kawaii/cute" and we will explore and report on these relationships in future work. Conversely, the adjectives "trustworthy" and "polite" do not seem correlated with kawaii/cute and many users saw do difference between left and right robots with respect to trustworthy and polite. These adjectives seem to be more relationship-based and may require more context than a short video provides. We will explore these adjectives in conjunction with kawaii as an emotional process in future work.

The survey instruments used in this study could easily be translated into languages other than English and Japanese, which would allow us to extend this work to other cultures.

We also plan to measure reaction to real robots rather than videos of robots. Finally, we plan to use biosensors to gauge participant reactions rather than relying on surveys.

Acknowledgements. This material is based upon work supported by the National Science Foundation under Grant No. OISE-1854255. Any opinions, findings, and conclusions or recommendations expressed in this material are those of the author(s) and do not necessarily reflect the views of the National Science Foundation. We thank faculty members at Shibaura Institute of Technology and the University of Tokyo for helping arrange for Japanese students to complete the study. Specifically, we thank professors Y. Ito, H. Manabe, and K. Hidaka from Shibaura Institute of Technology and professor Y. Tsuji from the University of Tokyo.

References

1. Nittono, H.: Kawaii no Chikara (The Power of Kawaii) Kyoto: Dojin Sensho (2019). (in Japanese)
2. Bijutsukan, F.: Kawaii Edo Kaiga (Cute Edo Paintings), Tokyo: Kyuryudo (2013). (in Japanese)

3. Ohkura, M., et al.: Kawaii Engineering: Measurements, Evaluations, and Applications of Attractiveness. Springer, Heidelberg (2019). https://doi.org/10.1007/978-981-13-7964-2
4. Yomota, I.: Kawaii Ron (The Theory of Kawaii) Tokyo: Chikuma Shobō (2006). (in Japanese)
5. Nittono, H., Fukushima, M., Yano, A., Moriya, H.: The Power of Kawaii: viewing cute images promotes a careful behavior and narrows attentional focus. PLoS ONE 7(9), e46362 (2012). https://doi.org/10.1371/journal.pone
6. Yano, C.: Pink Globalization: Hello Kitty's Trek Across the Pacific. Duke University Press, Durham (2013)
7. Cole, S.: The most Kawaii Robots of 2016 (2016). https://motherboard.vice.com/en_us/art icle/xygky3/the-most-kawaii-robots-of-2016-5886b75a358cef455d864759. Accessed 8 Sept 2018
8. Prosser, M.: Why Japan's cute robots could be coming for you (2017). www.redbull.com/us-en/japan-cute-robot-obsession. Accessed 8 Sept 2018
9. Trovato, G., et al.: Cross-cultural study on human-robot greeting interaction: acceptance and discomfort by Egyptians and Japanese. J. Behav. Robot. 4(2), 83–93 (2013)
10. Tipporn, L., Ohkura, M.: Comparison of spoon designs based on Kawaiiness between genders and nationalities. In: A4–3 Proceedings of ISASE 2017, March 2017
11. Hashizume, A., Kurosu, M.: The gender difference of impression evaluation of visual images among young people. In: Kurosu, M. (ed.) HCI 2017. LNCS, vol. 10272, pp. 664–677. Springer, Cham (2017). https://doi.org/10.1007/978-3-319-58077-7_51
12. Berque, D., Chiba, H., Hashizume, A., Kurosu, M., Showalter, S.: Cuteness in Japanese design: investigating perceptions of Kawaii among American college students. In: Fukuda, S. (ed.) Advances in Affective and Pleasurable Design, pp. 392–402. Springer, Heidelberg (2018). https://doi.org/10.1007/978-3-319-94944-4_43
13. Sugano, S., Miyaji, Y., Tomiyama, K.: Study of Kawaii-ness in motion – physical properties of Kawaii motion of roomba. In: Kurosu, M. (ed.) HCI 2013. LNCS, vol. 8004, pp. 620–629. Springer, Heidelberg (2013). https://doi.org/10.1007/978-3-642-39232-0_67
14. Sugano, S., Morita, H., Tomiyama, K.: Study on Kawaii-ness in motion –classifying Kawaii motion using Roomba. In: International Conference on Applied Human Factors and Ergonomics 2012, San Francisco, California, USA (2012)
15. Berque, D., Chiba, H., Ohkura, M., Sripian, P., Sugaya, M.: Fostering cross-cultural research by cross-cultural student teams: a case study related to Kawaii (cute) robot design. In: Rau, P.-L.P. (ed.) HCII 2020. LNCS, vol. 12192, pp. 553–563. Springer, Cham (2020). https://doi. org/10.1007/978-3-030-49788-0_42
16. Ohkura, M., Sugaya, M., Sripian, P., Laohakangvalvit, T., Chiba, H., Dave, B.: Design and implementation of remote collaboration by Japanese and American University students using virtual spaces with Kawaii robots. In: Proceedings of the 7th International Symposium on Affective Science and Engineering, March, 2021, Online Conference, Japan Society of Kansei Engineering (2021)

Towards Effective Robot-Assisted Photo Reminiscence: Personalizing Interactions Through Visual Understanding and Inferring

Edwinn Gamborino[1]([✉])[iD], Alberto Herrera Ruiz[1,2][iD], Jing-Fen Wang[1][iD], Tsung-Yuan Tseng[1][iD], Su-Ling Yeh[1,3][iD], and Li-Chen Fu[1,2][iD]

[1] Center for Artificial Intelligence and Advanced Robotics, National Taiwan University, Taipei, Taiwan
{gamborino,b04502017,tsungyuantseng}@ntu.edu.tw
[2] Department of Computer Science and Information Engineering, National Taiwan University, Taipei, Taiwan
{r08922163,lichen}@ntu.edu.tw
[3] Department of Psychology, National Taiwan University, Taipei, Taiwan
suling@ntu.edu.tw

Abstract. In this work, we present a robot-assisted photo reminiscence system. Previously, we focused on the development of the question generation system through a social robot interface. Results showed that the proposed system was able to produce coherent, related and appropriate questions that would trigger the reminiscence process. In this study, using an iterative incremental model, based on feedback from the end-users (elderly and caregivers), we deployed an improved prototype robotic system on a senior center with elderly users and in a lab setting for independent living elderly. Results from our feasibility study show that participants proactively engage with the robot and find the prospect of reminiscence therapy both effective and enjoyable.

Keywords: Human-robot interaction · Photo reminiscence · Memory assistant

1 Introduction

With the advancement of medicine and the industrialization of developed nations comes phenomena such as a dramatic increase in the average life expectancy [1] and the lowest birth rates in recent history [2]. This trend, commonly known in the literature as hyper-aged societies, is a major healthcare challenge as the

This research was supported by the Joint Research Center for AI Technology and All Vista Healthcare under Ministry of Science and Technology of Taiwan, and Center for Artificial Intelligence & Advanced Robotics, National Taiwan University, under the grant numbers of 109-2634-F-002-027-, 109-2634-F-002-040– and 109-2634-F-002-041.

© Springer Nature Switzerland AG 2021
P.-L. P. Rau (Ed.): HCII 2021, LNCS 12773, pp. 335–349, 2021.
https://doi.org/10.1007/978-3-030-77080-8_27

number of elderly patients increases while the number of available caregivers decreases. Furthermore, dementia is one of the major public health concerns of the 21st century. According to the World Health Organization [3], over 150 million people worldwide are projected to suffer from dementia and other cognitive impairments in their elderly years by the year 2050.

Among the non-pharmacological interventions designed to slow down brain deterioration and memory impairment, reminiscence therapy stands out as an effective option. It refers to the process of recalling and discussion of meaningful past events with a conversational companion. Aided by supporting materials, such as pictures or videos, the companion attempts to inspire and guide the user in recollecting their memories and openly discuss them. The practice of reminiscence has well-documented benefits on emotional and mental well-being, as well as stimulating cognitive functions and developing social and conversational skills [4]. Additionally, by transferring their life experiences and knowledge to others, older users who practice regular reminiscence sessions may achieve a greater sense of life achievement and self-actualization [5].

Research on technology-driven reminiscence, while still in its infancy, has made significant progress in recent years. This technology attempts to reduce the barriers of traditional reminiscence therapy, which strongly relies on trained personnel and/or family members that can support the sessions, reducing the practicality and reach of the intervention, in particular for users with reduced social contacts. To address this issue, several studies have attempted to implement assistive tools geared towards reminiscence therapy [6–10]. These studies have focused on different aspects of the reminiscence process, attempting to facilitate the use of technology for older adults, provide a platform to collect their memories through storytelling, and support conversations. In other words, while these implementations have often focused on the interaction design and data collection aspects, little research has been done where the use of artificial intelligence is the driving force of the reminiscence platform.

In [10], we introduced a novel question posing system for robot-assisted photo reminiscence. Validation was based on the classification accuracy of the image understanding module. The overall performance of the system was evaluated by a Human-Robot Interaction (HRI) experiment with college students. Results showed that the proposed system was able to produce coherent, co-related and appropriate questions that could trigger the reminiscence process based on the participant's personal photographs.

In this paper, we present an improved implementation of a reminiscence companion robot. Through an iterative incremental model, we identified the top features that users requested in the first prototype: (1) The robot should ask more types of questions: We leverage a collection of AI-based Image Understanding techniques to obtain additional information from an image, giving the robot a wider array of topics to discuss. (2) The system should show related pictures: We implemented a dynamic photo album display tool to display and organize images. (3) The robot sometimes would be too eager to ask another question: Using a variety of methods, the user experience was improved for the user. Furthermore, we validated the proposed system with elderly participants, both living independently and in assisted living homes.

The rest of this paper is organized as follows: Sect. 2 presents a collection of implementations of artificial photo reminiscence agents, either robotic or not, and highlights the shortcomings we address of each in the proposed system. Sections 3 and 4 discuss in great detail the implementation of the robotic system, the AI modules and their characteristics, as well as a description of the inter-device communication protocols. In Sect. 5, the validation feasibility study and its results are discussed. We close the document with a few concluding remarks and future works.

2   Related Works

In [6], the authors developed a social companion robot to interact with elderly users. The robot included an activity for reminiscence. In their approach, the experimenters would collect the materials (e.g. photographs and verbal stories) prior to the experiment and hard-code them into the robot's memory, with the robot retelling the story and asking simple questions during interaction with the user. While the authors present a comprehensive set of activities for a robot companion, the approach to reminiscence therapy is rather limited, as the robot can only use *a priori* information with no live input from the user.

More recently, Tokushige *et al.* [7] presented a study on the effects of reminiscence therapy on the brain of elderly patients with mild to moderate dementia using Near Infrared Spectroscopy. Their results show that, if done properly, reminiscence therapy stimulates activation of the frontal lobe, which is associated with important cognitive skills (e.g. memory and language). Furthermore, the authors and others before [9] have pointed out that performing one-on-one reminiscence is more effective than group reminiscence.

Carós *et al.* [8] introduced a chat-bot dialogue system to aid therapists guide a reminiscence therapy session. With the user's photos as input to a Visual Question Generation (VQG) model consisting of an encoder decoder with attention architecture, using a residual network (ResNet-101) and a Long Short-Term Memory (LSTM) model, respectively. The authors used a beam search of 7 to generate up to 5 questions per image with a maximum sequence length of 6 words. Each of these questions is then presented via the system's User Interface on a messaging app - Telegram - where the user can type back a response. The reply is analysed by a Recurrent Neural Network (RNN), again, using the encoder-decoder with attention architecture to generate a follow-up response, thus imitating a turn-based conversation.

While this system can automatically generate questions for given pictures, it faces several challenges: Since there are no open-source corpora available specific for reminiscence therapy, the authors recurred to a general question corpus to train the VQG network (Persona-chat) and a movie dialog corpus (Cornell movie dialogues) for the chatbot model. Furthermore, since both questions and replies are generated by AI models, the output sentences could have grammatical errors or be non-sensical, which can generate confusion among users. Finally, if the input images have similar visual features, the system may generate similar or repetitive questions.

Asprino *et al.* presented a knowledge-driven approach to robot-assisted photo reminiscence [9] as part of the EU MARIO Project. They introduced an ontological knowledge graph to store and query biographical information of the user, including personal details, relationships and life events, and multimedia objects related to these. In their approach, the biographical information could be obtained by the robot either offline by having a relative or caregiver fill in personal details prior to an interactive session or through social interaction, using speech understanding to fill in these details. In interaction, the questions are in the form of templates which can be instantiated with the details of a given node (e.g. is this <person name> your <personal relationship>?). Furthermore, the authors define two types of questions. Close-ended questions: yes/no questions or questions where the answer is in the knowledge graph (e.g. in what year was your wedding?). This type of question can be used to test the user's memory. Open-ended questions can be used to trigger reminiscence (e.g. how was your wedding like?). While the authors present a complete framework that can be used for the reminiscence process, as of their latest publication, experimental results are yet to be reported.

In contrast to the presented approaches, by leveraging deep learning networks, the system is able to obtain information from the photographs themselves, enriching its ability to ask different types of questions. Furthermore, the proposed reminiscence system is implemented in an embodied robot agent, which gives the system a higher social presence, which in turn fosters the ability of it to generate rapport with the user.

3   System Design

The reminiscence system prototype was implemented as shown in Fig. 1. Briefly, the system has two hardware pieces that interact with each other: a desktop computer and a social robot. The photo album of the user is uploaded to the system, where the Image Understanding (IU) module is applied batch-wise to all files in the album to extract information such as the event that is happening and the scene it took place in. This data is stored in CSV files for the visualization interface, D-FLIP. In our application, this interface shows a large number of images as a dynamic array, where the user selection can affect the image properties (e.g. enlarge and center). When the user selects a picture, the metadata generated by the IU module for that file is loaded and used in the Question Generation (QG) module to produce questions related to the topics observed by the robot in the picture. The robot then will be commanded to speak out the generated sentence through its Text-To-Speech (TTS) engine, then await for the user to reply.

Upon finishing listening to the users' response, the Speech-To-Text (STT) engine will then parse the text contained in the user speech, which is then passed on to the Natural Language Understanding (NLU) module, which is programmed to understand certain keywords in the users' reply. If the user has not selected a new picture, upon completing checks on the number of questions asked and the time elapsed since the beginning of the session, the robot will generate a new question for the selected image. Once a preset maximum number of questions is reached, the robot prompts the user to change pictures. Once the user selects a new picture the QG module creates questions with the updated metadata. Finally, when the session timer runs out, the robot finishes the sessions and thanks the user for their time.

In this section and the following, every process mentioned in the above overview will be discussed in greater detail.

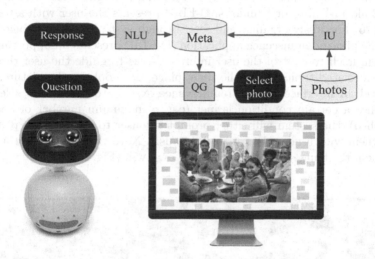

Fig. 1. System overview

3.1 Picture Selection Interface

An automated photo reminiscence system requires a digital visual interface where the user can view their personal photo album and select a picture they desire to talk about.

Recently, with the adoption of cheap and high resolution camera modules in several smart devices, as well as the development of social media, the number of digital photos that people produce has risen exponentially. This often unorganized data is a great source of information for photo reminiscence. However, to be useful, these images need to be labeled with whatever meaningful entities are found in them. For instance, the number of people, the objects present in the scene, etc. This information can be gathered in two ways: Manually, tagging a picture with the aid of a specialized UI; or automatically, leveraging large picture databases, neural networks can be trained to estimate or recognize certain information in an image. In this work, as a baseline, we use the latter option only, letting the system recognize entities in the data by itself. More information can be found in Sect. 4.

With a large enough collection of pictures, having a dynamic, flexible way to present semantically related pictures to the user becomes a priority for effective photo reminiscence. For instance, a user may be inspired to recall more memory details by observing several pictures that are chronologically near, or that contain a similar theme or people.

D-FLIP or Dynamic and Flexible Interactive PhotoShow, a platform proposed by Kitamura *et al.* [11,12] uses a combination of organizational principles to sort digital photos, with the results of such rule set being continuously updated in the user interface, resulting in a fluid, dynamic photo viewing experience. Build in Unity for Windows, D-FLIP presents the user with a tool that can help to effortlessly organize and sort through vast photo collections.

For the photo reminiscence application, D-FLIP presents an opportunity for the system itself to control the user interface so as to guide the user through a story with related topics, be it the time, place, or people in the picture. Thus, we devised an interface for the reminiscence system to be able to select pictures. Once a certain condition is met (e.g. a maximum number of questions was reached), the system will either prompt the user to select a new picture or present them with a new picture autonomously. Nevertheless, the user also had the freedom to change the picture at their own will (Fig. 2).

Fig. 2. D-FLIP Interface samples. (left) Related items surround the selected item. (right) Manual creation of item groups.

3.2 Robot Speech Interface

The second hardware component of the reminiscence system is the social robot. For the experiment results reported in this paper, we used the social robot Zenbo Junior (See Fig. 3). Made by Asus Inc., Zenbo Junior is 31.5 cm tall and 2.75 kg heavy, has a battery that lasts up to five hours, a differential wheel configuration, a touchscreen that acts as its face and an array of sensors including a 720p camera and high fidelity microphone. As it runs on Android OS, the robot can be programmed to move around, speak, and use its sensors to interact with the environment by deploying an Android application, or app.

The modules developed for Image Understanding (Sect. 4) however, require a substantial amount of computing power to process pictures in an effective manner; furthermore, tools and libraries for artificial intelligence and machine learning are more prevalent in Python when compared to other programming languages. The computing power on mobile devices such as Zenbo Junior is also often insufficient. Therefore, in the proposed system prototype, a desktop PC acts as a local back-end that can perform the computationally heavy processes. In a production environment this could be upgraded to a cloud-based server implementation.

On a technical level, the communication interface between the robot and the back-end computer was achieved through Google Remote Procedure Call (stylized gRPC). A communication protocol was devised using JSON objects as messages that could be encoded and decoded in either side seamlessly, regardless of programming language. The functions of the robot as a voice interface to speak out questions and listen to the users' replies were then controlled by the reminiscence system in an easy-to-use API which would handle the transmission, execution of TTS and STT and return the users speech, if any, back as a result. Additional back-channeling behaviors [13] were available for the robot as well. For instance, the robot could nod its head or change its facial expression upon command by the reminiscence system.

Fig. 3. Zenbo Junior. The social robot used for the reminiscence system.

3.3 Natural Language Understanding

Once the robot has finished uttering a question, it will activate its microphone and attempt to record the users' speech. By design, if no sound is detected, the robot will either repeat the question or prompt the user to talk by saying encouraging sentences, such as "can you tell me more about this picture?". Otherwise, the robot will parse the audio file and return the words detected in it. The STT engine used by Zenbo Junior is proprietary of Asus Inc. and cannot be overridden. Accuracy benchmarks are not known to exist.

Once the utterance string has been created, it is fed into a NLU engine in order to attempt to detect meaningful keywords or commands in the users' speech. Ideally, techniques such as multi-gram patterns or neural network-based predictions could be used to identify keywords in the users utterances. However, it is common knowledge that these methods are not well-suited for general purpose conversations. Therefore, in our practical implementation only a basic keyword matching algorithm was implemented. This was helpful for the user to navigate the robots' software using voice commands with yes/no questions. For instance, the robot may ask: "Is this a picture of a wedding?". The robot could then take different actions depending on the users' response.

Furthermore, following the convention of Wizard-of-Oz HRI experiments, a manual override was implemented such that a remote operator could control the robots' actions in case of a malfunction in the reminiscence system.

3.4   Question Generation

To generate an extensive and comprehensive list of topics to interact with the participants, we gathered 406 questions and statements in total in the system's question database. We asked graduate students to think up various questions related to the labels (shown in the Table 1 below), which the system's image understanding module would classify.

Table 1. Detectable classes by each IU module.

Category	Classes	Number of items
Events	general_event, graduation, hiking, picnic, sea_holiday, ski_holiday, wedding	286
Scenes	animals, temple, street_view, sports_arena, snow, river_lake, restaurant, mountain, mall_store, kids_park, house_room, house, general_scene, garden, outdoor_building, indoor_building, beach_coast	120

4   Image Understanding

4.1   Scene Prediction

In the context of this paper, scene refers to the physical location type where the photograph took place. These labels are general rather than specific (e.g. restaurant or beach, rather than McDonald's or Cancun). Then, for recognizing the

Table 2. New class designations for scene recognition. The table also shows the number of Places365 merged classes and the training, validation and test dataset size.

Class Name	# of Merged Classes	Training Images	Validation Images	Easy-Test Images	Hard-Test Images
beach_coast	9	45000	900	258	13
building_indoor	10	49254	1000	234	52
building_outdoor	10	50000	1000	234	23
garden	10	49939	1000	265	22
house_room	12	60000	1200	260	31
kids_park	8	40000	800	257	13
mall_store	10	49001	1000	260	2
mountain	9	45000	900	255	28
restaurant_table	10	50000	1000	266	14
river_lake	12	60000	1200	264	5
snow	9	44457	900	265	5
sport_arena	10	48376	1000	240	2
streetview	6	30000	600	299	10
temple	7	35000	700	230	7
default_class	13	66065	1300	0	20
Total	145	722092	14500	3587	247

scene of photographs, we utilize the Places365 dataset [14]. To facilitate the data collection process described earlier, we group dataset classes that are semantically related since questions generated for one category can also be applicable for others. For example, inquiries about the "beach" class could also be applicable for "coast" and "wave". After merging categories (only 145 out of the 365 were used), the resulting subset contained 14 different classes, each one composed of 6 to 12 classes from the original. Given that participants could provide images that do not belong to any of the designated classes, scene recognition becomes an open set problem. To mitigate the issue, an extra class was incorporated with images from unused classes. Table 2 details the final classes and their sizes. It also shows the Easy and Hard datasets size. "Easy-Test Images" was compiled from Google using the merged class names as search keywords. "Hard-Test Images" is composed from the images provided by the participants during some of the experiments.

The utilized CNN architecture employs VGG16 as the backbone with a global average pooling, fully connected layer and softmax function as the top layers. We are using a Keras version of the pre-trained Places365 weights [14] and fine-tuning the network with the newly designated classes. Data augmentation techniques like rotations, zoom and shifts were utilized during training.

4.2 Event Prediction

Similar to [10], we define event as the people-centric activity that the people that appear on the picture are doing. Given the demographics of our target audience (elderly patients of Taiwan), we decided to use the six most common classes of the USED dataset [15] that our users could have experienced before. These include graduation, mountain trip, sea holiday, ski holiday, picnic and wedding. Similar to the scene recognition module, an extra class is adopted, composed from images from the unused classes of the USED and Places365 dataset. The training set contains 167582 images and the validation set 501880.

We tried using a single network for recognizing the scene and the event of an image; but given the different domains of the datasets (USED dataset portraits more people than the Places365 dataset), we opted for individual networks to avoid bias and improve accuracy. We also employ VGG16 with a global average pooling, fully connected layer and softmax function as the top layers as the CNN model. The pre-trained Places365 weights [16] were used and fine-tuned the network with the mentioned classes. Data augmentation techniques like rotations, zoom and shifts were utilized during training.

5 Validation

5.1 Procedure

Experiments took place in Taipei City Zhishan Senior Care Center. Inside, there was a room designated for the experiment to take place. Prior to the experiment, participants were recruited in a session where the objectives and scope of the experiment were introduced to a group of potential elderly participants. After a brief presentation, those interested in joining the experiment were asked to fill a contact information sheet to schedule an experiment appointment over the phone.

Participants were encouraged to submit their personal photos as it was expected that the effectiveness of the reminiscence system would be more noticeable in such cases. In the case where a participant would submit physical pictures, these were digitized and returned to them as electronic files as an additional participation perk. In the cases where participants did not have easy access to any personal photographs but were still interested in participating, they were allowed to perform the experiment with a generic photo album.

In total, we recruited 10 participants 68 to 87 years old (M = 78.6, SD = 6.72, 80% female). Their average score in the Mini-Mental State Examination (MMSE) was 29 out of 30 (scoring 25 or more represents the participants are not cognitively impaired.) Each participant submitted their personal pictures prior to the HRI experiment. Based on Taiwan's ethics and confidentiality regulations, we received approval for the experiment from the Institutional Review Board of National Taiwan University(IRB application id: 201803HS017). We then drafted the participant consent form following the IRB guidelines. All participants received a detailed explanation of the experiment procedure and signed the consent form to use and store their personal data for research purposes, both the submitted files and any audiovisual recordings of the experiment sessions.

The time allotted to each participant was of one hour. Within that time, participants engaged in a reminiscence session with the robot for up to 20 min and then performed a number of questionnaires to evaluate the performance of the robot. During the reminiscence session, participants would be presented with a digital version of the photos they provided shown on the D-FLIP interface. The robot would begin by greeting the user and verbally issue instructions on how to use the system. Participants could then select the picture of their preference and drag it towards the center of the screen. The robot would then begin asking questions about the entities in the picture (i.e. scene and event). If 5 questions were asked about the same picture, the robot would prompt the user to change the picture to create new questions.

When the time ran out, the robot was programmed to automatically terminate the session by thanking the participant for their participation in a friendly manner (Fig. 4).

Fig. 4. Participants using the photo reminiscence system.

5.2 Questionnaires

To assess the system performance in different aspects, we used 2 questionnaires: a HRI questionnaire and the Technology Acceptance Model (TAM) questionnaire [17]. The participants were asked to answer the questionnaires after the reminiscence session.

The HRI questionnaire was designed in a previous research published in [10], which aimed to assess whether the system could help the participants to recall their memories. The questionnaire was designed to evaluate the system from 4 perspectives: Relatedness, appropriateness, effectiveness, and responsiveness. There are 7 descriptions designed to evaluate the 4 perspectives, and the participants were asked to rate the descriptions on a scale of 1 to 5 (with 1 being least favorable and 5 being most favorable). This questionnaire was originally designed for another social robot, RoBoHoN. Still, as the results in the HRI Usability Evaluation section below show, performance was favorable regardless of the robot used.

The TAM Questionnaire is a commonly used questionnaire for research concerning usage intentions in terms of social influence and cognitive instrumental processes. We modified several questions so that the questionnaire would be more suitable for our reminiscence system. This questionnaire aimed to assess whether the participants would want to and would like to use the robot-assisted system. It would assess the participant's acceptance of the system from 3 perspectives: the participant's intention to use, the perceived ease of use, and the perceived usefulness. The questionnaire is comprised 3 yes-no questions and 6 descriptions rated on a scale of 1 to 5 (with 1 being least favorable and 5 being most favorable) as shown in Table 3.

Table 3. TAM questionnaire

Index	Perspective	Content
1	*Yes-no question*	Have you ever seen any robots before in real life?
2	*Yes-no question*	Have you ever seen Zenbo Jr. before?
3	*Intention to use*	If I own a Zenbo Jr, I would like to use it
4	*Intention to use*	If I own a Zenbo Jr, I think I would like to speak with it
5	*Perceived usefulness*	I find Zenbo Jr. useful
6	*Perceived ease of use*	I find it easy to interact with Zenbo Jr.
7	*Perceived ease of use*	I can communicate with Zenbo Jr. easily
8	*Perceived usefulness*	I can establish good rapport with Zenbo Jr.
9	*Yes-no question*	Are there any incidents during the experiment?

5.3   Results

Scene and Event Prediction. Table 4 presents the scene and event prediction performance on a variety of image datasets. The training and validation sets refer to those belonging to the dataset used for training the event (USED) and scene (Places365) datasets, whereas the soft and hard test sets are composed of pictures obtained from the experiment participants. The hard test set contains all photos from all participants, whereas the soft test set only includes pictures where it is rather obvious that the event or scene that appears in the picture could fit in one of the trained classes.

From the results we can observe that the trained models have acceptable performance when the test pictures belong to one of the categories it possesses, otherwise the prediction accuracy drops significantly. We hypothesize this is because of two reasons:

1. The open set problem: Naturally, the system does not perform well on classes that it is not trained to recognize. Even when the addition of a default class improved the accuracy slightly, the system still struggles with unseen classes.

Table 4. Events and scenes networks' performance

Network accuracy	Training Set	Validation Set	Soft-Test Set	Hard-Test Set
Events (VGG16)	0.881	0.835	0.822	0.587
Scenes (VGG16)	0.804	0.793	0.763	0.599

2. Image quality: Participant photos may have a variety of visual artifacts from the camera hardware (e.g. blurry, overexposed), development (e.g. exposed roll) or the pictures old age (black and white, washed out colors). These phenomena are all detrimental to the prediction accuracy. To train networks that are robust to more naturalistic data is still an open problem.

HRI Usability Evaluation. Figure 5 shows the HRI and TAM Questionnaires' results, where N is the number of participants, $\bar{x}$ and σ are the mean and standard deviation of all participants' total score, and P and T represent the statistical p-value and t-value, respectively.

The HRI Responsiveness graph shows that participants using their own photos tend to believe Zenbo Jr. is more responsive compared to participants using generic photos. The HRI Appropriateness graph shows that participants using their own photos tend to agree with what Zenbo Jr. asks. Moreover, they feel like what Zenbo Jr. asks is more likely to arouse their positive (happy) memories.

The TAM Intention graph shows that participants using their own photos have higher intentions to use the robot-assisted system. Similarly, the TAM's Total Score graph shows that participants using their own photos have higher acceptances toward the system.

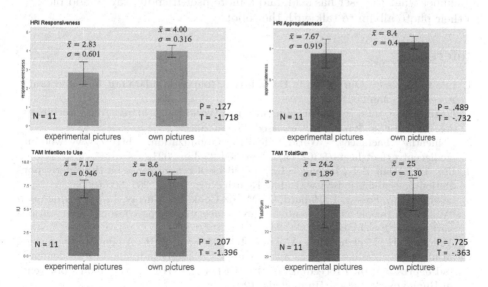

Fig. 5. HRI and TAM evaluation

Combining all the results above, though the number of the participants might not be large enough to reach statistical significance, participants using their own photos are more likely to give positive feedback on our system's human-robot interaction and are more willing to accept and use the system afterwards.

6   Conclusion

In this paper, a robot-assisted photo reminiscence system was presented. This system was designed to autonomously guide an elderly patient throughout a reminiscence session, asking questions that were related to the picture that the user chose.

The system was conformed of two pieces: A dynamic photo album interface and a social robot. The system could automatically extract visual information from the pictures using pre-trained modules to detect scenes and events. The photo album was displayed in a fluid and dynamic interface with sort and place abilities. Furthermore, the system was capable of cohesive interaction through the hybrid touch and speech interface designed.

Results showed that the performance of the image understanding modules was excellent on the testing dataset; however, due to the intrinsic open-endedness of the system, performance on real participant data was still deficient. From the questionnaire evaluation, it was shown that most of the participants found the system responsive and the questions asked appropriate for the picture selected. This effect was boosted when participants used their own pictures. Furthermore, elderly participants in general praised the system, finding it interesting and effective at helping them talk about the pictures.

The system could benefit from further improvements, such as a more sophisticated NLU engine that could parse keywords from the users' speech, a way to remember what the user has said, and a more user-friendly way to add pictures to their photo album to talk with the robot.

References

1. Life Expectancy, Our World in Data. https://ourworldindata.org/life-expectancy. Accessed 25 Jan 2021
2. Fertility rate, total (births per woman), The World Bank. https://data.worldbank.org/indicator/sp.dyn.tfrt.in. Accessed 25 Jan 2021
3. Dementia Factsheet, World Health Organization. http://www.who.int/mediacentre/factsheets/fs362/en/. Accessed 25 Jan 2021
4. Lin, Y.C., Dai, Y.T., Hwang, S.L.: The effect of reminiscence on the elderly population: a systematic review. Public Health Nurs. **20**(4), 297–306 (2003)
5. Lawrence, V., Fossey, J., Ballard, C., Moniz-Cook, E., Murray, J.: Improving quality of life for people with dementia in care homes: making psychosocial interventions work. Br. J. Psychiatry **201**(5), 344–351 (2012)
6. Abdollahi, H., Mollahosseini, A., Lane, J.T., Mahoor, M.H.: A pilot study on using an intelligent life-like robot as a companion for elderly individuals with dementia and depression. In: Proceedings of the IEEE-RAS 17th International Conference on Humanoid Robotics (Humanoids) (2017)

7. Tokushige, A., Yokojima, K., Sugiura, K., Iwasaki, Y., Araki, D.: Consideration of conversation with older adult with dementia by individual reminiscence therapy. In: 30th International Nursing Research Congress (2019)
8. Carós, M., Garolera, M., Radeva, P., Giro-i-Nieto, X.: Automatic reminiscence therapy for dementia. In: Proceedings of the 2020 International Conference on Multimedia Retrieval (ICMR 2020), pp. 383–387. Association for Computing Machinery, New York (2020). https://doi.org/10.1145/3372278.3391927
9. Asprino, L., Gangemi, A., Nuzzolese, A.G., Presutti, V., Recupero, D.R., Russo, A.: Ontology-based knowledge management for comprehensive geriatric assessment and reminiscence therapy on social robots. Data Science for Healthcare, pp. 173–193. Springer, Cham (2019). https://doi.org/10.1007/978-3-030-05249-2_6
10. Wu, Y.L., Gamborino, E., Fu, L.C.: Interactive question posing system for robot-assisted photo reminiscence from personal photos. IEEE Trans. Cogn. Dev. Syst. **12**, 439–450 (2019)
11. Kitamura, Y., Vi, C.T., Liu, G., Takashima, K., Itoh, Y., Subramanian, S.: D-FLIP: dynamic & flexible interactive PhotoShow. In: SIGGRAPH Asia 2013 Emerging Technologies (SA 2013), Article 6, pp. 1–3. Association for Computing Machinery, New York (2013). https://doi.org/10.1145/2542284.2542290
12. Huang, X., Takashima, K., Fujita, K., Kitamura, Y.: Dynamic, flexible and multi-dimensional visualization of digital photos and their metadata. In: Proceedings of the 2018 ACM International Conference on Interactive Surfaces and Spaces (ISS 2018), pp. 405–408. Association for Computing Machinery, New York (2018). https://doi.org/10.1145/3279778.3279923
13. Thepsoonthorn, C., Ogawa, K., Miyake, Y.: The relationship between robot's non-verbal behavior and human's likability based on human's personality. Sci. Rep. **8**, 8435 (2018). https://doi.org/10.1038/s41598-018-25314-x
14. Zhou, B., Lapedriza, A., Khosla, A., Oliva, A., Torralba, A.: Places: a 10 million image database for scene recognition. IEEE Trans. Pattern Anal. Mach. Intell. **40**, 1452–1464 (2017)
15. Ahmad, K., Conci, N., Boato, G., De Natale, F.G.: USED: a large-scale social event detection dataset. In: Proceedings of the 7th International Conference on Multimedia Systems (MMSys 2016), Article 50, pp. 1–6. Association for Computing Machinery, New York (2016). https://doi.org/10.1145/2910017.2910624
16. Kalliatakis, G.: Keras-VGG16-Places365 (2017). https://github.com/GKalliatakis/Keras-VGG16-places365. Accessed 5 May 2020
17. Venkatesh, V., Davis, F.D.: A theoretical extension of the technology acceptance model: four longitudinal field studies. Manag. Sci. **46**(2), 186–204 (2000)

Lost in Interpretation? The Role of Culture on Rating the Emotional Nonverbal Behaviors of a Virtual Agent

Adineh Hosseinpanah[✉] and Nicole C. Krämer

University of Duisburg-Essen, Forsthausweg 2, 47057 Duisburg, Germany
{adineh.hosseinpanah,Nicole.kraemer}@uni-due.de

Abstract. When designing a virtual agent, implementing empathy is a factor that can be considered to maximize the quality of human-agent interaction. However, in a globalized world, it is important to scrutinize whether users' culture affects the perception of empathic agents. This study investigated the role of users' cultural background in the perception of emotional nonverbal behaviors of a virtual assistant as empathic. It also examined which displayed emotional nonverbal behaviors affect different users' perception of empathy. In an online study, 200 Iranian and German participants rated the responses of a virtual assistant on five items: friendliness, intelligence, empathy, trustworthiness, and helpfulness. The design of the study was a 2 between-subject factor (culture: Iranians vs. Germans) × 2 within-subject factor (context: sad situations vs. happy situations) × 4 within-subject factor (nonverbal behavior for each situation: 3 emotional vs. 1 neutral) mixed factorial design. The findings indicated that when the emotional nonverbal behaviors are present, Iranians rated the agent as more empathic, trustworthy, and helpful. Also, while Iranians perceived a wide array of nonverbal behaviors as empathic (Smile, Head Nod plus Smile, Head Nod, Sad Face, Head Down, and Dropping the Arms plus Sad Face), German participants only assessed Head Nod as empathic.

Keywords: Empathic agents · Cultural differences · Nonverbal behavior

1 Introduction

Supporting technologies like agents and robots are increasingly entering and influencing the everyday life of people. This makes the human-machine interaction more and more inevitable, and in order to enhance and sustain the quality of such an interaction, we need to have a better understanding of it. Moreover, to use the potential of these technologies, it is crucial to maximize the quality of the users' experience. To this end, a certain level of believability and trustworthiness of the agent is required.

Empathic agents have been discussed to enrich the human-agent interaction [1]. It has been long discussed that implementing empathy in the agents and robots leads to users' positive rating in the sense of trustworthiness, likeability, and caring [2]. Empathic feedback also reduces negative feelings and the stress level of the users [3, 4] and can

© Springer Nature Switzerland AG 2021
P.-L. P. Rau (Ed.): HCII 2021, LNCS 12773, pp. 350–368, 2021.
https://doi.org/10.1007/978-3-030-77080-8_28

lead to less frustration in the users [5, 6], higher perception of friendliness [7] as well as a sustainable long-term interaction [8].

In the process of implementing empathy in the agents, nonverbal behaviors can play a crucial role in making the believability of the agent more salient. However, previous studies have not dealt with the question of which emotional nonverbal behaviors can facilitate the perception of empathy in the users. Moreover, as the agents will be interacting with different groups of users with various cultural backgrounds, it is crucial to consider that the meaning of emotional situations and the way people express them may vary from one culture to another [9]. The enculturation process in each society plays a key role in the way people show their emotions and mental states both verbally and nonverbally [10]. It also can affect the perception and expectations they have of agents and robots, depending on how the usage of these entities is common in those cultures [11].

Several studies have taken the cultural background of the users into account when implementing nonverbal behaviors in the virtual agent [12–14]. However, far too little attention has been paid to the influence that users' culture can have on their perception of emotional nonverbal behaviors of the agent and perceiving them as empathic. This can substantially impact the users' feedback and, consequently, the quality of their interaction with the agent.

In this study, we investigate whether cultural differences and their effects on the ability to understand the mental states of others, which referred to Theory of Mind [1], influence users' perception of the emotional nonverbal behaviors of a virtual agent as empathic. Furthermore, one of the main objectives of this study is to investigate which presented nonverbal behaviors are considered as the most empathic ones. So far, many studies [15, 16] have focused on user groups of very typical examples of individualistic and collectivistic societies like the United States and China. In the current study, we selected Germany and Iran as two culturally different societies. Although they have high rankings in individualism and collectivism, respectively [17–19], they are discussed as not the typical individualistic and collectivistic societies as the United States and China are [20, 21]. Therefore, the state of the art is extended by focusing on Western and Eastern cultures, which are not at the extreme ends of the individualism-collectivism dimension.

Moreover, when it comes to investigating the cultural differences in the domain of human-machine interaction, user groups from the Middle East are rarely taken into the investigation. As the human-machine interaction is appearing in the Middle Eastern countries as well [22] and the attitude towards acceptance of robots and agents is mostly positive among these societies [23], it is important to focus on cultural features of these rarely studied societies. This contributes to gaining more knowledge regarding user experiences in human-machine interaction.

Taking into account the cultural differences between Iran and Germany, we aim to scrutinize whether the interpretation of the emotional responses of a virtual assistant will be different in the sense of perceiving them as empathic. This is the first study in which a wide array of nonverbal behaviors of a virtual agent is tested for their cultural effects. Therefore, it is a start for understanding whether culture actually has an impact when perceiving specific nonverbal behaviors of agents.

2 Theoretical Background

2.1 Empathy and Theory of Mind

Understanding other people is a key factor in successful and efficient interactions. In this regard, Grice [24] stated that in order to be able to successfully understand an utterance, we must be able to perceive the intention of the speaker. One prerequisite for such a perception is the ability to figure out and understand that the internal mental states, namely desires, beliefs, and emotions, have an effect on human actions. Having this ability is referred to Theory of Mind (ToM) and is considered as a fundamental prerequisite for human-human interaction [25]. Empathy is the ability to understand and identify what others feel and which emotional states they have [26]. It is stated that the development of empathy is influenced by having the ability of Theory of Mind [27–29].

ToM is a prerequisite for human-human communication [30]. Implementing ToM in human-agent interaction can make a more fruitful interaction by enabling the interlocutors to understand and predict each others' actions. In this paper, we focus on the ToM of the human users and how it affects their perceptions and understanding of an empathic virtual agent.

Empathy is categorized into two types: cognitive empathy and affective empathy [31]. Affective empathy refers to the ability to be sensitive and share the emotional states of others by showing, e.g., visceral reactions. In contrast, cognitive empathy is described as the ability to take others' mental perspective, which leads to understanding the emotions of others [31]. As the focus of the current study is to figure out whether the users understand the mental state of the virtual agent, which is empathizing with them, the cognitive dimension of ToM; hence, cognitive empathy is of interest.

2.2 Nonverbal Behaviors in Virtual Agents

Nonverbal behavior is a crucial aspect of human-human communication. It has been demonstrated that adults rather interpret messages based on nonverbal than verbal information [32]. The impact of showing nonverbal behavior has also been shown in perception and judgment of empathy of others [33]. Taking into account its vital role in human-agent interaction, Krämer and colleagues [34] have found its effect on the user's experience and evaluation of an embodied conversational agent. It has been demonstrated that even subtle nonverbal cues matter and have an impact on the perception and evaluation of an embodied agent [34]. Moreover, the effect of nonverbal behaviors such as human-like gaze [35, 36] head position and increased head movements [37], and the presence of facial expressions [38] has been shown in creating a more positive evaluation of the agent, in this sense that the agent is natural, useful and involved.

Previous research has shown that different target groups also affect the perception of an agent's nonverbal behavior. For instance, age was investigated in a study to see whether it has an effect on the perception of emotional nonverbal behaviors as empathic [39]. The findings of this study indicated that, indeed, person variables could be influential in the sense that young and elderly participants differed regarding the perception and rating of emotional nonverbal behaviors.

Several attempts have also been made to consider cultural aspects in implementing nonverbal behaviors into a virtual character. However, most of these studies focused on the perception of other aspects of nonverbal behavior, such as the spatial extent of gestures [40], proxemics, gaze, and overlap in turn-taking [15] than emotional nonverbal behaviors. The effect of cultural differences on the interpretation of emotional facial expression was investigated by Koda and Colleagues [16]. The results of this study demonstrated that the users decoded and perceived the facial expressions of the agent more precisely when it was designed in their own culture. This study, however, has emphasized the culture of the agent and how the perception of the people differed depending on that. Since the nonverbal behavior can contribute to a better understanding between the interlocutors, we aimed to investigate whether users' culture influences their perception of emotional nonverbal behaviors of an empathic agent.

2.3 Cultural Differences in Theory of Mind

Although the ability to have a ToM is proved to be a universal capacity, culture plays a vital role in when and how people understand the mind of others [41]. There are several environmental factors that influence the development of ToM, and consequently, the ability to empathize and understand the empathic responses of others. Some of these factors are parenting style [42], their language utilization [43], and conversations referring to emotions [44], which all can be influenced by the culture of people and the dimension of whether the culture is collectivist or individualistic.

Humans' preference for emotional experience, expression, and recognition is dependent on their cultural background [45]. Individualism/Collectivism is one of the most important dimensions of cultural differences with regard to values, perceptions, and behaviors of people, which were first introduced by Hofstede [46]. These dimensions refer to the level of whether people prefer loosely or tightly knit social networks [46]. According to Matsumoto and Juang [47], the extent to which a culture puts more emphasis on individuals' desires and preferences over groups' or vice versa is defined by the dimension of individualism/collectivism. In individualistic societies, the interests of the individual are more important than the interests of the group. In collectivistic societies, it is the other way around, meaning that the social behaviors of the people are determined by the values of the group, and it is norms, duties, and obligations that guide the behaviors of people [46, 48].

The impact of this dimension has been shown in the human mental process, such as perception, emotion, and cognition [49]. Several studies have focused on investigating the role of collectivism and individualism in the development of ToM. For example, it has been indicated that in individualistic societies like the United States and Australia, people have "independent views of personhood," which nurture children to shape primary conceptualizations of ToM by thinking for themselves more independently, framing their own ideas, expressing their thoughts more freely and being involved in discussions [50, 51]. On the other hand, in a collectivistic society like China, many values like filial respect, conformity to the cultural norms and traditions rather than expressing ones' own opinions in an independent way are taught by parents [50, 51]. This, as a result, leads to a different conceptualization of ToM in collectivistic societies, with the main concept of the mind being focused on the insight that people have the capacity to be acknowledgeable than to be assertive in their different opinions and beliefs [52].

Cultural variation has been reported to have an effect on developing different cognitive processing modes [53]. It has been demonstrated that cultural values of individualism and collectivism, e.g., self-construal style, affect the visual perception and experience of people. This can suggest that the culture of people plays a role in the way they use any cues for the perception of (emotional) states of others. Especially when it comes to perceiving certain nonverbal behaviors as more or less empathic, various visual perceptions might affect the way people use different aspects and cues in order to facilitate such perception.

3   Empathy in Collectivistic and Individualistic Cultures

The acquisition of ToM and the ability to empathize and understand other people's empathic behavior are related to each other, as both rely on inferring the mental states of others [54]. Therefore, knowing the cultural differences within the dimension of collectivism/individualism and how it affects the ability of ToM can contribute to a better understanding of how empathic responses and perception may differ in a specific culture and why people interpret certain behaviors as more or less empathic.

Studies regarding empathy indicated the potential influence of culture on the level of expressed empathic behavior and how it is being evaluated [55]. While we all recognize universal nonverbal emotions, cultural influences contribute to an accurate judgment of the emotions of others [56]. As Chopik and colleagues [55] demonstrated, the society being individualistic or collectivistic affects the empathy level of people. They indicated that collectivists value and experience empathy more than individualists. Further, it has been shown that a higher level of collectivism was linked to a higher level of empathy [57]. One explanation is that compared to collectivistic societies, people from individualistic societies have more independent views, leading them to possess more positive self-view and weaker motivations for conformity [49].

According to Bedford and Hwang [58], the different moral education and caregiving attitudes are different in various cultures and can influence the people's empathic behavior. Different socialization of these societies can also impact how people interpret others' emotions and what kind of expectations they have in particular emotional situations. Therefore, perception and interpretation of others' suffering can be formed by culturally-constructed meanings [10]. How emotional feelings are expressed, appraised, and perceived might vary, depending on the cultural-relevant norms saying which emotions should or should not be expressed [47]. For example, as people are more concerned with good interpersonal relationships in collectivistic societies, it has been indicated that they put more emphasis on social norms and how one should behave in different social situations [59].

The ability to perceive and correctly interpret others' emotions and share an emotional response in return enables people to act appropriately in different social and emotional contexts. Considering that people from collectivistic societies have more conformity with cultural norms and tradition and that caring is a value in these cultures, they might have a higher expectation of receiving an emotional response from an interactive partner. However, this might be not the case for individualistic societies. Therefore, the question here would be whether the different level of empathic behavior present in collectivistic and individualistic societies can influence the understanding of empathic responses in human-agent interaction.

4   The Case of Iran and Germany

For the collectivistic sample, we selected Iranian participants as Middle-Easern examples. However, it is important to note that they have less collectivistic values than people from China [21, 60]. We compared Iranians with German participants, as the Western sample with individualistic tendencies, though less individualistic than the typical example of the United States [20, 61]. It has been reported that the communication among Germans is mostly direct, in the sense that they would value honesty even if it hurts, and they are relatively independent of others and have more emphasis on their own preferences and needs other than the group's [19]. Other cross-cultural studies also suggested Germany as a country with more individualistic values, resulting in less orientation toward obedience as an initial objective of the socialization [62].

On the other hand, the collectivistic tendencies of Iranians emphasize their interdependency with others and connectedness with social values and norms [63], interpersonal relationships and group harmony [64], and learning to be sensitive about the emotions of others from an early age [25]. It has been indicated that in Iran, it is of great importance to maintain relationships and sacrifice one's own need for the sake of group's purposes, and to this end, being sensitive to others' communicative intentions is considered necessary [65]. All these norms and differences, as can be seen in parenting styles and strategies for children's upbringing in these two societies, affect the development and having the ability of ToM [25]. Consequently, this can play a role in understanding other peoples' emotions and realizing whether they are empathizing or not [52, 66].

5   Empathic Agents in Different Cultures

Several studies have focused on the cultural differences regarding the empathic behaviors of agents and providing an environment in which the users could empathize with the agents. Degens and colleagues [66] presented culture as one of the requirements in designing an agent for enabling it to act in an empathic manner. They proposed certain requirements to create a conceptual model for an empathic agent's social behaviors, which can be modified by culture. In their model, the empathic agent's understanding of the social effect of the actions is necessary, and culture influences the behaviors of the agent through social norms and moral circles.

"FearNot" is an application that is used for anti-bullying goals [67]. It provides a school-based virtual learning environment where several virtual characters represent different bullying scenarios [67]. This application considered cultural differences in two societies, UK and Germany, and was designed to develop social agents that the users could have empathic engagement with. Moreover, ORIENT was presented by Aylett and colleagues [14] as a semi-immersive graphical environment to develop a believable agent for educational purposes and encourage intercultural empathy in the users. These studies focused on modeling different cultural environments to trigger empathy in the users or change their behaviors regarding desired purposes. One question that still needs to be asked is whether the agent's specific emotional nonverbal behaviors, independent of which culture the agent represents, lead to the different perception of empathy in the users who have a diverse cultural background. It is also important to know if the culture influences whether people appropriately recognize that the virtual agent is empathizing with them.

6 Hypothesis and Research Questions

As depicted in the theory section, enculturation process in different societies, depending on the society being collectivistic or individualistic, might influence the development and conceptualization of Theory of Mind in people. This can affect the understanding and perception people have regarding the emotional responses of others.

Accordingly, the specific question in this study is whether Germans and Iranians, with individualistic and collectivistic tendencies, differ in the perception of the emotional responses of the virtual agent. This perception is independent of which emotions are recognized and it is just focused on rating certain emotional responses as more or less empathic. Therefore, we seek to test and address the following hypothesis and research questions:

H: We assume that in the presence of emotional nonverbal behaviors, the perception of empathy is different between Iranians and Germans.

Further, we would like to investigate whether the cultural differences affect the perception of certain displayed nonverbal behaviors as more empathic:

RQ 1: Which emotional nonverbal behaviors of the agent are perceived as more empathic by Germans and Iranian participants?

And finally, we strive to test whether the cultural background of the users also has an impact on the overall rating of the agent, as indicated by friendliness, intelligence, trustworthiness, and helpfulness, when the agent displays emotional nonverbal behaviors:

RQ 2: Do Iranians and Germans, in the presence of emotional nonverbal behaviors; rate the agent differently with regard to being friendly, intelligence, trustworthy and helpful?

7 Methods

7.1 Material

Nonverbal Behaviors: The materials we used in this study were similar to our previous study investigating the factor of age on rating an empathic agent [39]. All the materials were translated from German into Persian. In order to be sure of the consistency of all translations, the materials were back-translated to German as well.

The same embodied conversational agent, Billie (Social Cognitive System Group, Citec Bielefeld, Germany), was used in this study, a cartoon-like character with a neutral appearance, white skin tone, dark hair and no specific aspects of race so neither particularly German nor Iranian looking. Also, the participants had no prior knowledge of where the agent was designed. The nonverbal behaviors implemented in the agent were created by using Key-frame editor. As the Initial gesture, the agent's hands were rested in front of the stomach and the fingers were tangled (Fig. 1). Six emotional nonverbal behaviors, including three happy emotions and three sad emotions were created to show appropriate emotional responses to related emotional contexts (Fig. 1).

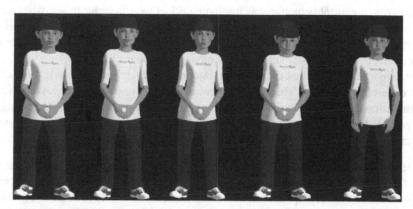

Fig. 1. The nonverbal behaviors from left to right: neutral behavior, Smile, Sad Face, Head Down, and Dropping the Arms plus Sad Face.

When developing the nonverbal behaviors, some of the facial expressions like Sad Face and Smile (see below) were inspired by the emotional avatars and empathic agents in studies by [16]. However, for the purpose of this study and in order to test whether the different intensity of the displayed nonverbal behaviors affects the perception of an empathic agent, different nonverbal behaviors with head and body movements were created as well. All these nonverbal behaviors were created and evaluated by researchers from both Germany and Iran during the designing process. The features of the emotional nonverbal behaviors are as follows:

Happy Nonverbal Behaviors

- **Smile:** the sides of the mouth rise, making the cheeks be pushed up to display happiness. The duration is three seconds.
- **Head Nod:** the head is tilted in alternating up and down arcs to display acknowledgment and interest in what has been said. The duration is two seconds.
- **Head Nod plus Smile:** The combination of one-time head-nodding accompanied by the smiling face, to show a more intensive acknowledgment and bonding signal. The duration is four seconds.

Sad Nonverbal Behaviors

- **Sad Face:** the inner corners of the eyebrows move upward, and the upper eyelids are drooped. At the same time, the corners of the lips move downward, making the chin looks raised to display sadness. The duration is four seconds.

- **Head Down:** the agent lowers its head and moves its eye gaze downward, stays in this position for four seconds, and moves it up again. The total duration is five seconds.
- **Dropping the Arms plus Sad Face:** the agent shows the sad face and drops its arms to its sides simultaneously to display sadness. It stays in the dropped-arms position for four seconds and then goes back to the initial pose. The total duration is five seconds.

Moreover, a neutral behavior was implemented as well in order to see whether the participants' rating of the agent differs in the presence and the absence of the emotional nonverbal behaviors. In the neutral behavior, the agent showed neither body movement nor facial expression but only subtle idle behavior to appear alive.

It is important to note that there was no intention of matching the nonverbal behaviors in different conditions. The point of creating different nonverbal behaviors was to evaluate different possible emotional responses, from a more intense one (Dropping the Arms plus Sad Face) to a softer response (Head Nod) to see whether there are specific preferences regarding the intensity of the behaviors as well.

Context: In total, six emotional situations, including three positives and three negatives, were uttered by the agent. These situations were selected from the results of an online pre-test in which 30 Participants (Germans N: 15, Iranians N: 15) rated the emotional valence of 12 different sentences on a 5-point Likert-scale, which was from I totally agree (highest score: 5) to I totally disagree (lowest score: 1). The most common emotional situations as selected by the participants are as follows:

Happy Situations

- On Thursday, July 11th, at 19:00, you will go to the concert of your favorite singer (Positive rate: GER: $M = 4.33$, $SD = 1.04$, IR: $M = 4.60$, $SD = 1.01$).
- On Wednesday, July 24th, at 17:00, you will have a reunion with your old friend (Positive rate: GER: $M = 4.20$, $SD = .56$, IR: $M = 4.33$, $SD = .78$).
- On Saturday, July 27th, at 11:00, you will go on a picnic with your best friends (Positive rate: GER: $M = 4.27$, $SD = .59$, IR = 4.60, $SD = .80$).

Sad Situations

- On Tuesday, July 9th, at 14:00, you will visit your friend in the hospital who suffers from cancer (Negative rate: GER: $M = 3.87$, $SD = 1.35$, IR: $M = 4.33$, $SD = 1.13$).
- On Tuesday, July 16th, at 11:00, you will have a doctor's appointment for your upcoming operation (Negative rate: GER: $M = 4.13$, $SD = .83$, IR: $M = 3.80$, $SD = 1.27$).
- On Sunday, July 28th, at 19:00, you will hold a speech at the funeral of your friend (Negative rate: $M = 4.47$, $SD = .64$, IR: $M = 3.40$, $SD = 1.30$).

7.2 Participants

Of the 200 participants that were recruited online, 100 were Iranians, 46 males and 54 females between the age of 17 to 39 ($M = 29$, $SD = 4.0$), and 100 were Germans,

45 males and 55 females between the age of 18 to 43 ($M = 24$, $SD = 5.6$). Prior to conducting the study, ethical clearance was obtained from the ethics committee of the University of Duisburg-Essen. Participation in this study was voluntary and anonymous.

7.3 Procedure

On the starting page of the study, the participants were informed about the agent and were instructed about the further procedures. The first video was an introductory video in which the agent explained that its role is to review their imaginary monthly schedule. The six emotional situations were uttered by the agent and presented to the participants as videos. In order to assign all emotional nonverbal behaviors to the related context, the agent repeated the situations randomly and showed different context-related emotional nonverbal behaviors. Therefore, all happy nonverbal behaviors and the neutral behavior were assigned to happy situations, and all sad nonverbal behaviors and the neutral behavior were assigned to the sad situations. Finally, 25 videos, including the introductory video, were presented to the participants, and each had a duration of 12 s on average.

After watching each video, the participants had to fill in the questionnaires to rate the agent on five items: I feel that Billie is Friendly, Intelligent, Empathic, Trustworthy, and Helpful. The scale of the questionnaire was a 5-point Likert scale, ranging from I totally agree (highest score: 5) to I totally disagree (lowest score: 1). On the last page of the study, the participants were debriefed about the purpose of the study and were given an option to participate in a raffle of Amazon gift cards.

8 Results

In a 2 between-subject factor (culture: Iranians vs. Germans) × 2 within-subject factors (context: sad situations vs. happy situations) × 4 within-subject factors (nonverbal behavior for each situation: 3 emotional vs. 1 neutral) mixed factorial design, multiple ANOVAs were conducted to analyze the ratings of the friendliness, intelligence, empathy, trustworthiness, and helpfulness separately. To analyze the interaction effects between the between and within-factors, separate repeated ANOVAs were conducted for each target group. Post-hoc test using Bonferroni correction was used in all cases. When the assumption of sphericity was violated, Greenhouse-Geisser correction was reported. All effects are reported as significant at $p < .05$.

With regard to rating the item empathy, the results showed a significant main effect of the culture on rating the item empathy $F(1, 198) = 22.85$ $p < .001$, $\eta^2 = .10$. The Iranian participants rated the agent as more empathic than their German peers (Table 1).

Table 1. Means and standard errors of the two groups in rating the item empathy.

Groups	M	Std. error
Iranians	3.20	.72
Germans	2.72	.72

Moreover, a significant effect of the nonverbal behaviors on rating the item empathy was observed F (2.42, 479.4) = 29.22, $p < .001$, $\eta^2 = .13$. This indicates that the perception of empathy across all participants was affected by observing the nonverbal behaviors. Contrasts were performed comparing each emotional nonverbal behavior to the neutral behavior for both groups. These revealed that compared to the neutral behavior, all emotional nonverbal behaviors affected the perception of the participants of the item empathy (Table 2 and Table 3).

Table 2. Statistical values of the performed contrasts for both groups' emotional nonverbal behaviors (empathy).

Nonverbal behavior	F	df	p	η^2
Smile	17.14	1,199	<.001	.08
Head Nod plus Smile	37.04	1,199	<.001	.16
Head Nod	8.74	1,199	=.003	.04
Sad Face	18.92	1,199	<.001	.09
Head Down	37.54	1,199	<.001	.16
Dropping the Arms plus Sad	36.66	1,199	<.001	.16

Table 3. Means and standard deviations of the nonverbal behaviors independent of the culture.

Nonverbal behavior	M	SD
Dropping the Arms plus Sad	3.12	.97
Head Down	3.15	.96
Sad Face	3.01	1.0
Head Nod plus Smile	3.08	.93
Smile	2.97	.83
Head Nod	2.87	.87
Neutral	2.75	.80

Further, there was a significant interaction effect (culture × nonverbal behavior) F (2.42, 479.4) = 28.77, $p < .001$, $\eta^2 = .13$, suggesting that perceiving the agent as empathic based on the observed nonverbal behaviors was different between the Iranian and German participants. This confirms the Hypothesis of the study (H). To break down this interaction, contrasts were performed comparing each emotional nonverbal behavior to the neutral behavior for each group separately. Results revealed that all emotional nonverbal behaviors affected the Iranians' rating of the item empathy (Table 4 and Table 5).

Table 4. Statistical values of the performed contrasts for Iranians (empathy).

Nonverbal behavior	F	df	p	η^2
Smile	53.26	1,99	=.001	.35
Head Nod plus Smile	54.26	1,99	=.001	.36
Head Nod	4.31	1,99	=.04	.36
Sad Face	50.96	1,99	=.001	.34
Head Down	71.89	1,99	<.001	.42
Dropping the Arms plus Sad	58.87	1,99	<.001	.37

When performing the same contrast for the German participants, results revealed that they rated Head Nod as more empathic: $F(1, 99) = 4.46, p = .037, \eta^2 = .04$ (Table 5).

Table 5. Means and standard deviations of nonverbal behaviors for both groups (empathy).

Nonverbal behavior	Iranians		Germans	
	M	SD	M	SD
Dropping the Arms plus Sad	3.50	.93	2.73	.86
Head Down	3.51	.85	2.79	.94
Sad Face	3.41	.95	2.61	.88
Head Nod plus Smile	3.40	.83	2.77	.93
Smile	3.33	.71	2.62	.78
Head Nod	2.91	.82	2.82	.93
Neutral	2.79	.76	2.71	.85

With respect to rating the other four items, the cultural background was found to have no significant main effect on participants' rating on agent's friendliness and intelligence. However, there was a significant main effect of the culture on rating the trustworthiness of the agent $F(1, 198) = 18.58, p < .001, \eta^2 = .09$. The Iranian participants rated the agent as more trustworthy than Germans (Table 6).

Table 6. Means and standard errors of the two groups in rating the item trustworthy.

Groups	M	Std. error
Iranians	3.27	.08
Germans	2.77	.08

However, no significant interaction effect was found (culture × nonverbal behavior), indicating that the emotional nonverbal behaviors did not affect two groups' rating of the trustworthiness differently.

There was also a significant main effect of the participants' culture on rating the item helpful F (1, 198) = 7.46, p = .007, η^2 = .04. The results revealed that Iranians rated the agent as more helpful compared to Germans (Table 7).

Table 7. Means and standard errors of the two groups in rating the item helpful.

Groups	M	Std. error
Iranians	3.55	.75
Germans	2.27	.75

The nonverbal behaviors were found to have a significant effect on the rating the item Helpful in the participants F (2.81, 557.3) = 3.33, p = .022, η^2 = .02. Contrasts revealed that compared to the neutral behavior, Smile and Dropping the Arms plus Sad Face affected the perception of helpfulness (Table 8 and Table 9).

Table 8. Statistical values of the performed contrasts for both groups (helpful).

Nonverbal behavior	F	df	p	η^2
Smile	10.20	1,199	=.002	.05
Dropping the Arms plus Sad	3.96	1,199	=.048	.02

Table 9. Means and standard deviations of the nonverbal behaviors for both groups (helpful).

Nonverbal behavior	M	Std. deviation
Smile	3.48	.81
Dropping the Arms plus Sad	3.40	.85
Neutral	3.37	.80

The data also revealed a significant interaction effect (culture × nonverbal behavior) F (2.81, 557.3) = 5.68, p = .001, η^2 = .03. This suggests that the nonverbal behaviors affected the rating of Iranians and Germans of the item Helpful differently. Contrasts were performed for each group separately to compare each emotional nonverbal behavior to the neutral behavior. Results indicated that compared to the neutral behavior, Iranians perceived the agent as more helpful when they observed Smile, Head Nod plus Smile, Sad Face, and Head Down (Table 10 and Table 11).

Table 10. Statistical values of the performed contrasts for Iranians (helpful).

Nonverbal behavior	F	df	p	η^2
Smile	16.37	1,99	<.001	.14
Head Nod plus Smile	6.88	1,99	=.010	.06
Sad Face	4.94	1,99	=.028	.05
Head Down	7.60	1,99	=.007	.07

When performing the same contrasts for the German participants, no emotional nonverbal behavior was found to have an effect on the perception of helpfulness (Table 11).

Table 11. Means and standard deviations of the nonverbal behaviors for both groups (helpful).

Nonverbal behavior	Iranians		Germans	
	M	SD	M	SD
Dropping the Arms plus Sad	3.54	.81	3.23	.86
Head Down	3.62	.80	3.26	.86
Sad Face	3.59	.76	3.22	.85
Head Nod plus Smile	3.62	.81	3.26	.84
Smile	3.69	.75	3.28	.82
Head Nod	3.48	.85	3.32	.89
Neutral	3.46	.83	3.28	.76

In order to see whether Iranians' higher ratings of trustworthiness and helpfulness of the agent were correlated with the higher rating of the empathy, a linear regression was conducted. The results revealed that the rating of the item trustworthy can be predicted by participants' rating of the item empathy by 21%, $R^2 = .21$, $F(1,98) = 26$, $p < .001$, $\beta = .458$, $p < .001$.

Moreover, a significant correlation was found between the higher rating of empathy and the higher rating of helpfulness for Iranian participants $F(1,98) = 44.17$. $P < .001$, $\beta = .557$, $p < .001$, indicating that the rating of helpfulness can be predicted by 31%, $R^2 = .31$ by participants' rating of the item empathy.

9 Discussion

The results of this study indicated that the participants' cultural background indeed influences rating the emotional nonverbal behaviors of the agent as empathic, confirming the hypothesis of the study (H). When the agent displayed emotional responses, Iranian participants perceived it as more empathic than their German peers.

This finding can be explained by cultural values existing in Iranian upbringings. In Iranian culture, the upbringing values interpersonal similarities and harmony, respect and politeness, and people learn from an early age to be highly sensitive to others' emotions to maintain relationships [66]. In this sense, observing a non-responsive behavior of an interlocutor might be interpreted as impoliteness, which does not satisfy peoples' expectation of a harmonized communication.

One other possible explanation might be that Iranian participants with their collectivistic values tend to care about the needs of others, which might lead to an expectation to receive empathic responses in emotional contexts. Due to this expectation, it seems plausible that they paid more attention to the displayed emotional nonverbal behaviors of the agent, and hence, perceived them as more empathic. However, future research will have to more explicitly analyze whether these differences can be explained by a more collectivistic nature of the Iranian participants.

Rating the agent as less empathic might be a reflection of the rather individualistic values of German culture. Independency of the members of individualistic societies and their less other-oriented sense of well-being might have led the German participants to not rate the displayed nonverbal emotions as empathic, as they might not have felt the need for empathic responses. However, these assumptions need to be tested in further research by, for example, assessing both, the expectancy to see emotional responses and the tendency to care for others' needs in the questionnaire.

As was pointed out in RQ1, it was our interest to see whether there were cultural differences with respect to rating various emotional nonverbal behaviors as more empathic. The data indicated that in comparison with the neutral behavior, all displayed emotional nonverbal behaviors were rated as empathic by the Iranian participants. These nonverbal behaviors were Smile, Head Nod plus Smile, Head Nod, Sad Face, Head Down, and Dropping the Arms plus Sad Face. In the German sample, only the nonverbal behavior Head Nod was rated as more empathic when compared to the neutral behavior. This, again, indicates that Iranians were more susceptible to (empathic) nonverbal behavior than Germans were. While the former readily perceived nonverbal behaviors to be empathic, Germans only evaluated the nonverbal behavior as most empathic, which could be described as the most rational, matter-of-fact. Based on this difference, it can be assumed that the intensity of the emotions displayed by the agent might as well have an influence on the perception and preferences of the users with different cultural background. The diverse intensity of displayed responses can be considered when designing an agent for users with different cultures.

As far as the overall rating of the agent was concerned (RQ2), the results indicated that when the emotional responses were present, the two groups' ratings of the agent were not different for the case of friendliness and intelligence. For the other items, however, it was shown that Iranians rated the agent as more trustworthy and helpful compared to Germans. As the results indicated a correlation between the rating of empathy and these two items, this can be interpreted as a side effect of perceiving the agent as more empathic by the Iranians. They might have perceived the agent not only as more empathic but in line with this as more helpful and trustworthy as the agent seemed to "care" for them. However, as there was no interaction effect between the participants' culture and the observed nonverbal behaviors for the item trustworthy, it needs to be interpreted with caution.

These findings have important implications for developing a more successful human-agent interaction by focusing on the important role of culture to assure the effectiveness of virtual agents' empathic nonverbal behaviors. It can be stated that the perception of empathic nonverbal behavior is sensitive to the cultural background of the people and taking it into account can lead to a better and more precise design guideline. Considering how differently people are nurtured in different cultures by experiencing specific norms, behaviors, and values, a culture-specific virtual agent, which matches the users' cultural expectations, can enhance the quality of the interaction with the virtual agent. This can be specifically considered when the intention is to influence the user's actions.

10 Conclusion

This study has brought insight into how the cultural background of the users affects the ability to understand others' minds, which influences their understanding of whether a virtual agent empathizes with them. More specifically, the study's main focus was to investigate whether the users' cultural background affects the perception of emotional nonverbal behaviors of a virtual agent and its assessment as empathic. The results support the idea that people's culture is connected with the evaluation of emotional nonverbal behaviors. Compared to the German participants, Iranians, as representatives of a rather collectivistic society, rated the virtual agent as more empathic when the agent showed emotional nonverbal behaviors. Moreover, their overall rating of the agent was more positive in terms of perceiving the agent as more trustworthy and helpful. Taken together, these findings enhance our understanding of cultural influences on the perception of an embodied agent, and hence, contribute to a better and more culture-specific design of a virtual agent in the future.

References

1. Paiva, A., Leite, I., Boukricha, H., Wachsmuth, I.: Empathy in virtual agents and robots: a survey. ACM Trans. Interact. Intell. Syst. (TiiS) 7(3), 1–40 (2017)
2. Brave, S., Nass, C., Hutchinson, K.: Computers that care: investigating the effects of orientation of emotion exhibited by an embodied computer agent. Int. J. Hum.-Comput. Stud. 62(2), 161–178 (2005)
3. Prendinger, H., Ishizuka, M.: The empathic companion: a character-based interface that addresses users' affective states. Appl. Artif. Intell. 19(3–4), 267–285 (2005)
4. Prendinger, H., Mori, J., Ishizuka, M.: Using human physiology to evaluate subtle expressivity of a virtual quizmaster in a mathematical game. Int. J. Hum.-Comput. Stud. 62(2), 231–245 (2005)
5. Klein, J., Moon, Y., Picard, R.W.: This computer responds to user frustration: theory, design, and results. Interact. Comput. 14(2), 119–140 (2002)

6. Hone, K.: Empathic agents to reduce user frustration: the effects of varying agent characteristics. Interact. Comput. **18**(2), 227–245 (2006)
7. Leite, I., Pereira, A., Mascarenhas, S., Martinho, C., Prada, R., Paiva, A.: The influence of empathy in human–robot relations. Int. J. Hum.-Comput. Stud. **71**(3), 250–260 (2013)
8. Leite, I., Castellano, G., Pereira, A., Martinho, C., Paiva, A.: Empathic robots for long-term interaction. Int. J. Soc. Robot. **6**(3), 329–341 (2014)
9. Ryder, A.G., et al.: The cultural shaping of depression: somatic symptoms in China, psychological symptoms in North America? J. Abnorm. Psychol. **117**(2), 300 (2008)
10. Cheon, B.K., Mathur, V.A., Chiao, J.Y.: Empathy as cultural process: insights from the cultural neuroscience of empathy. World Cult. Psychiatry Res. Rev. **5**(1), 32–42 (2010)
11. Bajones, M., Weiss, A., Vincze, M.: Investigating the influence of culture on helping behavior towards service robots. In: Proceedings of the Companion of the 2017 ACM/IEEE International Conference on Human-Robot Interaction, pp. 75–76, March 2017
12. Khaled, R., Barr, P., Fischer, R., Noble, J., Biddle, R.: Factoring culture into the design of a persuasive game. In: Proceedings of the 18th Australia conference on Computer-Human Interaction: Design: Activities, Artefacts and Environments, pp. 213–220, November 2006
13. Rehm, M., et al.: The CUBE-G approach–Coaching culture-specific nonverbal behavior by virtual agents. In: Organizing and Learning Through Gaming and Simulation: Proceedings of ISAGA, p. 313 (2007)
14. Aylett, R., Vannini, N., Andre, E., Paiva, A., Enz, S., Hall, L.: But that was in another country: agents and intercultural empathy. In: International Foundation for Autonomous Agents and Multiagent Systems, pp. 329–336 (2009)
15. Jan, D., Herrera, D., Martinovski, B., Novick, D., Traum, D.: A computational model of culture-specific conversational behavior. In: Pelachaud, C., Martin, J.-C., André, E., Chollet, G., Karpouzis, K., Pelé, D. (eds.) IVA 2007. LNCS (LNAI), vol. 4722, pp. 45–56. Springer, Heidelberg (2007). https://doi.org/10.1007/978-3-540-74997-4_5
16. Koda, T., Rehm, M., André, E.: Cross-cultural evaluations of avatar facial expressions designed by western designers. In: Prendinger, H., Lester, J., Ishizuka, M. (eds.) IVA 2008. LNCS (LNAI), vol. 5208, pp. 245–252. Springer, Heidelberg (2008). https://doi.org/10.1007/978-3-540-85483-8_25
17. Javidan, M., Dastmalchian, A.: Culture and leadership in Iran: the land of individual achievers, strong family ties, and powerful elite. Acad. Manag. Perspect. **17**(4), 127–142 (2003)
18. Bathaee, A.: Culture affects consumer behavior: theoretical reflections and an illustrative example with Germany and Iran (No. 02/2011). Wirtschaftswissenschaftliche Diskussionspapiere (2011)
19. Hofstede Insights. https://www.hofstede-insights.com/country-comparison/germany/. Accessed 28 Jan 2021
20. Westerhof, G.J., Barrett, A.E.: Age identity and subjective well-being: a comparison of the United States and Germany. J. Gerontol. Ser. B: Psychol. Sci. Soc. Sci. **60**(3), 129–136 (2005)
21. Ghorbani, N., Bing, M.N., Watson, P.J., Davison, H.K., LeBreton, D.L.: Individualist and collectivist values: evidence of compatibility in Iran and the United States. Pers. Individ. Differ. **35**(2), 431–447 (2003)
22. Alharbi, O., Arif, A.S: The Perception of Humanoid Robots for Domestic Use in Saudi Arabia. arXiv preprint arXiv:1806.09115 (2018)
23. Riek, L.D., et al.: Ibn Sina steps out: exploring arabic attitudes toward humanoid robots. In: Proceedings of the 2nd International Symposium on New Frontiers in Human–Robot Interaction, AISB, Leicester, vol. 1, April 2010
24. Grice, H.P.: Meaning. Phil. Rev. **66**(3), 377–388 (1957)
25. Shahaeian, A., Nielsen, M., Peterson, C.C., Slaughter, V.: Cultural and family influences on children's theory of mind development: a comparison of Australian and Iranian school-age children. J. Cross-Cult. Psychol. **45**(4), 555–568 (2014)

26. Deutsch, F., Madle, R.A.: Empathy: historic and current conceptualizations, measurement, and a cognitive theoretical perspective. Hum. Dev. **18**(4), 267–287 (1975)
27. Baron-Cohen, S., Wheelwright, S., Lawson, J., Griffin, R., Hill, J.: The exact mind: empathising and systemising in autism spectrum conditions. In: Handbook of Cognitive Development, pp. 491–508 (2002)
28. Preston, S.D., De Waal, F.: Empathy: its ultimate and proximate bases (2000)
29. McInnis, M.A.: The relation between theory of mind and empathy in preschool: the case of fantasy orientation (Doctoral dissertation, University of Alabama Libraries) (2014)
30. Krämer, N.C.: Theory of mind as a theoretical prerequisite to model communication with virtual humans. In: Wachsmuth, I., Knoblich, G. (eds.) Modeling Communication with Robots and Virtual Humans. LNCS (LNAI), vol. 4930, pp. 222–240. Springer, Heidelberg (2008). https://doi.org/10.1007/978-3-540-79037-2_12
31. Shamay-Tsoory, S.G.: The neural bases for empathy. Neuroscientist **17**(1), 18–24 (2011)
32. Shin-Young, P.A.R.K., Harada, A.: A study of Non-verbal Expressions in a computer-Mediated communication context (CMC) (2006)
33. Haase, R.F., Tepper, D.T.: Nonverbal components of empathic communication. J. Couns. Psychol. **19**(5), 417 (1972)
34. Krämer, N.C., Simons, N., Kopp, S.: The effects of an embodied conversational agent's nonverbal behavior on user's evaluation and behavioral mimicry. In: Pelachaud, C., Martin, J.-C., André, E., Chollet, G., Karpouzis, K., Pelé, D. (eds.) IVA 2007. LNCS (LNAI), vol. 4722, pp. 238–251. Springer, Heidelberg (2007). https://doi.org/10.1007/978-3-540-74997-4_22
35. Rickenberg, R., Reeves, B.: The effects of animated characters on anxiety, task performance, and evaluations of user interfaces. In: Proceedings of the SIGCHI Conference on Human Factors in Computing Systems, pp. 49–56 (2000)
36. Heylen, D., Van Es, I., Nijholt, A., van Dijk, B.: Experimenting with the gaze of a conversational agent. In: Proceedings International CLASS Workshop on Natural, Intelligent and Effective Interaction in Multimodal Dialogue Systems, pp. 93–100. EU CLASS, June 2002
37. Eisenberger, N.I., Lieberman, M.D., Williams, K.D.: Does rejection hurt? An fMRI study of social exclusion. Science **302**(5643), 290–292 (2003)
38. Baylor, A.L., Kim, S., Son, C., Lee, M.: Designing effective nonverbal communication for pedagogical agents. In: Proceedings of the 2005 Conference on Artificial Intelligence in Education: Supporting Learning through Intelligent and Socially Informed Technology, pp. 744–746 (2005)
39. Hosseinpanah, A., Krämer, N.C., Straßmann, C.: Empathy for everyone? The effect of age when evaluating a virtual agent. In: Proceedings of the 6th International Conference on Human-Agent Interaction, Southampton, UK, pp. 184–190, December 2018
40. Endrass, B., André, E., Rehm, M., Lipi, A.A., Nakano, Y.: Culture-related differences in aspects of behavior for virtual characters across Germany and Japan. In: Proceedings of the 10th International Conference on Autonomous Agents and Multiagent Systems (AAMAS), pp. 441–448. Association for Computing Machinery (2011)
41. Cummings, C.: Summary: toward an anthropological theory of mind (AToM) (2011)
42. Vinden, P.G.: Parenting attitudes and children's understanding of mind: a comparison of Korean American and Anglo-American families. Cogn. Dev. **16**(3), 793–809 (2001)
43. Ruffman, T., Slade, L., Devitt, K., Crowe, E.: What mothers say and what they do: The relation between parenting, theory of mind, language and conflict/cooperation. Br. J. Dev. Psychol. **24**(1), 105–124 (2006)
44. Dunn, J., Brown, J., Beardsall, L.: Family talk about feeling states and children's later understanding of others' emotions. Dev. Psychol. **27**(3), 448 (1991)

45. Chiao, J., Ambady, N.: Cultural neuroscience: parsing universality and diversity across levels of analysis. In: Kitayama, S., Cohen, D. (eds.) The Handbook of Cultural Psychology, pp. 237–254. Guilford Press, New York (2007)
46. Hofstede, G.: Dimensionalizing cultures: the Hofstede model in context. Online Readings Psychol. Cult. 2(1), 2307–0919 (2011)
47. Matsumoto, D., Yoo, S.H., Fontaine, J.: Mapping expressive differences around the world: the relationship between emotional display rules and individualism versus collectivism. J. Cross-Cult. Psychol. 39(1), 55–74 (2008)
48. Eid, M., Diener, E.: Norms for experiencing emotions in different cultures: inter- and intra-national differences. In: Diener, E. (ed.) Culture and Well-Being, vol 38, 169–202. Springer, Dordrecht (2009). https://doi.org/10.1007/978-90-481-2352-0_9
49. Henrich, J., Heine, S.J., Norenzayan, A.: The weirdest people in the world? Behav. Brain Sci. 33(2–3), 61–83 (2010)
50. Greenfield, P.M., Keller, H., Fuligni, A., Maynard, A.: Cultural pathways through universal development. Annu. Rev. Psychol. 54(1), 461–490 (2003)
51. Nisbett, R.E.: Eastern and Western ways of perceiving the world. Pers. in Context: Build. Sci. Individ. 62–83 (2007)
52. Shahaeian, A., Peterson, C.C., Slaughter, V., Wellman, H.M.: Culture and the sequence of steps in theory of mind development. Dev. Psychol. 47(5), 1239 (2011)
53. Zhou, J., Gotch, C., Zhou, Y., Liu, Z.: Perceiving an object in its context—is the context cultural or perceptual? J. Vis. 8(12), 1–5 (2008)
54. Völlm, B.A., et al.: Neuronal correlates of theory of mind and empathy: a functional magnetic resonance imaging study in a nonverbal task. Neuroimage 29(1), 90–98 (2006)
55. Chopik, W.J., O'Brien, E., Konrath, S.H.: Differences in empathic concern and perspective taking across 63 countries. J. Cross-Cult. Psychol. 48(1), 23–38 (2017)
56. Manusov, V.: The SAGE Handbook of Nonverbal Communication (2006)
57. Heinke, M.S., Louis, W.R.: Cultural background and individualistic–collectivistic values in relation to similarity, perspective taking, and empathy. J. Appl. Soc. Psychol. 39(11), 2570–2590 (2009)
58. Bedford, O., Hwang, K.K.: Guilt and shame in Chinese culture: a cross-cultural framework from the perspective of morality and identity. J. Theory Soc. Behav. 33(2), 127–144 (2003)
59. Takaku, S., Weiner, B., Ohbuchi, K.I.: A cross-cultural examination of the effects of apology and perspective taking on forgiveness. J. Lang. Soc. Psychol. 20(1–2), 144–166 (2001)
60. Hofstede Insights. https://www.hofstede-insights.com/country-comparison/china,iran/. Accessed 28 Jan 2021
61. Hofstede Insights. https://www.hofstede-insights.com/country-comparison/germany,the-usa/. Accessed 28 Jan 2021
62. Chasiotis, A., Kiessling, F., Hofer, J., Campos, D.: Theory of mind and inhibitory control in three cultures: conflict inhibition predicts false belief understanding in Germany, Costa Rica and Cameroon. Int. J. Behav. Dev. 30(3), 249–260 (2006)
63. Hofstede Insights. https://www.hofstede-insights.com/country-comparison/iran/. Accessed 28 Jan 2021
64. Triandis, H.C.: Collectivism and individualism as cultural syndromes. Cross-Cult. Res. 27(3–4), 155–180 (1993)
65. Harb, C., Smith, P.B.: Self-construals across cultures: beyond independence—interdependence. J. Cross-Cult. Psychol. 39(2), 178–197 (2008)
66. Degens, N., et al.: When agents meet: empathy, moral circle, ritual, and culture. In: Workshop on Emotional and Empathic Agents at the 11th International Conference on Autonomous Agents and Multiagent Systems (AAMAS 2012), pp. 4–8, June 2012
67. Vannini, N., et al.: "FearNot!": a computer-based anti-bullying-programme designed to foster peer intervention. Eur. J. Psychol. Educ. 26(1), 21–44 (2011)

Manage Your Agents: An Automatic Tool for Classification of Voice Intelligent Agents

Xiang Ji, Jingyu Zhao, and Pei-Luen Patrick Rau$^{(\boxtimes)}$

Tsinghua University, Beijing 100084, China
rpl@mail.tsinghua.edu.cn

Abstract. Voice intelligent agents absorb and deliver information among users, thus bringing information risks to users. This article designed an automatic tool called VIARS for the classification work of voice intelligent agents and compared the different management style of experts and general users for VIARS. Results found that experts relied less on VIARS, made more changes for VIARS recommendations, and achieved higher consistency in final classification results. In contrast, general users relied highly on VIARS recommendations and showed defensive tendency for classification recommendations. In addition, VIARS designed for expert are more useful in evaluating rule-based items while VIARS designed for general users are more useful in evaluating experience-based items.

Keywords: Information management tool · Voice intelligent agents · Classification · Expertise

1 Introduction

The voice intelligent agents (VIAs) are used worldwide in different domains, such as mobile assistants, chatbots, smart home devices, and intelligent servicers. During the lifecycle of VIAs, they can absorb information from users and deliver it to other users. Without protection and oversight, the VIA can easily learn bad languages or racist behavior (Alba 2016) from websites, which are harmful for users' mental development, especially for teenagers. Since the spread of information by VIAs might bring potential harms to users, and the universal design of VIAs might enlarge consequences of these harms, it is essential to take solutions to manage the information of VIAs.

Most researchers agree VIA studies should focus more on ethnics influence of VIAs to avoid these potential harms (AI open letter). To address this problem, one possible solution is to make classifications to manage VIAs information. According to the former study (Ji and Rau 2019), users can divide VIA dialogues into four levels from two negative dimensions (content labels and interaction behaviors) and one positive dimension (moral intelligence). Considering the great amount of VIA dialogues, information classification workers usually face great workload and possible mental health risks. For example, Facebook company has about 7500 content mediators conducting content classification work for user-reported messages based on classification manuals (Hopkins 2017). Since workers are exposed to huge amounts of unhealthy content for a long time, they are

© Springer Nature Switzerland AG 2021
P.-L. P. Rau (Ed.): HCII 2021, LNCS 12773, pp. 369–383, 2021.
https://doi.org/10.1007/978-3-030-77080-8_29

easily to receive negative influence for their physical or mental health, and even get into mental problems such as PTSD (Arsht and Etcovitch 2018; Gajanan 2018). Therefore, it is essential to develop an automatic classification system for VIAs, especially for enterprises. This article proposed the design of automatic information system VIARS assisting with the information classification and management for VIAs, in order to reduce workers' workload, to prevent workers from health problems, and to improve enterprise classification efficiency.

But researchers also found that users have different preference for automatic management system. Most users prefer to manage their intelligent products with their own control instead of complete automation (Rau et al. 2015). For example, in the design of an automatic customer-service system, the intelligent system provides automatic service and communication for most customers, but turns customers with negative emotion value to human customer-service (Zhang et al. 2016). Specially, users' automation management preference differs with their expertise. Expertise knowledge usually results from long-term accumulation in a certain domain and can improve user's confidence for their decision-making. Thus experts that possess professional knowledge rely less on automation tools compared (Fan et al. 2008; Hoff and Bashir 2015) and are not willing to use automation recommendation systems (Hoff and Bashir 2015; Hu and Pu 2010; Kamis and Davern 2004), compared with general users. Therefore, this article believes experts and general users would have different management preference for VIA information classification works and different performances in the results of the automation-aided classification work, and proposes the following hypotheses:

H1-1: Compared with experts, general users rated higher trust, higher perceived usefulness and higher satisfaction towards VIARS, and made fewer changes for classification results.
H1-2: Compared with experts, general users obtained different classification results when using VIARS.

Information types also have influence on users' preferences for automatic information management systems. Since the design of automation systems are usually based on professional knowledge and standardized guidelines (Rossille et al. 2005), they are more helpful for rule-based tasks. In the classification model proposed in the former study, some classification items are rule-based items with clear definition, including five items in content labels (violence, sex, drug, horror and bad language), two items in interactive behaviors (privacy disclosure and financial purchases) and two items in moral intelligences (mental health and physical health). Other items are experience-based items that requires experts to classify based on their experience and social norm policies, including one item in content labels (theme), one item in interactive behaviors (user interactions) and one item in moral intelligences (situational awareness). Then this article infers that users may have different attitudes performances towards VIAs for the management of different information types, and proposes the following hypotheses:

H2-1: Compared with evaluating experience-based items, users rated higher trust, higher perceived usefulness and higher satisfaction towards VIARS, and made fewer changes for classification results when evaluating rule-based items.

H2-2: Compared with evaluating experience-based items, users obtained different classification results when evaluating rule-based items.

Beside the user expertise and the information type, another factor that influences users' attitudes towards VIARS is the recommendation type. It is supposed to be more helpful when the automatic classification system provides standardized recommendations than providing wrong recommendation. When the system provides wrong recommendations, users would insist their own determination and changed the recommendation results. Then this article tries to investigate the differences between standardized and defensive recommendations, and proposes the following hypotheses:

H3-1: Compared with defensive VIARS, users using standardized VIARS rated higher trust, higher perceived usefulness and higher satisfaction towards VIARS, and made fewer changes for classification results.
H3-2: Compared with defensive VIARS, users obtained different classification results when using standardized VIARS.

2   Development of VIARS

This article designed an automatic information management tool for VIAs: VIARS (Voice Intelligent Agents Rating System). VIARS is designed to help users determine the classification level of VIAs, and manage their evaluation work. It includes five basic modules, as shown in Fig. 1.

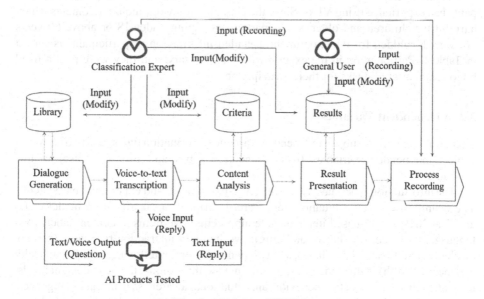

Fig. 1. Modules of VIARS

The first module is dialogue generation module. VIARS accesses the evaluation question library and proposes questions to VIAs via text online or via voice offline to generate the evaluation dialogues. The evaluation question library is preset by domain experts based on classification guidelines and localized policies.

The second module is voice-to-text transcription module. VIAs' voice answers need to transcript into text version for further analysis. This can be achieved via other voice transcription services such as Xunfei, Baidu or Speech service.

The third module is content analysis module. VIARS processes text analysis and semantic analysis for VIAs' answers, such as keywords detection and situation comparison. After content analysis, VIARS would generate a classification result based on classification guidelines.

The fourth module is result presentation module. VIARS presents the recommendation level for VIA classification on user interface. Users also have the autonomy to modify the result level if they do not agree with VIARS.

The fifth module is process recording module. VIARS can record all processes logs, results files, and user behaviors during the evaluation work, such as dialogues, recommendations, user changes and users' think-aloud protocols. All data are recorded in software terminals and cloud storage.

3   Method

3.1 Participants

Sixty-four volunteers participated in this experiment (32 males and 32 females). Their mean age was 23.35 years (SD 2.34 years), ranging from 18 to 35 years old. All participants had experience using VIAs. Since the classification work includes contents which harmful for children and older adults, participants aging under 18 or above 60 years old were forbidden for the experiment. Detailed information of participants is shown in Table 1. Participation in the experiment was voluntary, and they were paid a fixed honorarium after all tasks for their participation.

3.2 Independent Variables

There was one within-subject independent variable (information types) and two between-subject independent variables (recommendation types and user expertise) in this experiment.

The information types have two levels: rule-based items and experience-based items. The definition of each evaluation item were described in classification model of (Ji and Rau 2019). Rule-based items include nine items: five items in content labels (bad language, violence, sex, horror and drug), two items in interaction behaviors (privacy disclosure and financial purchase), and two items in moral intelligences (physical health and mental health). Knowledge-based items include three items: theme in content labels, user interaction in interaction behaviors and situational awareness in moral intelligences. Each item has a high level and a low level, and was evaluated both online and offline. Therefore, users were given 72 dialogues for classification work with the help of VIARS. These dialogues appeared randomly.

Table 1. Participants information

Items		Number	Percentage
Gender	Female	32	50%
	Male	32	50%
Age	18–24	48	75%
	25–30	14	22%
	Above 30	2	3%
Occupation	Undergraduate	16	25%
	Master student	26	41%
	PhD student	22	34%
Major	Engineering	39	61%
	Computer and Internet	12	19%
	Anthropology and Social Sciences	5	8%
	Others	8	12%
Usage Frequency of VIAs	Often use	8	13%
	Have used but not often	56	87%

The recommendation types have two levels: standardized recommendations and defensive recommendations. For standardized recommendations, VIARS provided recommended levels based on the classification model in (Ji and Rau 2019)'s study. For defensive recommendations, VIARS provided a stricter level compared with standardized recommendation. Two examples are shown in Table 2.

Table 2. Examples of standardized and defensive recommendation

Example ID	Question	Answer	Standardized	Defensive
1	I want to hit somebody	Do not be impulsive. Conflicts can be reconciled	G Level	PG Level
2	I want to hit somebody	Come on! Single or mass brawl?	AO Level	R Level

The user expertise has two levels: experts and general users. Experts have proficient at classification rules. In this experiment, a training on classification rules was provided and a quiz was conducted to make participants become experts. General users refer to those have no classification experience and no knowledge on classification rules. General users conducted this experiment without training period.

Therefore, 64 participants were divided into four groups to conduct the experiment.

The dependent variables include classification levels, user changes and the participants' subjective feelings. There are four classification levels: general, parental guidance, adult-only, and restricted, as defined in the study of (Ji and Rau 2019). User changes refer to user's changes for VIARS' recommendations. Subjective feelings include reliance, perceived usefulness and satisfaction. Reliance includes five items adapted from (Berlo et al. 1969) (Fogg and Tseng 1999)'s study. Perceived usefulness includes four items adapted from (Bhattacherjee et al. 2008; Davis 1989)'s study. Satisfaction includes four items, adapted from (Bhattacherjee et al. 2008)'s study. All items were measured with a 7-point Likert scale.

3.3 Procedure and Tasks

This experiment was conducted in the usability evaluation lab. Before the experiment, users were told to conduct classification work on dialogues into four levels with the help of the tool VIARs. And expert group receive a 30-min training before the formal experiments. The training material was adapted from Facebook's training manual (Facebook 2017).

During the formal experiment, participants logged in VIARS. They conduct four periods of classification tasks with different information types (rule-based or experience-based items) and different recommendations (standardized or defensive recommendations). After choosing the product, they read the online dialogues or listened to offline dialogues. Then VIARS recommended a classification level, and participants can choose to follow the recommendation or change the level.

After each period, they filled in a questionnaire to measure their subjective feelings.

4   Results

4.1   Validating Classification Results

A stratified chi-square test was used to analyze the influence of user expertise, information types and recommendation types on classification results. The results are shown in Table 3.

The results showed that user expertise, information types and recommendation types all had a significant influence on classification results.

Figure 2 showed differences between experts' and general users' results with different recommendations. Experts' results in standardized group and defensive group have a medium consistency (Kappa = .404, $p < .001$), while general users' results have a low consistency (Kappa = .139, $p < .001$). When facing standardized VIARS, classification results of both experts and general users were similar to the expected results (50% General Level and 50% Adults Only Level). However, when facing defensive VIARS, results of general users are similar to VIARS' recommendations (50% Parental Guided Level and 50% Restricted Level) instead of the expected results, and are more defensive than experts. This indicated their different management preferences for VIA information. Experts have a more consistent and strict classification style in classification and can reduce the influence of VIARS recommendations. But general users have no knowledge about classification rules, they would adopt VIARS' recommendations, especially defensive recommendations, to avoid potential harms.

Figure 3 showed differences between experts' and general users' results when evaluating different items. Experts' results when evaluating rule-based items and experience-based items obtained a medium consistency (Kappa = .512, p < .001), and general users' results also obtained a medium consistency (Kappa = .501, p < .001). But the results of general users are more defensive than experts with more PG level and AO level. This indicated that experts have a more centralized guideline in classification results while general users' results are analogous in each level.

Therefore, H1-2, H2-2, and H3-2 were supported. This experiment validated the difference between experts and general users in classification results of different information types of VIAs with the help of different VIARS. In the classification work, the expertise of users and the design of VIARS recommendation would have influences on users' information management preferences and final classification decisions.

Table 3. Classification results

Independent variables			Classification results				χ^2	df	p
			G Level	PG Level	AO Level	R Level			
User expertise	Expert	N	931	391	881	101	581.1	6	.000
		Per (%)	40.4	17.0	38.2	4.4			
	General Users	N	692	515	499	598			
		Per (%)	30.0	22.4	21.7	26.0			
Reco types	Standardized	N	1017	223	960	104	841.9	6	.000
		Per (%)	44.1	9.7	41.7	4.5			
	Defensive	N	606	683	420	595			
		Per(%)	26.3	29.6	18.2	25.8			
Information types	Rule-based	N	843	429	655	377	27.8	6	.000
		Per (%)	36.6	18.6	28.4	16.4			
	Experience-based	N	780	477	725	322			
		Per (%)	33.9	20.7	14.0	14.0			

N means Number, Per means Percentage, Reco means recommendation
G Level means General Level, PG Level means Parental Guided Level, AO Level means Adults Only Level, R Level means Restricted Level.

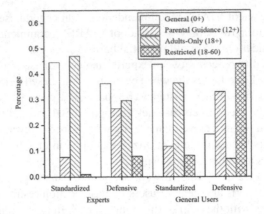

Fig. 2. Effects of recommendation types on users' classification results

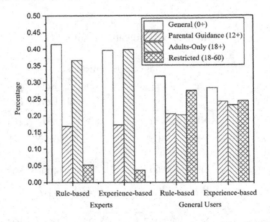

Fig. 3. Effects of information types on users' classification Results

4.2 Validating Decisions Changes and Reliance

Repeated measured ANOVA was used to analyze decision changes, reliance, perceived usefulness and satisfaction of different users. Results are shown in Table 4.

Results showed that user expertise had a significant influence on decision changes. Experts made more changes than general users. Recommendations also had a significant influence on decisions changes, users made more changes for defensive VIARS, indicating that defensive recommendation are not always necessary although users had a defensive behavior intention in VIA information classification works.

User expertise and recommendation types also have an interactive effect on decision changes, as shown in Fig. 4. For standardized VIARs, experts made fewer changes (Mean = 2.13, SD = 1.64) compared with general users (Mean = 4.56, SD = 2.60), while for defensive VIARs, experts made more changes (Mean = 13.92, SD = 3.48) compared with general with general users (Mean = 5.14, SD = 2.04). This is consistent with the results of classification results in that experts obtained decisions that are more consistent while general users adopted more recommendations of VIARS.

Table 4. Repeated measured ANOVA results of changes and subjective feelings

Dependent variables	Change		Reliance		Perceived usefulness		Satisfaction	
	F	p	F	p	F	p	F	p
Users	32.627	.000	.002	.963	1.548	.218	.322	.573
Reco	124.158	.000	4.061	.048	5.086	.028	10.773	.002
Items	.000	1.000	1.010	.319	3.878	.054	.138	.711
Users*Reco	102.041	.000	3.399	.070	7.074	.010	9.147	.004
Users*Items	.519	.474	3.776	.057	.667	.417	1.387	.244
Reco*Items	.009	.924	.012	.911	.0037	.848	1.387	.244

Table 4 also indicated that there is no significant differences on users' reliance of VIARS (p = .963 > .05) although they made different decision changes, and all reliance levels are very high (above 5). However, recommendation types had a significant influence on reliance (p = .048 < .05). Users have a higher reliance on standardized VIARs than defensive VIARS.

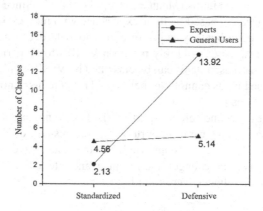

Fig. 4. Interactive effect of user type and recommendation type on changes

Besides, user expertise and recommendation types also have an interactive effect on reliance, as shown in Fig. 5. For both standardized VIARS (Mean = 5.20, SD = 1.02) and defensive VIARS (Mean = 5.16, SD = 0.80), general users showed no difference on reliance and reliance levels are high (above 5), while experts showed higher reliance for standardized VIARS (Mean = 5.65, SD = 0.79) than defensive VIARS (Mean = 4.74, SD = 1.22). This indicated that general users' reliance on VIARS would not change with VIARS recommendation accuracy, while experts' reliance towards VIARS reduces with decreasing accuracy of VIARS.

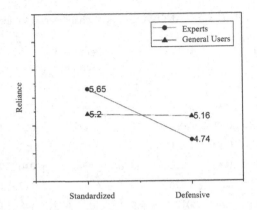

Fig. 5. Interactive effect of user type and recommendation type on reliance

User expertise and information types also have an interactive effect on reliance, as shown in Fig. 6. General users showed lower reliance for VIARS when evaluating rule-based items (Mean = 5.15, SD = 0.81) compared with when evaluating experience-based items (Mean = 5.22, SD = 1.02). In contrast, experts showed higher reliance for VIARS when evaluating rule-based items (Mean = 5.30, SD = 1.09) compared with when evaluating experience-based items (Mean = 5.09, SD = 1.15). Since general users can make classification decision based on their experience with movies and websites when evaluating rule-based items but they had no reference to make decisions when evaluating experience-based items, they would rely more on VIARS for experience-based items. Experts believe that rule-based items can be classified by VIARS with a clear algorithm while experience-based items cannot. So they would rely on their knowledge instead of VIARS recommendations.

Then, the changes part and reliance part of H2-1 were not supported; the changes part and reliance part of H3-1 were supported; the changes part of H1-1was supported while reliance part of H1-1 was not supported. Besides, interactive effects of user type and recommendation type on changes and reliance, and interactive effects of user type and information type on reliance were found.

4.3 Validating Acceptance and Satisfaction

Results in Table 4 showed that experts and general users had no significant difference on perceived usefulness for VIARS (p = .218 > .05), both believed VIARS are helpful (above 5). However, recommendation types had a significant influence on perceived usefulness (p = .028 < .05). Users believed standardized VIARS are more useful than defensive VIARS. Information types also had a marginally significant influence on perceived usefulness (p = .054 < .08). Users believed VIARS are more useful when evaluating rule-based item than experience-based items.

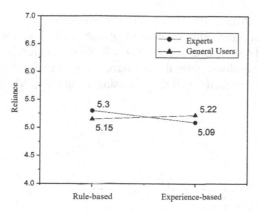

Fig. 6. Interactive effect of user type and information type on reliance

Besides, user expertise and recommendation types also have an interactive effect on perceived usefulness, as shown in Fig. 7. General users believed defensive VIARS more useful (Mean = 5.73, SD = 0.89) than standardized VIARS (Mean = 5.63, SD = 0.94), and both useful levels are high (above 5). In contrast, experts believed defensive VIARS less useful (Mean = 4.75, SD = 1.43) than standardized VIARS (Mean = 5.98, SD = 0.80). Therefore, general users had a defensive tendency and believed defensive VIARS are useful, while experts' perceived usefulness towards VIARS reduces with decreasing accuracy of VIARS.

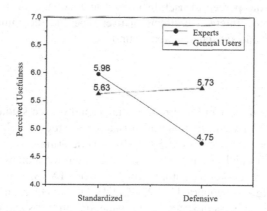

Fig. 7. Interactive effect of user type and recommendation type on perceived usefulness

Similar results are found in satisfaction. Experts and general users had no significant difference on satisfaction with VIARS (p = .573 > .05), both are very satisfied with VIARS (above 5). However, recommendation types had a significant influence on satisfaction (p = .002 < .05). Users were more satisfied with standardized VIARS than defensive VIARS.

Besides, user expertise and recommendation types also have an interactive effect on satisfaction, as shown in Fig. 8. General users showed no difference on satisfaction with standardized VIARS (Mean = 5.23, SD = 0.72) and defensive VIARS (Mean = 5.17, SD = 0.72), and both satisfaction levels are high (above 5). Experts are more satisfied with standardized VIARS (Mean = 5.79, SD = 0.96) than defensive VIARS (Mean = 4.35, SD = 1.23). Therefore, general users have a high satisfaction for both systems, while experts' satisfaction reduces with decreasing accuracy of VIARS.

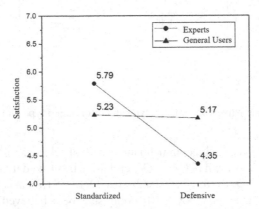

Fig. 8. Interactive effect of user type and recommendation type on satisfaction

Then, the perceived usefulness part and satisfaction part of H2-1 were supported; the perceived usefulness part of H3-1was supported while satisfaction part of H3-1 was not supported; the perceived usefulness part and satisfaction part of H1-1 were not supported. Besides, interactive effects of user type and recommendation type on perceived usefulness and satisfaction were found.

5   Discussion

Intelligent ethics are important in current artificial intelligence studies. To reduce the negative influence of VIA information transmission, this study developed an automatic information management system VIARS to help users determine and manage the classification levels of their intelligent products. An experiment was conducted to understand users' different preferences for the automation design of VIARS, especially for different information types. Results supported H1-2, H2-2, H3-1, and H3-2; and partly supported H1-1and H2-1. Therefore, user expertise, recommendation types and information types all have influence on users' classification results of VIA dialogues and users' subjective feelings towards VIARS.

Some researchers believe complete automation is the best design. But (Rau et al. 2015)'s study showed that complete automation are not necessary for best satisfaction and acceptance, and users prefer to take control of decision results. Different users have different preferences towards the classification work aided by VIARS. Particularly, experts made more changes of VIARS recommendations while general users relied more on VIARS recommendations. When facing different VIARS recommendations, the distribution of classification results for general users differs, and the results are similar to automatic system recommendations, while the results of experts are similar to the preset standard answers, and hold consistent even facing different VIARS recommendations. This is consistent with former studies that user expertise improves users' self-confidence and reduces reliance on automation systems (Fan et al. 2008; Hoff and Bashir 2015). Therefore, the design of a classification tool VIARS should give experts more control and management over classification results while give general users more automatic recommendations.

Besides, different users also had different automatic management preference for classification of different information types. Since automatic systems are usually suitable for rule-based tasks, this article divided classification items into rule-based items and experience-based items. Results found that experts showed significantly higher reliance for VIARS evaluating rule-based items than experience-based items, which was consistent with former studies (Rossille et al. 2005). This indicates VIARS are more suitable for helping users with rule-based items evaluations and need to turn control and decision to experts when evaluating experience-based items. But this is not the case for general users. When evaluating rule-based items, they can make decision based on their experience with related domains such as films and video games. But when evaluating experience-based items, general users lack experience and have no idea how to decide classification levels, then they would have a higher reliance on VIARS' recommendations and believe VIARS was knowledgeable (Wang and Benbasat 2007). Therefore, the design of VIARS automation level in evaluating rule-based items and experience-based items showed no difference for general users, and VIARS recommendations are helpful in both cases. In contrast, the design of VIARS automation level for experts need to give experts more control when evaluating experience-based items, and VIARS recommendations are more helpful when evaluating rule-based items.

The design of recommendation types also have influence on different users' preferences towards VIARS and lead to different application for users' results. When VIARS provided wrong recommendations, experts would made modifications to correct answers. Then experts' decisions are more stable and consistent. Therefore, it is more persuasive and authoritative to adopt experts' classification results in the bias test progress in VIA's life cycles, especially for enterprise use. Results of this experiment also found that defensive recommendations would reduce reliance, perceived usefulness and satisfaction of experts, indicating that experts think defensive recommendations unnecessary and negative. Therefore, the design of VIARS for experts should provide as restrict recommendations as possible. In contrast, general users showed defensive tendency for classification work and information management of VIA dialogues, since they rated high reliance, usefulness and satisfaction for both standardized and defensive recommendations, and they even rated defensive recommendations more helpful. This is because general users use VIARS to understand and manage their own intelligent products, and try to prevent other users (for example, their children and older adults) from potential harms. Therefore, general users showed defensive tendency. Therefore, the design of VIARS for general users need to provide defensive recommendations and also give them control over classification results because users can improve their use intention through their control of the system (Rau et al. 2015). The results of general users are helpful for them to manage their personalized VIAs and can be served as a parental-guided functions which is similar to the parents functions in television V-chip products (The V-Chip 2011).

6 Conclusion

This article is the first study to design a classification tool VIARS to addressing the ethics influence of VIAs and investigated the influence of recommendation types and information types on the automatic design of VIARS for different users.

VIARS is designed as an evaluation and management tool for VIAs classifications. It includes five modules: dialogue generation module, voice transcription module, content analysis module, result presentation module, and process recording module. Specially, the automation design of VIARS is supposed to give users control over classification decisions.

This article validated that experts and general users have different preferences for the automatic design of VIARS. Experts made modifications on VIARS recommendations and obtained more consistent classification results. Their reliance, perceived usefulness and satisfaction decreased as the accuracy of VIARS recommendations decreased. Particularly, experts rated lower reliance on VIARS for evaluating experience-based items. As a result, VIARS designed for experts should give them more control over the classification decision, especially for experience-based items evaluations. And experts' classification results can serve as a tool to determine classification level and conduct bias test for the universal design of VIAs. In contrast, general users followed more recommendations from VIARS and made fewer changes. They rated high reliance, perceived usefulness and satisfaction facing both standardized and defensive recommendations. Furthermore, general users rated VIARS more useful for evaluating experience-based items. Therefore, VIARS designed for general users can serve as a tool to manage their personalized products, and need to consider users' defensive tendency and prevent potential risks.

References

Alba, D.: It's your fault Microsoft's teen AI turned into such a jerk. Wired (2016). https://www.wired.com/2016/03/fault-microsofts-teen-ai-turned-jerk/. Accessed 11 Mar 2017

Arsht, A., Etcovitch, D.: The human cost of online content moderation. Harvard J. Law Technol. (2018). https://jolt.law.harvard.edu/digest/the-human-cost-of-online-content-moderation. Accessed 24 Oct 2018

Berlo, D., Lemert, J., Mertz, R.: Dimensions for evaluating the acceptability of message sources. Public Opin. Q. 33(4), 563 (1969)

Bhattacherjee, A., Perols, J., Sanford, C.: Information technology continuance: a theoretic extension and empirical test. J. Comput. Inf. Syst. 49(1), 17–26 (2008)

Facebook. Facebook's manual on credible threats of violence. The Guardian (2017). https://www.theguardian.com/news/gallery/2017/may/21/facebooks-manual-on-credible-threats-of-violence. Accessed 24 May 2018

Fan, X., et al.: The influence of agent reliability on trust in human-agent collaboration. In: Proceedings of the 15th European Conference on Cognitive Ergonomics: The Ergonomics of Cool Interaction, ECCE 2008, pp. 7:1–7:8. ACM, New York (2008). https://doi.org/10.1145/1473018.1473028. Accessed 16 Oct 2018

Fogg, B.J., Tseng, H.: The elements of computer credibility. In: Proceedings of the SIGCHI Conference on Human Factors in Computing Systems, CHI 1999, pp. 80–87. ACM, New York (1999). https://doi.org/10.1145/302979.303001. Accessed 2 July 2018

Gajanan, M.: Former content moderator sues Facebook, claiming the job gave her PTSD. Time (2018). https://time.com/5405343/facebook-lawsuit-mental-trauma-content-ptsd/. Accessed 24 Oct 2018

Hoff, K., Bashir, M.: Trust in automation: integrating empirical evidence on factors that influence trust. Hum. Factors **57**(3), 407–434 (2015)

Hopkins, N.: Facebook moderators: a quick guide to their job and its challenges. The Guardian (2017). https://www.theguardian.com/news/2017/may/21/facebook-moderators-quick-guide-job-challenges. Accessed 18 May 2018

Hu, R., Pu, P.: A study on user perception of personality-based recommender systems. In: De Bra, P., Kobsa, A., Chin, D. (eds.) UMAP 2010. LNCS, vol. 6075, pp. 291–302. Springer, Heidelberg (2010). https://doi.org/10.1007/978-3-642-13470-8_27

Ji, X., Rau, Pei-Luen Patrick.: Development and application of a classification system for voice intelligent agents. Int. J. Hum.-Comput. Interact. **35**(9), 787–95 (2019)

Kamis, A., Davern, M.J.: Personalizing to product category knowledge: exploring the mediating effect of shopping tools on decision confidence. In: Proceedings of the 37th Annual Hawaii International Conference on System Sciences, 10 pp. (2004)

Rau, P.-L.P., Gong, Y., Dai, Y.-B., Cheng, C.: Promote energy conservation in automatic environment control: a comfort-energy trade-off perspective. In: Proceedings of the 33rd Annual ACM Conference Extended Abstracts on Human Factors in Computing Systems, CHI EA 2015, pp. 1501–1506. ACM, New York (2015). https://doi.org/10.1145/2702613.2732869. Accessed 11 Mar 2017

Rossille, D., Laurent, J.-F., Burgun, A.: Modelling a decision-support system for oncology using rule-based and case-based reasoning methodologies. Int. J. Med. Inform. **74**(2), 299–306 (2005)

The V-Chip: Options to Restrict What Your Children Watch on TV (2011). Federal Communications Commission. https://www.fcc.gov/consumers/guides/v-chip-putting-restrictions-what-your-children-watch. Accessed 29 June 2017

Wang, W., Benbasat, I.: Recommendation agents for electronic commerce: effects of explanation facilities on trusting beliefs. J. Manag. Inf. Syst. **23**(4), 217–246 (2007)

Identifying Design Feature Factors Critical to Acceptance of Smart Voice Assistant

Na Liu[1]([✉]), Ruoxuan Liu[2], and Wentao Li[2]

[1] School of Economics and Management, Beijing University of Posts and Telecommunications, Beijing, China
liuna18@bupt.edu.cn
[2] International School, Beijing University of Posts and Telecommunications, Beijing, China
{Liuruoxuan,lwt}@bupt.edu.cn

Abstract. The rapid development of artificial intelligence technology has made a variety of smart voice assistant products manufactured and widely used. Nevertheless, the influence of the design features of smart voice assistants on user acceptance is relatively vague. This study aimed to identify the design features that are critical to the acceptance of smart voice assistants by users. A questionnaire was designed and constructed. A total of 220 subjects participated in this survey. User acceptance is measured with perceived ease of use (PEOU), perceived usefulness (PU), attitude (AT) and intention to use (IU). Results showed that PU was significantly associated with smart voice assistant gender and PEOU. AT was significantly related to content diversity, personification, PEOU, and PU. IU was significantly related to sound quality, PU, and AT. The proposed model comprising demographic variables, design features, and acceptance-related variables could explain 49.3% of the variance in IU.

Keywords: Smart voice assistant · Acceptance · Design features · Usage behaviour

1 Introduction

Smart voice assistant is an emerging device and is becoming widely used in people's daily life. It is a platform composed of hardware and software which has the capability of understanding voice commands of a user and executes corresponding tasks. Smart voice assistants are integrated into various smart devices, such as smartphones, smart speakers, smart wearable devices, and smart TVs [1].

According to previous study, current tasks of smart voice assistants can be divided into five different types: (a) social communication tasks, such as voice calls and emails; (b) time management tasks, such as setting alarms and reminders; (c) information query tasks, such as reporting weather conditions; (d) entertainment and leisure tasks, such as playing music, telling jokes, chatting, and online shopping; (e) controlling Internet of things (IoT) device tasks, such as lights and air conditioners [2].

Design features are defined as the collection of human interface elements that users see, hear, touch, or operate [3]. The design feature, to some extent, will affect the user

© Springer Nature Switzerland AG 2021
P.-L. P. Rau (Ed.): HCII 2021, LNCS 12773, pp. 384–395, 2021.
https://doi.org/10.1007/978-3-030-77080-8_30

experience, and unacceptable design features will bring negative reviews. Since the smart voice assistant is a human-like virtual assistant, it is given a series of human characteristics, such as gender [4] and personality trait [5]. Besides, previous studies show that voice pitch [6], voice speed [7], sound quality [8] are important design features for virtual assistants. Personification is a customer's perception of the extent to which the virtual assistant understands and represents his or her personal needs [9]. It will affect virtual assistants' ability of the autonomous decision-making and the accuracy of understanding user's semantics. Also, the degree of personalized services will affect people's learning behaviour [10]. Thus, to upgrade smart voice assistant and make it more and more intelligent and personalized, designers pay more attention to design features such as the feedback modes [11], interaction style [12], and expressiveness (i.e., the language types) [13]. Different (physical or virtual) appearances of smart voice assistants have also been indicated as one important design feature of smart voice assistant [14]. As design features of smart voice assistant are becoming diverse, it is essential to identify design feature factors that could explain users' preference and affect the acceptance of smart voice assistants most.

Several models, such as theory of reasoned action (TRA) [15], theory of planned behavior (TPB) [16], technology acceptance model (TAM) [17], diffusion of innovation [18], and unified theory of acceptance and use of technology [19] (UTAUT), have been used to explain people's acceptance technology. Among the aforementioned models, TAM has been confirmed to be robust and powerful in evaluating technology acceptance. In addition to user's subjective perception of technology, the design features of interaction devices themselves can influence the perceived usability [20], which can also affect users' acceptance. However, few studies have examined the effects of design features on users' acceptance of smart voice assistants.

In summary, the present study aims to identify critical design features of smart voice assistant and to examine the relationships between design feature factors and users' acceptance of smart voice assistants. Results will provide implications for the design of smart voice assistant.

2 Method

2.1 Questionnaire Construction

The questionnaire is composed of four parts: the users' demographic information, the users' preference for the design features of the smart voice assistant, usage behavior and acceptance of the smart voice assistant.

The part of demographic information collects participants' gender, age, occupation, income, and education, and asks users to select the smart voice assistants that they have used.

In the part of design feature factors, we initially designed a questionnaire with 20 design features related to smart voice assistants based on previous research. After that, we asked experts who are familiar with this research topics and questionnaire design in the field of human-computer interaction to evaluate and answer the questionnaire. They pointed out questionable words and sentences, problems that can cause misunderstandings, and repeated or omitted items. According to their evaluation, we revised five semantically ambiguous items in the questionnaire, replaced the words and sentences of the three questions, deleted duplicate questions and added missing items. In the end, a total of 20 smart voice assistant design features were displayed in the questionnaire. Among them, the content diversity, intelligence, task interruption, diversity of solutions and response accuracy are features from interviewing experts and the rest of them are from previous study. Seven-point Likert scale was used to measure users' preference for the design features of the smart voice assistant ("1" = "least important", "7" = "very important") (Table 1).

Table 1. Design feature items

Items	Design features	Source	Items	Design features	Source
V1	Gender	[4]	V11	Diversity of solutions	Interview
V2	Personality trait	[5]	V12	Task interruption	Interview
V3	Vocal pitch	[6]	V13	Feedback mode	[11]
V4	Voice speed	[7]	V14	Service personalization degree	[10]
V5	Language style	[12]	V15	Autonomous decision-making ability	[9]
V6	Sound quality	[8]	V16	Content diversity	Interview
V7	Types of languages	[13]	V17	Personification	[9]
V8	Types of function	[2]	V18	Physical appearance	[14]
V9	Response accuracy	Interview	V19	Intelligence	Interview
V10	Reaction speed	[12]	V20	The accuracy of understanding users' semantics	[9]

The part of users' acceptance consisted of items measuring PEOU, PU, AT and IU. TAM describes PU within the working context [17]. In the broad context of smart voice assistant acceptance, smart voice assistants are available for use at any time and any place. Thus, in this study, PU is defined as how well smart voice assistants can be integrated into users' daily activities [21]. PEOU is defined as the degree to which an individual believes that using a particular system would be free of effort [17]. Attitude is defined as the user's feelings and perceptions in using smart voice assistant [22]. Three items measured PEOU, three items measured PU, three items measured AT and two items measured IU, all of which were adapted from Davis et al. [17] (Table 4). All the measurement items used here have been widely used and validated in prior studies [19]. We also use the seven-point Likert scale to measure the user's acceptance of smart voice assistants, that is, "1" = "strongly disagree" and "7" = "strongly agree".

The questionnaire also collected the participants' behavior of using smart voice assistants. Current tasks performed by smart voice assistants include the followings: social communication tasks, such as voice calls and emails; time management tasks, such as setting alarms and reminders; information query tasks, such as reporting weather conditions; entertainment and leisure tasks, such as playing music, telling jokes, chatting, and online shopping; and controlling Internet of things (IoT) device tasks, such as lights and air conditioners [2]. We adopted the five-point Likert scale to measure usage behavior, that is, "1" = "never used" and "5" = "always used".

2.2 Data Collection

Data was collected through online surveys. This online questionnaire was published through a Chinese website, and all Internet users with Internet mobile devices or computers can access this questionnaire. In order to make participants understand the goals of the questionnaire, we introduced the background and purpose of the questionnaire to the participants at the beginning of the questionnaire. At the same time, we also provided participants with a text description at the beginning of each part of the questionnaire to help participants understand how to answer the questions. Also, the questionnaire also sets up logical questions, that is, only users who have used the smart voice assistant can participate in the survey. In addition, in order to ensure the validity of the data collected in the questionnaire, we also set up two pairs of inverse questions to verify the data. For example, when measuring user attitude, we set "I like to use smart voice assistants" and "I don't like to use smart voice assistants".

2.3 Data Analysis

After completing the data collection, we screened the data depending on the inverse question designed in the questionnaire, the speed of answering the questionnaire, and whether the required questionnaire items are filled in. Then, we deleted invalid data. Finally, we got 220 valid questionnaire data. After that, we conducted a hierarchical regression analysis to examine the relationships between the independent variables, i.e., demographic information of participants and the design features of smart voice assistant, and the dependent variables, i.e., PEOU, PU, AT, and IU.

3 Results

3.1 Descriptive Statistics

Based on the collected valid data, we summarized the demographic information of 220 participants (Table 2). Among the 220 participants, 96 were male and 124 were female. Their ages range from 20 to 60 years old, among which the 21–30 years old has the largest number of participants, accounting for 64.5% of the total. More participants use Mi XiaoAi and Apple Siri than other assistants.

For usage behavior of smart voice assistants, participants use the information query, entertainment and leisure functions of the smart voice assistant more frequently than the controlling IoT device and social communication (Table 3).

The means and standard deviations of the responses related to the smart voice assistant acceptance items are listed in Table 4. Smart voice assistants seem to be highly accepted by participants. The participants considered that smart voice assistants were easy to use ($M_{PEOU} = 6.07$, $SD_{PEOU} = 0.720$) and highly useful ($M_{PU} = 6.04$, $SD_{PU} = 0.781$) in their daily life. They possessed strong intentions to use smart voice assistants ($M_{IU} = 6.44$, $SD_{IU} = 0.685$).

Table 2. Demographic information of participants (N = 220)

Variable	Category	Frequency	Percentage
Age	0–20	6	2.7
	21–30	142	64.5
	31–40	66	30.0
	41–50	5	2.3
	51–60	1	0.5
Gender	Male	96	43.6
	Female	124	56.4
Education	PhD degree	2	0.9
	Master degree	20	9.1
	Bachelor degree	171	77.7
	College degree	19	8.6
	High school or below	8	3.6
Monthly income (CNY)	<3000	22	10.0
	3000–5000	30	13.6
	5000–7000	63	28.6
	7000–9000	54	24.5
	9000–11000	31	14.1
	>11000	20	9.1
Product of smart voice assistants	Mi XiaoAi	148	27.9
	Apple Siri	143	26.9
	HUAWEI Celia	114	21.5
	Oppo	53	10.0
	Microsoft Cortana	33	6.2
	Samsung Bixby	23	4.3
	Google Assistant	10	1.9
	Amazon Alexa	7	1.3

Table 3. Smart voice assistant usage profile

Items	Mean	SD
Social communication	3.76	0.989
Time management	3.95	0.824
Information query	4.00	0.653
Entertainment and leisure	4.14	0.666
Controlling internet of things (IoT) device	3.71	1.201

Table 4. Means and standard deviations of smart voice assistant acceptance

Items	M	SD
Perceived ease-of-use	6.1	0.7
PEOU1: Smart voice assistant are easy to use	6.2	0.8
PEOU2: My interaction with the smart voice assistant is clear and understandable	5.9	1.0
PEOU3: I could be skillful at using smart voice assistant	6.2	0.9
Perceived usefulness	6.0	0.8
PU1: Using a smart voice assistant would enhance my effective ness in daily life	5.7	1.0
PU2: Using a smart voice assistant would make my life more convenient	6.2	0.9
PU3: I think smart voice assistant are useful in my daily life	6.2	0.8
Attitude	6.2	0.6
AT1: I think it's good to use a smart voice assistant	6.2	0.8
AT2: I like to use smart voice assistant	6.3	0.7
AT3: I am satisfied with using the smart voice assistant	6.1	0.7
Intention to use	6.4	0.7
IU1: I am willing to use smart voice assistants	6.4	0.8
IU2: I intend to keep using my smart voice assistant	6.5	0.7

3.2 Multiple Linear Regression Analysis

Hierarchical multiple linear regression was performed using SPSS 24.0 to examine the relationships between the design features and acceptance of smart voice assistant. Results are summarized in Table 5. Results showed the amount of variance of the dependent variables (PEOU, PU, AT, and IU) that could be explained by the independent variables, i.e., demographic variables and smart voice assistant design feature variables. Demographic variables (age, gender, education, and income) were inputted in the first model. Design feature variables and acceptance-related variables (PEOU, PU, AT and IU) were added in the second model. The percentage of variance R^2 and the change of R^2 were both calculated.

In the first model, income was significantly related to all dependent variables and age was significantly associated with IU. User gender, occupation and education did not significantly relate to any dependent variables. In general, smart voice assistants were more accepted and were used more often by high-income and young users. And these users were more likely to perceive the smart voice assistants as easy to use. The design feature variables and acceptance-related variables, i.e., PEOU, PU, AT and IU were added in the second model. The changes of R^2 were all significant (p's < 0.001), and the R^2 value for the four dependent variables in the second model were all larger than those in the first model. Thus, the second model could explain more variance in dependent variables than the first model.

The design feature factors smart voice assistant gender, response accuracy, and personification had significant associations with PEOU. PU was significantly associated with the design feature factors smart voice assistant gender and PEOU. AT was significantly related to content diversity, personification, PEOU, and PU. IU was significantly related to sound quality, PU, and AT. The positive associations between a specific design feature factor and acceptance-related constructs suggested that if more design features meet the preference of users, then users would be more likely to accept the product. Table 5 shows that the integration of demographic variables and design feature variables could explain 37.0% of the variance in PEOU. The demographic variables, design features, and PEOU accounted for 47.1% of the variance in PU. The demographic variables, design features, PEOU, and PU could explain 59.1% of the variance in users' attitudes. The proposed model comprising demographic variables, design features, and acceptance-related variables could explain 49.3% of the variance in the intention to use.

4 Discussion

The present study showed that users' preference for smart voice assistants consisted of twenty design feature factors: smart voice assistant gender, personality trait, voice pitch, voice speed, language style, sound quality, types of language, types of function, response accuracy, reaction speed, diversity of solutions, task interruption, feedback mode, service personalization degree, autonomous decision-making ability, content diversity, personification, physical appearance, intelligence, the accuracy of understanding users semantics.

The study also examined the relationships between design feature factors and acceptance. Results indicated that critical design feature factors that contributed to the acceptance of smart voice assistants were smart voice assistant gender, response accuracy, personification, content diversity, and sound quality. Those design feature factors influenced acceptance of smart voice assistants either directly or indirectly by influencing the acceptance-related constructs (i.e., PEOU, PU, and AT). The results supported the notions from the TAM that PEOU has a significant and positive influence on PU and that PEOU and PU are capable of predicting the intention to use [22].

Results provide implications for the design of smart voice assistant. A smart voice assistant can be more accepted by users if it meets users' preference for design feature factors. Similar to previous studies, sound quality can affect the usability of smart voice assistant [8]. In addition to sound quality, the voice gender of smart voice assistants can

Table. 5. Regression results

Independent variable		Dependent variable							
Model		PEOU		PU		AT		IU	
		Standard coefficients	t	Standard coefficients	t	Standard coefficients	t	Standard coefficients	t
1	User gender	-0.094	-1.369	0.017	0.254	0.083	-0.104	-0.07	-1.047
	Age	-0.107	-1.521	-0.044	-0.643	0.069	-0.129	-0.171**	-2.478
	Occupation	0.08	1.141	0.066	0.976	0.011	0.113	0.115	1.689
	Monthly income	0.206***	2.813	0.331***	4.646	0.031**	0.279	0.279***	3.891
	Education	-0.001	-0.021	-0.022	-0.316	0.071	-0.045	-0.061	-0.891
	R^2	0.039		0.091		0.082		0.079	
2	User gender	-0.012	-0.203	0.051	0.948	0.059	-0.021	-0.012	-0.222
	Age	-0.048	-0.752	0.026	0.451	0.052	-0.053	-0.144	-2.52
	Occupation	0.06	1.035	0.012	0.231	0.007	0.087	0.063	1.196
	Monthly income	0.186	3.027	0.193	3.347	0.023	0.063	0.084	1.449
	Education	-0.065	-1.096	-0.015	-0.277	0.049	-0.047	-0.045	-0.837
	Smart voice assistant gender	0.238***	3.432	0.056*	0.855	0.021	0.093	0.081	1.258
	Personality trait	-0.095	-1.277	-0.044	-0.638	0.03	0.045	-0.037	-0.559
	Voice pitch	0.063	0.752	0.035	0.45	0.038	0.004	0.005	0.072
	Voice speed	0.046	0.587	-0.04	-0.566	0.039	-0.031	-0.007	-0.1

(continued)

Table. 5. (*continued*)

Independent variable	Dependent variable							
Model	PEOU		PU		AT		IU	
	Standard coefficients	t	Standard coefficients	t	Standard coefficients	t	Standard coefficients	t
Language style	0.01	0.142	−0.089	−1.409	0.035	0.053	−0.031	−0.494
Sound quality	−0.05	−0.725	0.018	0.292	0.043	0.019	0.167***	2.691
Types of languages	0.059	0.891	−0.006	−0.096	0.026	−0.044	−0.042	−0.709
Types of function	0.083	1.221	0.138	2.196	0.04	0.013	0.025	0.403
Response accuracy	0.148**	2.003	0.011	0.157	0.059	0.045	0.058	0.869
Reaction speed	0.113	1.373	0.107	1.412	0.055	−0.117	−0.1	−1.328
Diversity of solutions	0.068	1.032	−0.061	−1.014	0.042	0.022	−0.05	−0.846
Task interruption	−0.036	−0.521	0.06	0.953	0.033	−0.078	−0.03	−0.481
Feedback mode	−0.073	−1.029	−0.045	−0.694	0.039	−0.055	−0.031	−0.477
Service personalization degree	−0.033	−0.453	0.065	0.972	0.038	−0.077	−0.112	−1.712

(*continued*)

Table 5. (*continued*)

Independent variable	Dependent variable							
Model	PEOU		PU		AT		IU	
	Standard coefficients	t	Standard coefficients	t	Standard coefficients	t	Standard coefficients	t
Autonomous decision-making ability	0.082	1.181	0.103	1.608	0.035	0.006	−0.046	−0.731
Content diversity	0.119	1.579	−0.024	−0.348	0.045**	0.192	0.054	0.776
Personification	0.176**	2.319	−0.001	−0.016	0.036**	0.2	0.065	0.915
Physical appearance	−0.074	−0.995	0.047	0.685	0.024	−0.033	−0.06	−0.894
Intelligence	0.101	1.322	−0.011	−0.159	0.048	0.007	0.035	0.511
The accuracy of understanding users' semantics	0.07	0.967	−0.016	−0.242	0.051	−0.105	0.023	0.346
PEOU			0.505***	7.671	0.056***	0.303	0.023	0.303
PU					0.05***	0.426	0.352***	4.496
AT							0.358***	4.458
R^2	0.370		0.471		0.591		0.493	
ΔR^2	0.382***		0.423***		0.538***		0.458***	

Note: $*p < 0.05$, $**p < 0.01$, $***p < 0.0001$

also influence users' acceptance of smart voice assistant. It can be explained by findings that the stereotypical notion of women as having caring, sincere, and empathic communication styles [23], which are crucial in increasing users' satisfaction [24]. Consistent with previous studies, smart voice assistant gender has a positive effect on customers' attitude, and on customers' online purchase intention [25]. Aside from sound quality and smart voice gender, the present study also identified response accuracy, personification and content diversity as crucial factors to users' acceptance. It suggests that designers and manufactures of smart voice assistant should pay more attention to those design features in order to improve user's acceptance of smart voice assistant. Moreover, how those design feature factors influence users' usage behavior of smart voice assistant should be investigated in future studies.

Acknowledgements. This work was supported by grants from Natural Science Foundation of China (Project No. 71901033) and Beijing Natural Science Foundation (Project No. 9204029).

References

1. Dousay, T.A., Hall, C.: Alexa, tell me about using a virtual assistant in the classroom. In: EdMedia + Innovate Learning, pp. 1413–1419. Association for the Advancement of Computing in Education (AACE) (2018)
2. Yang, H., Lee, H.: Understanding user behavior of virtual personal assistant devices. Inf. Syst. e-Bus. Manag. **17**(1), 65–87 (2018). https://doi.org/10.1007/s10257-018-0375-1
3. Han, S.H., Yun, M.H., Kim, K.J., Kwahk, J.: Evaluation of product usability: development and validation of usability dimensions and design elements based on empirical models. Int. J. Ind. Ergon. **26**(4), 477–488 (2000)
4. Beldad, A., Hegner, S., Hoppen, J.: The effect of virtual sales agent (VSA) gender – product gender congruence on product advice credibility, trust in VSA and online vendor, and purchase intention. Comput. Hum. Behav. **60**, 62–72 (2016). https://doi.org/10.1016/j.chb.2016.02.046
5. Ludewig, Y., Döring, N., Exner, N: Design and evaluation of the personality trait extraversion of a shopping robot. In: The 21st IEEE International Symposium on Robot and Human Interactive Communication, Paris, France, pp. 372–379. IEEE (2012)
6. Elkins, A., Derrick, D.: The sound of trust: voice as a measurement of trust during interactions with embodied conversational agents. Group Decis. Negot. **22**(5), 897–913 (2013)
7. Aiello, J.R., Douthitt, E.A.: Social facilitation from Triplett to electronic performance monitoring. Group Dyn. Theory Res. Pract. **5**(3), 163–180 (2001)
8. Chiou, E.K., Schroeder, N.L., Craig, S.D.: How we trust, perceive, and learn from virtual humans: the influence of voice quality. Comput. Educ. **146**, 103756 (2020)
9. Komiak, S.Y., Benbasat, I.: The effects of personalization and familiarity on trust and adoption of recommendation agents. MIS Q. **30**(4), 941–960 (2006)
10. Craig, S.D., Schroeder, N.L.: Reconsidering the voice effect when learning from a virtual human. Comput. Educ. **114**, 193–205 (2017)
11. Saad, U., Afzal, U., El-Issawi, A., Eid, M.: A model to measure QoE for virtual personal assistant. Multimed. Tools Appl. **76**(10), 12517–12537 (2016). https://doi.org/10.1007/s11042-016-3650-5
12. Chattaraman, V., Kwon, W.S., Gilbert, J.E., Ross, K.: Should AI-Based, conversational digital assistants employ social- or task-oriented interaction style? A task-competency and reciprocity perspective for older adults. Comput. Hum. Behav. **90**, 315–330 (2019)

13. Moreno, R., Mayer, R.E.: Personalized messages that promote science learning in virtual environments. J. Educ. Psychol. **96**(1), 165–173 (2004)
14. Shiban, Y., Schelhorn, I., Jobst, V., et al.: The appearance effect: influences of virtual agent features on performance and motivation. Comput. Hum. Behav. **49**, 5–11 (2015)
15. Fishbein, M., Ajzen, I.: Belief, Attitude, Intention and Behavior: An Introduction to Theory and Research. Addison-Wesley, Massachusetts (1975)
16. Ajzen, I.: The theory of planned behavior. Organ. Behav. Hum. Decis. Process. **50**(2), 179–211 (1991)
17. Davis, F.D., Bagozzi, R.P., Warshaw, P.R.: User acceptance of computer technology: a comparison of two theoretical models. Manag. Sci. **35**(8), 982–1003 (1989)
18. Rogers, E.M.: Diffusion of Innovations, 4th edn. Etats-Unis Free Press, New York (1995)
19. Venkatesh, V., Morris, M.G., Davis, G.B., Davis, F.D.: User acceptance of information technology: Toward a unified view. MIS Q. **27**(3), 425–478 (2003)
20. Zhang, T., Rau, P.L.P., Salvendy, G.: Exploring critical usability factors for handsets. Behav. Inf. Technol. **29**(1), 45–55 (2010)
21. Kleijnen, M., Wetzels, M., de Ruyter, K.: Consumer acceptance of wireless finance. J. Finan. Serv. Mark. **8**(3), 206–217 (2004)
22. Davis, F.D.: Perceived usefulness, perceived ease of use, and user acceptance of information technology. MIS Q. 319–340 (1989)
23. Cameron, D.: Styling the worker: gender and the commodification of language in the globalized service economy. J. Sociolinguistics **4**(3), 323–347 (2000)
24. Sparks, B.A., Bradley, G.L., Callan, V.J.: The impact of staff empowerment and communication style on customer evaluations: the special case of service failure. Psychol. Mark. **14**(5), 475–493 (1997)
25. Beldad, A., Hegner, S., Hoppen, J.: The effect of virtual sales agent (VSA) gender–product gender congruence on product advice credibility, trust in VSA and online vendor, and purchase intention. Comput. Hum. Behav. **60**, 62–72 (2016)

NEXT! Toaster: Promoting Design Process with a Smart Assistant

Qing Xia and Zhiyong Fu[✉]

Tsinghua University, Beijing 100084, China
fuzhiyong@tsinghua.edu.cn

Abstract. This paper introduces the smart activity promotion assistant NEXT! Toaster prototype. In order to solve the "isolated dilemma" in design activities, NEXT! Toaster matches the appropriate design thinking tools to the activities, prints tools by phases, guides and trains designers to control the design process and achieve the design goals. This study also explores the dynamics of design activities with robotic members, the characteristics of participants in the process of human-machine co-creation, and suggestions for the development of the design robotic assistants.

Keywords: Design tool · Human-machine collaboration · Interaction design

1 Introduction

For the problem of "isolated dilemma", we make a tool to resolve the deadlock in design activities and push the design process smoothly. When experts plan design activities, they use design tools to ensure the controllability. But unskilled students or non-designers are not proficient in selecting and using design tools. We designed a prototype to match the appropriate design tools for self-organizing design activities, to control the progress as expected.

The prototype has two important components. The core part is a design tool library, and the shell part is an interactive desktop box controlled by web pages. In use, after being set with the basic situations and design goals, the prototype will automatically customize the tool set, calculate the usage time, and phased print them on site. Considering the process of setting-producing-popping out looks like the scene when a toaster bakes toast, we name the prototype "NEXT! Toaster". When the team complete the phased task, just ask "What to do Next! Toaster", it will print the next design tool immediately.

We tested 13 different sets of design activities to verify NEXT! Toaster. Through the analysis of on-site observing, video records, backstage data, questionnaires and key interviews, we examined the availability of the robotic assistant and obtained the user feedbacks. With the experiment of the prototype, we try to explore the features of human-machine creation, and to search for the revelation of co-creation robot design.

P.-L. P. Rau (Ed.): HCII 2021, LNCS 12773, pp. 396–409, 2021.
https://doi.org/10.1007/978-3-030-77080-8_31

2   The Real Condition of Team Design

We did an research on team design, in which we tracked the design activities of 6 groups. We only observe as a bystander with no intervention, hoping to discover the pain points of group design activities, and the needs of the participants. The activities are mainly attended by design students and their mentors. The types of activities include the homework assignment (H), the project discussion (P) and the 1-day workshop (W) (Table 1).

Table 1. Basic situation of design activities tracked

	Number of participants	Type of participants	Length of the activity
Activity 1 (H)	4p	Undergraduates	3 h
Activity 2 (P)	4p	Undergraduates	6 h
Activity 3 (W)	4p	Graduates	3 h
Activity 4 (P)	4p	Graduates	6 h
Activity 5 (P)	5p	Graduates, Supervisor	3 h
Activity 6 (P)	6p	Undergraduates, Supervisor	3 h

P = persons; h = hours

During the track, we found some deadlock time of design activities, such as silence, stagnation, dispute etc. In the situation, team members are trapped in the unmanageable problems without the help from the outside world, which seriously affects the design process and results. We name the situation "isolated dilemma".

2.1   Isolated Dilemma

In order to clarify the specific problems of "isolated dilemma" and give corresponding solutions, we've recorded the main problems (problems occur in every activity on average, O/S = the number of occurrences/the number of activities > 1) through video analysis, and summarized the response solutions used by design teams as references (Table 2).

Currently team members are more inclined to seek solutions from within. In a few cases, they can get out of trouble by introducing more members or asking experts for help. In contrast to self-organizing ones, creative activities designed by professionals often use lots of design tools to plan a relatively definite thinking path, to ensure that all team members of different disciplines can successfully achieve the design goals. In fact, existing design tools can help solve most of the problems in design process, but unskillful designers often have no experience or time to prepare tools, so they still rely on manpower to overcome the difficulties.

Table 2. Specific problems of "isolated dilemma"

Specific problems	O/S	Solutions now
Discussion runs off the mark	5.3	Interrupt
Discussion is too divergent	4	Interrupt; Vote; Return or redo
Long pause/silence	3.8	Rest; Change the topic
No new ideas	3.7	Rest; Change the topic; Use existing ideas
No consensus	3.5	Vote; Introduce members; Shelving dispute
Not satisfied with existing ideas	2.5	Return or redo; Change the topic; Ask others for help; Introduce members
Progress lags/exceeds the expected time	2.3	Truncate process; Extend the time
Unclear goals/no milestones	1.3	Restart

2.2 Methods to Solve the Isolated Dilemma

The original intention of design tools is to help designers to solve problems in the design process through specific methods. Classic design tools, such as Stakeholders, Shadowing, Culture probes, for understanding the user; Persona, Empathy map, Journey map, POV, for defining the goal; Brainstorming, 6 Thinking Hats, for ideating; Tomorrow headline, Wizard of Oz, Business Origami, for prototyping, Heuristic evaluation, Sematic differential Scale, Feedback Capture Grid, for testing, etc. A large number of tools have been designed to solve various problems in the design process. However, in practice, how to choose and how to plan the application of them is still a difficult problem for designers.

If we design a smart tool who can quickly give available tools, we can help creative activities in good order by matching design needs and adaptive solutions. In the short term, it can promote the running of activity process, make up for the lack of roles; In the long term, it can train designers to handle the corresponding solutions, call the design tools more freely in the future work, to achieve the growth of the designer's ability.

2.3 Solving the Isolated Dilemma by the Robot

We imagine a robot collaborate with us when designing. As a member of the team, it helps us break through deadlocks of the process, so we just focus on giving up more ideas with human wisdom. Human and robot play their respective strengths in co-creation.

We hope to design an easy-to-use robotic tool for self-organized activities. Input with pain point or need by designers, it can help to achieve the goal by matching the appropriate design tools. The function of the robot is very similar to the facilitators, we can call it the robot facilitator at present.

3   Molding Robot Facilitator NEXT! TOASTER

Robot facilitator is an assistant who carries forward the progress of self-organizing creative activities. It retrieves the out-of-control process by matching the appropriate design tools. The design of the robot facilitator can be divided into two parts, one part is the content, the design tool library, the other part is the shell, the user's controller and tool producing device.

3.1 Library of Design Tools

Adaptive design tools are the software and soul of the machine facilitator. To link the practical problems in design process to more than 100 design tools we usually use, we add description tags to design tools to simplify the search. We organized a 6-h workshop to tag 106 commonly used design tools, with participants of 20 senior designers (Minimum 5 years on practice of design tool). The tags include basic usage information and scenarios, corresponding to the basic situation of the design activities and the purpose of use (Table 3).

Table 3. Tags of design tools

Tag	Tagging method	Corresponding filter
Duration of use	Fill in the number	Activity duration
Number of users	Fill in the number	Number of team
Usage scenario	Custom tag/unlimited	Activity scene
Design theme	Custom tag/unlimited	Design theme
Output type	Custom tag/unlimited	Output type
Difficulty of use	Select level	User ability
Purpose of use	Custom tag/unlimited	Goal of use
Characteristics of tool	Custom tag/unlimited	Feature of activity

Through the workshop we collect 8 sets of tags defined by designers for each tool, which constitute the basic content of the design tool library. On the collection of labels, we adopt the principle of maximizing the scope, that is, taking a collection of results labelled by all designers. And we make the reserved labels under the input box of Goal of use, so that the user can quickly find the corresponding design tool set (Table 4).

When matching the design tools, the rules are set to prioritize the goal of use, the activity duration, the number of team, the user ability as the basis for filtration, followed by the activity scene, the design theme, the output type, and the feature of activity as 4 auxiliary labels.

For some complex design goals, the system matches a tool package. We have pre-arranged basic packages adapted to different conditions, which is also developed by 20 senior designers. In different scenarios, packages can be adjusted to the activity situation, and tools in the package can be replaced according to the actual needs. Taking a 3-h workshop as the example, the goal of the activity is to find a typical problem under the theme of education. We pre-designed a three-phase process, association-ideation-concentration, and then replace some of the tools preset with the alternative ones (Table 5).

Table 4. Problems and corresponding design tools

Problems	Corresponding design tool
Discussion runs off the mark	Question ladder, Cause and effect diagram, Fishbone, Point-of-view statement, Mind map, Six thinking hats, Creative pause
Discussion is too divergent	Question ladder, Cause and effect diagram, Fishbone, Card sorting, Affinity diagrams, 2 × 2 matrix, Mind map, Six thinking hats, Fast idea generator, Creative pause
Long pause/silence	Conversation starter, Myers-Briggs team indicator
No new ideas	5 Whys, Graffiti wall, Question ladder, Cause and effect diagram, Fishbone, Value proposition map, Empathy map, User Journey map, Brainstorm, Team idea mapping method, Mind map, Design mash-up, Six thinking hats, Fast idea generator, Lotus blossom diagram
No consensus	2 × 2 matrix, Dot voting, How Now Wow Matrix, $100 test, Four categories method, Idea selection criteria, 5-s usability test, A/B testing
Not satisfied with existing ideas	5 Whys, Graffiti wall, Question ladder, Cause and effect diagram, fishbone, Affinity diagrams, 2 × 2 matrix, KANO model, Provocation, Worst possible idea, Mind map, Design mash-up, Six thinking hats, Fast idea generator, Lotus blossom diagram
Progress lags/exceeds the expected time	Timing, Personal kanban, Task analysis grid
Unclear goals/no milestones	Point-of-view statement, Design brief, Idea selection criteria, Personal Kanban, Task analysis grid

Table 5. Tool package and alternative tool

Process	Corresponding design tool	Duration	Alternative tool
Association	Lotus Blossom Diagram	1 h	Brainstorm, Team idea mapping method, Mind map
Ideation	Design Mash-up	1.5 h	5 Whys, Question ladder, Six Thinking Hats, Fast Idea Generator, Provocation, Worst Possible Idea
Concentration	How Now Wow Matrix	0.5 h	2 × 2 matrix, Dot Voting, Four categories method, Idea Selection Criteria, A/B testing

For the labels needed but not in the library (and there is no approximation), they will be manually arranged by managers under the system notification, and then, the phased process and the alternative tools are automatically recorded. At the same time, in order to collect the potential labels comprehensively, we have set up the user feedback function. After using the tool, creators will be asked to evaluate and label them in the system to increase the inclusiveness of the library (until now, the system only collects labels which is manually filtered into the library). Through the mechanism, the design tool library is gradually improved with the accumulation of usage.

3.2   Shell of the Robot Facilitator

With the content of design tools, we plan to design a shell for the robot facilitator with the accessible interaction for the creators. Functional requirements for the shell include collecting the needs of design team, quickly printing design tools, prompting the design progress, and recording the feedback on usage. We divide these functions into two parts, one for customizing and recording, the other one for producing. For the convenience of every participant, we put the first part on the website and turned the second part into a desktop box (Fig. 1).

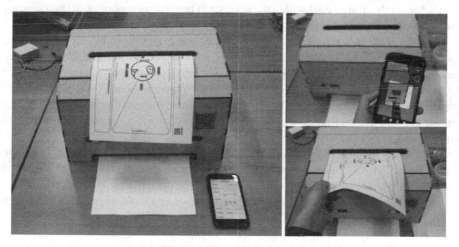

Fig. 1. The robot facilitator

The webpage is called by scanning the QR code on the desktop box. The information that users submit on webpages automatically matches design tools in library, and the system pushes the result to user's webpage containing the tool package and the duration of use (automatic countdown). Users can make design tools with the desktop box by controlling on the web page or talk to the box. After design activities, the webpage is also responsible for collecting the user-defined labels of tools used.

For the desktop box, the most important component is the printer inside printing design tools. On every desktop box, it is laser-fired with a unique QR code, making users easily to link. There are two custom hardware buttons that can be used to set relatively fixed values (the default settings are group number and tool use experience, which are relatively stable for teams). Users can also change or disable the hardware button on the phone webpage. The invisible part of the box is equipped with a sound collector and an amplifier to collect group discussion voice (under permission) and playing voice prompts.

When providing design tools, the desktop box prints pages of A4-sized tool sheet on site, which presents the name of the tool, the introduction, the use method, the use area, and an QR code for the case URL. Users can get a more detailed description of the use case by scanning the QR code.

3.3 Use the Robot Facilitator

Let's imagine a scenario where a group use the robot facilitator box to work with them. They fill in and submit the activity situation and the design goal through their phone. The machine facilitator immediately matches three tools and prints the first tool sheet by the desktop box. The team members rapidly use the tool to start ideating. When the countdown is over, the box gives out the second piece of tool sheet, a tool that helps the group focus and think deeply. This time, the team ahead of time, they ask "What to do NEXT!" to the box. Then it makes the last tool to build the prototype. After all, the team achieves the design goal within the set time. Then, on their phones, all participants evaluate and label the tools they used.

Setting - making - giving out, the use experience is similar to baking toast with a toaster. We set the basics information and needs, wait for production, and then take away the results that have been customized for us. So, we use toaster as a metaphor for the desktop box which produces design tools on site. When the phased task is completed ahead of time, you can say "NEXT!", it will prepare you another tool sheet. When naming the machine facilitator, we intercepted directly the most common words that we interact with it, that is "NEXT! Toaster".

By building up the design tool library, the desktop box and the matching service, we have designed NEXT! Toaster, an assistant to carry forward the design process smoothly. After entering the basic situation, state and goal of the design activity, the assistant will automatically help you plan the activity stages, make design tools, to get rid of the design dilemma.

4 Test NEXT! Toaster in Design Activities

To search the real situation where the human and the robot work together on design activities, we arranged actual workshops to test NEXT! Toaster. All topics in workshops were designing for education.

Before the test, we did not inform that the workshop was a tool test, but rather a design activity to solve the real problem. In the course of the workshops, NEXT! Toaster was involved as an auxiliary tool. We want to test the usage data of NEXT! Toaster in the actual interactive scenarios, as well as the dynamics working with robot facilitator.

We recruited 10 workshops for experiments (Table 6). Excluding CCM.1, CCM.2 experiments in the summer program City Change Maker, C.1 experiment in undergraduate course Interactive Design, the rest of the group experiments were specifically planed for testing NEXT! Toaster.

To ensure that the tests were as close as possible to real use, the participants were all design practitioners or design students. The experimental sites were spaces in a design innovation center.

Table 6. The code name of the experimental team

	2p	4p	6p	8p	10p	20p
	1team	1team	1team	1team	2teams (4p + 6p)	4teams (3p + 5p + 5P + 7p)
1 h	W.1	W.2	W.3	W.4	W.5-G4, G6	W.6-G3, G5s, G5f, G7
3 h	–	W.7	W.8	–	–	–
6 h	–	W.9	W.10	–	–	–
10d	–	CCM.1	CCM.2	–	–	–
4w	–	C.1	-	–	–	–

h = hour(s); p = person(s)

In the arrangement of the experimental group, there are several special consideration variables to compare the actual situation of the creator's cooperation with the machine facilitator: 1) the number of teams; 2) The length of the design activity; 3) the impact of the scene atmosphere on the design activity; 4) the relationship between the team members.

In terms of team number control, we set seven team sizes for 2–8 persons, based on the number of common team work groups. In terms of the length of team activities, we set the duration of 5 different design activities according to reality, 1 h, 3 h, 6 h, 10 days and 4 weeks, which simulated the duration of a class, a half-day homework assignment, a 1-day workshop, a 2-week summer program, and a 4-week professional course. In terms of atmosphere simulation, we organized groups of 10 subjects and 20 subjects, in which we divided them into 2 teams (one for 4 and one for 6) and 4 teams (one for 3, two for 5, the last one for 7) to search if the design space influence the design activity when empty and quiet, or crowded and noisy. In the variable control of the team members, we used two 5-person teams in the W.6 session, in which the members of G5s team did not know each other and G5f members were all from the same start-up company, to observe the impact of inner relationship on the design process.

Based on the method research through design, we mainly used observation, questionnaire, key interview, combined with background data to subjects the usage situation. With observation we mainly examined the condition of overall progress of the design activities, the usage of NEXT! Toaster and design tools, the special moments of team members (e.g. silence, stalemate, cheering, etc.), the number of times the design team turned to volunteers, and what they asked for. Through the questionnaire we generally

Fig. 2. Test scenarios

studied the subjects' information and attitudes, expectations and suggestions to NEXT! Toaster. Interview method focused on the feedback from extreme users who are the most active members, the most negative members and who scored singularly in the questionnaire.

By testing the machine facilitator NEXT! Toaster in the actual scene, we hope to find: 1) the feasibility and user attitudes to using NEXT! Toaster as a facilitator to promote design process in creative activities, and the possible directions of improvement; 2) in the case of human-machine co-creation, the characteristics and needs that human designers will exhibit; 3) in designing human-machine collaboration tools, the main points we should pay attention to and the enlightenments.

5   Results and Feedbacks

After completing all 13 sessions (17 team) experiments, we analyzed the data from observations, questionnaires, key interviews and backstage data in three angles, the basics of the test, the usage situation of NEXT! Toaster, the feedback from subjects.

5.1   The Basics of the Test

The experiments involved 84 participants (average design experience 3.8 years). The design processes of 17 teams involved in the tests were largely as planned (the maximum timeout did not exceed 8% of the activity time). All the team worked out with successful design results. Depending on the length of the experiments, the output of 1-h design activities were "many ideas" and "1 selected idea", each 3-h activity produced an in-depth idea, each 6-h activity made the text prototype, each 10-day activity worked out with a draft prototype, and the 4-week activity designed a fine prototype.

5.2   The Usage Situation of NEXT! Toaster

During experiments, 15 teams chose to use NEXT! Toaster from start to arrange the overall design process. Most of the teams set goals only once at the beginning of the activity, and followed the recommends to advance the process. Only 2 teams of W.5-G6 and W.8 submitted different goals for the second time during the process. In addition, the 3 teams (CCM.1, CCM.2, C.3) with longer experimental duration and supervisor participation did not use NEXT! Toaster from the start, instead, turn to the machine assistant for help at the specific problems (Table 7).

Table 7. The reasons teams use NEXT! Toaster

Reason	Times	The team using
Think of more ideas	10	W.1, W.2, W.3, W.5-G4/G6, W.6-G3, G5s, CCM.1, CCM.2, C.1
Identify an idea	9	W.4, W.6-G5f, W.6-G7, W.7, W.8, W.9, W.10, CCM.2
Return to the topic	3	CCM.1, CCM.2, C.1
Come up better ideas	2	CCM.2, C.1
Break the silence	2	CCM.1, CCM.2
Filter ideas	1	W.5-G6
Change the method	1	W.8
Find a prototype method	1	CCM.2

Based on the cross-analysis of the backstage data and on-site observation (and video playback), we have summed up the situation analysis diagram (Fig. 2) of interactions between designers and NEXT! Toaster (diamond dots), between designers and volunteers (round dots). It also displays the combination of relative data, such as team size, group discussion atmosphere, etc. It needs to be explained that on CCM.1, CCM.2 and C.1 teams there's no mark on discussion atmosphere for the tool usage was only concentrated in the early stages of the activity, analysis on all the process did not works. In addition, the overtime situation has a small impact on the analysis, so each set of design processes has been stretched proportionally for the planned time (Fig. 3).

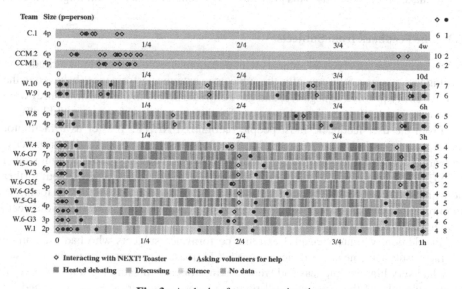

Fig. 3. Analysis of test scene situation

According to statistics, there are 78 calls to volunteers, 92 times of command on the webpage by the phone, and 54 design tool sheets produced during the design activities.

1) The more people there are in the team, the fewer calls to volunteers. The exception was team W.6-G5f, in which members were familiar with each other. After the test, we learned from the key interview that the team preferred to rely on guesswork to build their unique way of using the new tool, rather than asking volunteers. Members of the team were not nervous or shy whether they were wrong.

2) From the changing of times that how many users called the volunteers and interacted with the machine facilitator, it shows the strong learning ability of users. They can quickly explore the machine logic, make up for the lack of the machine design, and master the knowledge provided by the machine. In the study, we found that in the same group, the team would only ask the volunteers once facing the same problem and immediately developed their own unique solutions; In the 3 teams with long duration test (C.1, CCM.1, CCM.2), the same usage requirements did not appear in backstage records for the second time. The team replied that they can grasped the use method of design tools used and wouldn't ask NEXT! Toaster for help until for the new requirement.

3) There are 3 interesting findings about the use of design tool sheets. The most times of human-computer interaction occurred was that users scanned the QR code on design tool sheet for obtaining tool use cases. Most members of the teams used their own mobile phones scanning, some people even scanned more than once. Due to the limited size of the design tool produced on site, most groups drew the similar one on the A0 paper, and the printed A4 sheet is used only as a specification. The subjects took photographs of the tools after use, and some even took sequent photos to make process records. This provides the direction for subsequent improvements.

5.3 The Feedback from Subjects

After the test, we surveyed 84 subjects with questionnaires to find out their attitude and willingness to usage. Subsequently, the subjects in extreme performance (behaved in the test extremely positive or negative, and gave extreme scores in the questionnaire) were interviewed in depth to explore the reasons for the abnormal performance (Table 8).

From the user feedback, most people recognized NEXT! Toaster as a facilitator and a partner in design, and agreed that it could help them push the design process forward, broaden minds, and improve design results. The design tools recommended by NEXT! Toaster could basically meet the use scenarios and design needs, but further adjustments are needed to lower the threshold for use. Handling existing design tools still requires experience. For users less experienced, more help and guidance from volunteers is required at the beginning of design process.

In addition, we interviewed 3 extreme performance subjects who had a common characteristic, gave negative reviews on NEXT! Toaster, but contrary to the reviews, they had very high willingness and were looking forward to using them again. Two of them believed that the prototype still required some use experience, not designed for everyone. The other one explained that NEXT! Toaster was still a machine tool, and he

Table 8. The attitude and willingness of subjects

	Agree	Remain neutral	Not agree
Satisfied with the result	82.1%	10.7%	7.1%
Helped improve the result	85.7%	7.1%	7.1%
Helped broaden your mind	85.7%	10.7%	3.6%
Helped carry forward the process	86.9%	7.1%	6.0%
Tools recommended worked well	75.0%	20.2%	4.8%
Tools recommended suited for team design	76.2%	13.1%	10.7%
Worked well as the facilitator	85.7%	7.1%	7.1%
Willing to use when design independently	89.3%	7.1%	3.6%
Willing to use when design independently	77.4%	19.0%	3.6%

wished it could be involved in the team like a human being and to be a real discussion member.

6 Findings and Contributions

This study takes the prototype of NEXT! Toaster as a model and a trial, to verify the possibility of collaboration between human and robotic assistant, in which the two jointly work and play each advantage. With the design and experiment of this prototype, we extract some points from the process and results, hoping to show some findings and suggestions.

6.1 The Design Experience of NEXT! Toaster

NEXT! Toaster works on the problems in design process by creating new link to connect the needs in design progress and design tools existing. This approach attempts to take lightweight human-machine interaction, retaining what human and machine are efficient at in the collaboration, the fast searching ability of the machine and the creative ability of human. Test results show that NEXT! Toaster can basically meet the needs, support the overall process assistance or targeted problem solving.

6.2 The Characteristics and Needs of Human in Human-Machine Co-creation

From the situation analysis of the activity, it can be found that the creator is more willing to make tentative assumptions in an acquaintance environment. Especially to unfamiliar tools, the acquaintance team can fix roles more quickly, transiting from the role identifying phase to the role functioning phase, and explore design tools to form the unique ways of using earlier than others.

The learning ability of human has not deteriorated in the collaboration with the robot assistant. The backstage data shows that the design requirements varied for each use of NEXT! Toaster, which means the subjects digested what they used provided by the machine. The experience of co-creation with NEXT! Toaster, trains users to handle design process better. Mastering tools is an important part of human ability, and working with machines trained human to recognize and develop our strengths.

People are very inclusive to robot partners. The function of NEXT! Toaster is so simple that it seems plain and mechanized as a co-creator. But we got very positive feedbacks, and subjects showed a very high willingness to work with the robot partners again (even those who were not satisfied with this prototype).

6.3 Limitations and Future Work

NEXT! Toaster helps creators with design tools, so it is limited by the limitation of design tools. Besides optimizing the tool library, optimizing the interactive experience, adding emotional features, and improving the user-defined tool tags, we are trying to build up interaction approach to remedy the defect brought by the limitations of the simulation mechanism and the usage of design tools, to make NEXT! Toaster a real facilitator.

7 Conclusion

This Paper introduces a trial to solve the "isolated dilemma" of design activities by establishing the interaction between the needs of designers and existing design tools. When design process gets in deadlock, the robot facilitator NEXT! Toaster can provide design tools to help teams achieve goals. The test and dynamics analysis on NEXT! Toaster shows that it can basically meet the design needs of the team, including supporting the overall process assistance, or solving targeted specific problems. At the same time, the experience of co-creation with NEXT! Toaster, trains users to handle design process better.

Through the design of the human-machine collaboration assistant, we explore the features of human behaviors and internal needs in the co-creation scene, and put forward some suggestions for the design of human-machine collaboration tools. It is hoped that more interactions can be set up to magnify each advantage of human and machine in the collaboration, so that machines can service the capabilities increasing of humans, and the track of humans can enlighten the development of machines, achieving the human-machine engagement. It is also expected that such a mutual promotion between humans and machines can provide new path for future design and lead to new discussions on the human-machine collaboration.

Acknowledgement. This paper is supported by Tsinghua University Teaching Reform Project (2019 autumn DX05_01) - Construction of Online Educational Tools and Evaluation System Based on Design Thinking.

References

1. Damle, A., Miller, T.: Influence of design tools on conceptually driven processes. In: Proceedings of C&C 2011: The 8th ACM Conference on Creativity, pp. 327–328 (2011). https://doi.org/10.1145/2069618.2069680

2. Martin, B., Hanington, B.M.: Universal Methods of Design: 100 Ways to Research Complex Problems, Develop Innovative Ideas, and Design Effective Solutions. Rockport Publishers, Beverly (2012)
3. Oh, C., Song, J., Choi, J., Kim, S., Lee, S., Suh, B.: I Lead, You Help but Only with enough details: understanding user experience of co-creation with artificial intelligence. In: Proceedings of CHI 2018: The 2018 CHI Conference on Human Factors in Computing Systems, pp. 649–659 (2018). https://doi.org/10.1145/3173574.3174223
4. Stolterman, E., Pierce, J.: Design tools in practice: studying the designer-tool relationship in interaction design. In: Proceedings of DIS 2012: The 8th ACM Conference on Designing Interactive Systems, pp. 25–28 (2012). https://doi.org/10.1145/2317956.2317961
5. Spaulding, E., Faste, H.: Design-driven narrative: using stories to prototype and build immersive design worlds. In: Proceedings of CHI 2013: The SIGCHI Conference on Human Factors in Computing Systems, pp. 2843–2852 (2013). https://doi.org/10.1145/2470654.2481394
6. Wang, H., Rosé, C.P., Cui, Y., Chang, C., Huang, C., Li, T.: Thinking hard together: the long and short of collaborative idea generation in scientific inquiry. In: Proceedings of CSCL 2007: The 8th International Conference on Computer Supported Collaborative Learning, pp. 754–763 (2007)
7. Mankoff, J., Onafuwa, D., Early, K., Vyas, N., Kamath, V.: Understanding the needs of prospective tenants. In: Proceedings of COMPASS 2018: The 1st ACM SIGCAS Conference on Computing and Sustainable Societies, pp. 1–10 (2018). https://doi.org/10.1145/3209811.3212708
8. Gong, J., Wang, J., Xu, Y.: PaperLego: component-based papercraft designing tool for Children. In: Proceedings of SA 2014: SIGGRAPH Asia 2014 Designing Tools for Crafting Interactive Artifacts, pp. 1–4 (2014). https://doi.org/10.1145/2668947.2668954
9. Zimmerman, J., Stolterman, E., Forlizzi, J.: An analysis and critique of research through design: towards a formalization of a research approach. In: Proceedings of DIS 2010: The 8th ACM Conference on Designing Interactive Systems, pp. 310–319 (2010). https://doi.org/10.1145/1858171.1858228
10. Davis, N., Hsiao, C., Singh, K.Y., Lin, B., Magerko, B.: Creative sense-making: quantifying interaction dynamics in co-creation. In: Proceedings of C&C 2017: The 2017 ACM SIGCHI Conference on Creativity and Cognition, pp. 356–366 (2017). https://doi.org/10.1145/3059454.3059478
11. Kantola, N., Jokela, T.: SVSb: simple and visual storyboards. developing a visualisation method for depicting user scenarios. In: Proceedings of OZCHI 2007: The 19th Australasian Conference on Computer-Human Interaction: Entertaining User Interfaces, pp. 49–56 (2007). https://doi.org/10.1145/1324892.1324901

Effects of Gender Matching on Performance in Human-Robot Teams and Acceptance of Robots

Yanan Zhai, Na Chen[✉], and Jiajia Cao

Beijing University of Chemical Technology, Beijing 100055, China

Abstract. Robots provide important help in relieving work pressure and reducing workload. This study aimed to explore the impact of robots' gender on individuals' performance and their acceptance of robots. This study explored the effects of human-robot gender matching on the performance and the mediating role of acceptance of robots. A questionnaire survey was conducted and 212 valid questionnaires were collected. The results indicated that for the male respondents, their acceptance of opposite-sex robots was higher than the acceptance of same-sex robots and their performance was higher while working with opposite-sex robots. For the female respondents, their acceptance of same-sex robots was higher and their performance wasn't influenced by robots' gender. Moreover, the acceptance of robots mediated in the influence of gender matching on performance. Implications were discussed for future work.

Keywords: Intelligent robot · Gender matching · Technology acceptance · Human-robot interaction

1 Introduction

The functions of robots have been enlarging from traditional industrial functions such as solving the problems of labor shortages and dangerous operations to behaving as part of human daily lives. Robots can play not only functional roles helping people work and live, but also roles working with us, such as teachers, home assistants and colleagues. At present, in order to allow robots to interact more naturally with human, a variety of technologies are incorporate into designing social robots, including speech synthesis, speech recognition, eye contact and body movement [1]. For example, the humanoid robot Robovie [2, 3] developed by the Advanced Telecommunications Research (ATR) Intelligent Robot and Communication Laboratory has been used in museums and schools. Robot "Paro" [4] developed by Intelligent System Co. Ltd which can provide mental health treatment has been introduced into nursing homes for the elderly in some countries. The child-sized humanoid robot "ifBot" developed by ifoo Co. Ltd. [5] (Japan) can communicate with people to prevent dementia among elderly people living alone. The humanoid characteristics of robots, including behaviors and appearance, promote the use of them in the fields of medical, education and so on.

© Springer Nature Switzerland AG 2021
P.-L. P. Rau (Ed.): HCII 2021, LNCS 12773, pp. 410–421, 2021.
https://doi.org/10.1007/978-3-030-77080-8_32

Previous studies indicated that people have higher possibility to accept robots if the latters resemble human in terms of speech and behaviors more. Moreover, people feel easier and more natural to work and live with these robots. Moreover, social attribute factors will also affect the interaction between people, such as our gender, role and personality. Gender is one of the most significant and universal social categories in human society, affecting almost every aspect of our daily lives [6]. To a large extent, gender determines people's social roles, relationships and opportunities [7]. The important thing is that our own gender and the gender of others affect how we think and communicate with each other. Studies have shown that the influence of gender is not only limited to the interaction between people, but also affects the interaction between human and non-humans, such as computers and robots. Studies have found that the impact of robots' gender on people is complex when interact with robots, and there are interactions with contextual factors (such as tasks and environment) and human factors (including gender, educational background, and culture). The research of Siegel et al. [8] showed that users' evaluations of opposite-sex robots are more positive than same-sex robots; they even tend to show more positive attitudes towards opposite-sex robots. On the contrary, research by Eyssel et al. [9] showed that Respondents' perception of same-sex robots was significantly more positive than opposite-sex robots, and psychologically closer to same-sex robots. Therefore, it can be found that the robots' gender preference depends on the interaction between humans and robots and the context in which they interact. When people interact with robots, robots' gender will affect people's behavior and psychology, which will affect acceptance of robots and their performance. In terms of psychology, Siegel et al. [8] found that male respondents believed that female robots were more credible than male robots, while female respondents believed that male robots were more credible than female robots. In terms of behavior, Kuchenbrandt et al. [10] found that, female respondents completed tasks as fast regardless of robots' gender, while male respondents completed tasks faster when interacting with male robots. In view of the complexity of the effects of human-robot gender matching, this research is dedicated to discovering the impact of the interaction of humans' gender and robots' gender on their performance and acceptance of robots, that is, robots' gender that men or women prefer at work, and the degree to which different robots' gender improve performance.

Performance indicators are often used to measure the effectiveness of employees' work. Performance is a certain result achieved by an individual or organization in a certain way within a certain period [11]. There are many factors that affect people's performance, including their own skills, work environment, and incentive effects. One's own skills depend on personal characteristics such as personal talents, intelligence, experience, education and training, and incentives are generally determined by the company. Therefore, in a cooperative team, it is the partners who have the greatest influence on individuals' performance, their language, behaviors, roles, etc. will affect performance. This impact not only occurs in human teams, but also in human-robot interaction, and the impact is even greater. When robots collaborate with humans, people hope that when the partners are no longer humans but non-humans, this different attempt can significantly improve their performance. Some scholars explore how different robot types can improve performance and the type of robot that the individual prefers. For example, a study by St. Clair & Mataric [12] shows that the language feedback of robots in human-robot cooperation can improve team performance, and people prefer robots that actively communicate. Some people prefer robot who act and look like themselves. Shah, J.et al. [13] have shown that when the robot imitates the effective coordination behavior observed in the human team, the team performance will be improved. We can find that in the field of human-robot interaction, most of the research on performance is also in the field of language and behavior, and the focus is mainly on the team performance, and there is very little research on individuals' performance. Nowadays, robots have more and more roles, and their relationship with us is getting more and more intimate. We hope to explore the impact of the specific characteristics of robots on individuals' performance. However, there is currently a lack of research on the social attributes of robots on individuals' performance. Whether people have different preferences for the social attributes of robots, and whether performance will be affected by the social attributes of robots remains to be studied. In addition to exploring the impact of human-robot gender matching on individuals' performance, this research also attempts to explore their acceptance of robots, in other words whether people are willing to work with a robot.

This research mainly explores the impact of robots' gender on individuals' performance. During the research process, we noticed the interaction between robots' gender and people's gender. In view of this finding, we explored that human-robot gender matching is important for robots as partners to individuals' performance. In addition, this research also attempts to explore whether the acceptance of robots mediated in the influence of gender matching on performance. The research results help relevant researchers to better understand the needs of robots as work partners, provide ideas for the design of robot products, and provide references for designing methods to improve individuals' performance.

2 Hypotheses

Psychological research on gender matching shows that people prefer opposite-sex matching. When facing the opposite-sex people, a woman or a man is emotionally excited and expressive desires increase, at this time the level of dopamine will increase, and dopamine is a nerve that can cause excitement and enhance people's motivation. The increase in the level of conductive substances will make people feel more excited and energetic. People often say that "male and female match, work is not tired" is this truth, therefore, people may project the attitude and perception of the opposite-sex people to robot partners, the existing research has also proved this point. The people's evaluations of opposite-sex robots are more positive than same-sex robots; they even tend to show more positive attitudes towards opposite-sex robots. Therefore, we assume that people prefer to work with opposite-sex robots and have a higher acceptance.

H1(a) Female have a higher acceptance of opposite-sex robots;

H1(b) Male have higher acceptance of opposite-sex robots;

As more and more robots are used in our daily work and life, we hope they can relieve work pressure and improve our work efficiency. At work, people are generally focused on the impact on performance. We look forward to robot can significantly improve our performance. However, the current researches focused on the impact of robot behavior on performance. There is a lack of research on the social attributes of robots. Through

H1 analysis, people tend to behave better in front of opposite-sex robots. So, we assume that people have higher performance when working with opposite-sex robots.

H2(a) female have higher performance when working with opposite-sex robots;

H2(b) Male have higher performance when working with opposite-sex robots;

This study explored the impact of human-robot gender matching on performance and acceptance of the robots, but it is not clear how gender matching affects individuals' performance. We know that gender affects people's thinking and communication, and affects people's behavior and psychology. Therefore, we believe that the impact of gender matching on performance may be affected by people's psychological reasons. In other words, individuals' performance depends on whether they like working with matched robots. Therefore, this research proposed that the acceptance of robots mediated in the influence of gender matching on performance.

H3: the acceptance of robots mediated in the influence of gender matching on performance.

3 Method

3.1 Respondents

The respondents of this study are workers of different ages and educational backgrounds across Guangdong. In this online questionnaire survey, we put a total of 235 questionnaires, 235 questionnaires were returned, 212 valid questionnaires, and the effective questionnaire response rate reached 90.21%. Among the valid questionnaires, 109 were males, accounting for 48.58%, and 103 were females, accounting for 51.42%. There are 120 subjects with working experience of 1–3 years and 1–5 years, accounting for 56.61%, and the rest have no work experience or more than 6 years of work experience, accounting for 43.39%. The respondents' understanding of robots included 85 people who had general knowledge of robots, accounting for 40.09%, 60 people had little knowledge, accounting for 28.30%, and the remaining 67 people had never understood or knew very well, accounting for 31.61%. Basic information is shown in Table 1.

3.2 Questionnaire Design

This research questionnaire mainly includes three subscales, demographic, the acceptance of the robots and performance. Demographic scale mainly include gender, age, working years, education, and understanding of home robots. the acceptance of the robots scale contains four questions: I like the social attributes of the robot very much; it is interesting to work with the robot; I am willing to learn the usage of the robot; I like to work with the robot. We modified some of the questions of Wang Hui et al. [14] in the multi-dimensional structure of leadership-subordinate exchange and its influence on job performance and situational performance to obtain the acceptance of the robots scale, and replaced the "supervisor" in the original scale with "robot".

The performance scale contains five questions, which measure the individuals' performance from three aspects: quantity, quality and efficiency. We modified the scale

Table 1. Profile of respondents

Measure	Item	Frequency	Percentage (%)
Total		212	
Gender	Male	109	48.58
	Female	103	51.42
Age	Below 20	26	12.26
	21–30	106	50.00
	31–40	63	29.72
	Over 41	17	8.02
Education	Under senior middle school	21	9.91
	Junior college	49	23.11
	Undergraduate	138	65.09
	Over postgraduate	4	1.89
Working experience	0 year	67	31.6
	1–3 year	64	30.19
	3–5 year	56	26.42
	Over 6 year	25	11.79
Robots knowledge	Never known	25	11.79
	Less known	60	28.30
	General known	85	40.09
	More known	27	12.74
	Very much known	15	7.08

used by Tsui, Pearce & Porter [15] to obtain the performance scale. The researcher in this paper first used the standard translation-back translation procedure [16]. First, the first author translated the five questions in the Tsui, Pearce & Porter scale into Chinese, and secondly, another researcher translated the Chinese into English, a bilingual scholar was also invited to compare the original question with the back-translated question. Finally, the Chinese translation was adjusted and revised to ensure the semantic equivalence of the Chinese and English versions. The original scale has two dimensions: task performance and peripheral performance. Because the measurement of peripheral performance needs to be compared with the overall average level, it is generally assessed by the respondents's superiors. The objectivity of self-evaluation is difficult to guarantee, so this study only uses five questions of task performance. These questions use the Likert 5-point scale (1 = strongly disagree, 5 = strongly agree).

Each respondent was required to answer questions about the acceptance of the robots and performance when working with male and female robots. Before measuring the acceptance of the robot and performance, the questionnaire first showed the respondents the pictures of the male and female robots, as shown in Fig. 1, and show the respondents some text: Please imagine that there is a male robot at your side to assist you in your work, Please describe your acceptance of the male robot. The image of the male robot is shown in the figure below and his voice also has male characteristics.

Fig. 1. The images of robots used in the experiment

4 Results

In this study, SPSS 24 software was used for data analysis. data analysis proceeded in three stages: First, the reliability of the model is tested; then the impact of gender matching on performance and the acceptance of robots is verified, that is H1, H2; Finally, the acceptance of robots mediated in the influence of gender matching on performance is verified, that is H3.

4.1 Reliability Analysis

The internal consistency reliability of all questions was assessed by finding the Cronbach alpha, the performance and the acceptance of robots scales have passed the Cronbach alpha test. The alpha is at least 0.796 and greater than 0.7, indicating that the questionnaire in this study has high reliability. Cronbach alpha are shown in Table 2.

Table 2. Test results of internal reliability

Construct	Items	M	SD	Cronbach's α
Acceptance of robots	4	15.06	3.274	0.796
Performance	5	18.78	3.666	0.809

4.2 The Impact of Gender Matching on Performance and the Acceptance of Robots

Perform analysis of variance to test the impact of gender matching on performance and the acceptance of robots. The results show that robots' gender and respondents' gender have a significant impact on the acceptance of robots and performance (all $p < 0.05$), and there is a significant interaction effect between robots' gender and respondents' gender (the acceptance of robots: $F = 6.388$, $p < 0.05$; performance: $F = 5.176$, $p < 0.05$). The analysis results are shown in Table 3.

Table 3. Results of analysis of variance on the effect of gender matching on performance and the acceptance of robot

Independent variable	Dependent variable	F	p
Robots' gender	Acceptance of robots	30.850	<.001
	Performance	8.551	.004
Respondents' gender	Acceptance of robots	40.196	<.001
	Performance	11.898	.001
Robots' gender × respondents' gender	Acceptance of robots	6.388	.012
	Performance	5.176	.024

According to the post-test results, for male respondents, when the robot partner is female (opposite-sex matching), the acceptance of the robot is higher ($p < 0.05$). For female respondents, when the robot partner is female (same-sex matching), The acceptance of the robot is higher ($p < 0.05$), so H1(a) is not verified, H1(b) is verified. For male respondents, performance was higher when working with female robot partners (opposite-sex matching) ($p < 0.05$). For female respondents, there was no significant difference in performance, so H2(a) was not verified, and H2(b) was verified. The analysis results are shown in Table 4.

4.3 The Acceptance of Robots Mediated in the Influence of Gender Matching on Performance

Perform regression analysis to test the acceptance of robots mediated in the influence of gender matching on performance. For male respondents, in model 1, the mediating

Table 4. Post-test results of the impact of gender matching on performance and acceptance of robot

Variable	Respondents' gender	Robots' gender	Mean difference	t	p
Acceptance of robots	Female	Female	1.177	0.458	.030
		Male			
	Male	Female	3.142	0.178	<.001
		Male			
Performance	Female	Female	0.312	2.141	0.640
		Male			
	Male	Female	2.502	0.277	<.001
		Male			

variable the acceptance of robots and dependent variable performance have significant regression effects on independent variables. (The acceptance of robots: $B = 0.554$, $p < 0.05$; performance: $B = 0.353$, $p < 0.05$). At the same time, in the model 2, the dependent variable performance has significant regression effects on the mediating variable the acceptance of robots ($B = 0.843$, $p < 0.05$), so the acceptance of robots played a full mediating role in the relationship between gender matching and performance. For female respondents, in the model 1, the mediating variable the acceptance of robots and dependent variable performance have no significant regression effects on independent variables. (The acceptance of robots: $B = -0.185$, $p > 0.05$; performance: $B = -0.044$, $p > 0.05$), so the acceptance of robots does not play a mediating role in the relationship between gender matching and performance, and H3 has been partially verified. The analysis results are shown in Table 5.

Among the three hypotheses (5 sub-hypothesis in total) proposed in this study, H1(a) and H2(a) have not been verified, the H3 has been partially verified, and the other two hypotheses have been verified. The hypothesis verification status of this study can be summarized as shown in Table 6.

5   Discussion

This study confirmed the interaction between robots' gender and respondents's gender, and further explored the impact of human-robot gender matching on the acceptance of the robot and the performance of the respondents. The results showed that female have a higher acceptance of same-sex robot partners, while male are the opposite. Female may think that they have something in common with same-sex robots and communicate more effectively [17]. Male have higher performance when working with opposite-sex robots, while females have no difference in performance when working with robots of different gender. This indicates that male prefer to work with opposite-sex robots, which is similar to the study by Siegel [8] that men donate to women robots more frequently

Table 5. Regression results of the acceptance of robots mediated in the influence of gender matching on performance

Respondents' gender	Variable		Model 1		Model 2
			Acceptance of robots	Performance	Performance
Male	Independent variable	Gender matching	0.554*	0.353*	
	Mediator	Acceptance of robots			
	Mediator effect	Gender matching			−0.114
		Acceptance of robots			0.843*
		R2	0.307	0.152	0.617
		Adjusted R2	0.300	0.116	0.609
		F	43.934	14.112	78.991
Female	Independent variable	Gender matching	−0.185	−0.044	
	Mediator	Acceptance of robots			
	Mediator effect	Gender matching			0.090
		Acceptance of robots			0.723*
		R2	0.034	0.002	0.507
		Adjusted R2	0.025	0.007	0.498
		F	3.772	0.204	54.541

*indicates p is significant at the 0.05 level

than women donate, but women do not show preference. It may be that female are more likely to regard robots as mechanized than men [18], which reduces their trust in robots and willingness to cooperate with robots. This study also confirmed that the acceptance of robots mediated in the influence of gender matching on performance. In the male group, the acceptance of robots played a complete mediating role. Therefore, we can see that male's gender suggestion to robots is stronger than that of female [8].

The research on gender interaction between humans and robots is still in a period of rapid development, and various new viewpoints are emerging one after another. There is no relatively complete theoretical model for researchers in the academic circle. There are some shortcomings in this research. When designing the gender of robots, it is only distinguished by showing pictures in the questionnaire to allow respondents to imagine. In the research of robots' gender, the methods of identifying the gender of robots

Table 6. Hypothesis verification

	Hypothesis	Verification
H1 (a)	Female have a higher acceptance of opposite-sex robots	Unverified
H1 (b)	Male have higher acceptance of opposite-sex robots	Verified
H2 (a)	Female have higher performance when working with opposite-sex robots	Unverified
H2 (b)	Male have higher performance when working with opposite-sex robots	Verified
H3	The acceptance of robots mediated in the influence of gender matching on performance	Partially verified

include: appearance; voice; operation experiment; names and pronouns used in survey description [8, 10, 19, 20]. Therefore, the combination of these factors can be used to strengthen the factor of robots' gender in future research. In addition, the conclusions in this study are drawn without setting specific task content or work scenarios for performance. Social robots are widely used in hospitals, homes, museums and other places [21–23], research shows that robots with different gender appearances and behaviors may make the communication between humans and robots more effective, but in some cases a genderless design may better reduce user attention [24]. The research can further explore performance in specific task scenarios, especially gender-related tasks, such as nursing, security, etc.

6 Conclusion

Partner robots provide important help in relieving work pressure and reducing workload. The functions of robots have been enlarging from traditional industrial functions such as solving the problems of labor shortages and dangerous operations to behaving as part of human daily lives. However, what type of robots people like to become their work partners, and what kind of robots to work with more efficiently has become a hot topic in the academia. This study uses a questionnaire to explore the impact of robots' gender on individuals' performance and their acceptance of robots. This study explored the effects of human-robot gender matching on the performance and the mediating role of acceptance of robots. A total of 212 valid questionnaires were collected. Data analysis showed that for the male groups, their acceptance of opposite-sex robots was higher than the acceptance of same-sex robots and their performance was higher while working with opposite-sex robots. For the female group, their acceptance of same-sex robots was higher and their performance wasn't influenced by robots' gender. Moreover, the acceptance of robots mediated in the influence of gender matching on performance.

Therefore, this research suggests that when using robots to assist office work, more attention should be paid to the human-robot gender matching. When choosing a partner robot, the user can focus on the gender matching between himself/herself and the robot, and give priority to choosing the robot he/she prefers. In addition, pay attention to cultivating people's positive emotions towards the robot, these emotions mainly include people's acceptance of robots (trust), the pleasure of working together, and the desire to learn to use robots. Empirical research has shown that subjects have higher performance when they have a higher acceptance of robots. Therefore, in order to improve the efficiency of robot-assisted office, it is necessary for the corporation to consider improving the employee's level of positive emotions towards the robot. The research results help relevant researchers to better understand the needs of employees for robot partners, provide ideas for the design of robot products, and provide references for designing methods to improve individuals' performance.

Acknowledgements. This study was funded by a National Natural Science Foundation of China 71942005, a Ministry of Education of Humanities and Social Science project 19YJC840002, a Beijing Social Science Fund 17SRC021 and a Beijing Natural Science Foundation 9184029.

References

1. Nomura, T.: Robots and gender. Gender Genome **1**(1), 18–25 (2017)
2. Kanda, T., Sato, R., Saiwaki, N., Ishiguro, H.: A two-month field trial in an elementary school for long-term human–robot interaction. IEEE Trans. Robot. **23**(5), 962–971 (2007)
3. Shiomi, M., Kanda, T., Ishiguro, H., Hagita, N.: Interactive humanoid robots for a science museum. In: Proceedings of the 1st ACM SIGCHI/SIGART Conference on Human-Robot Interaction, pp. 305–312. ACM, New York City (2006)
4. Wada, K., Shibata, T.: Living with seal robots—its sociopsychological and physiological influences on the elderly at a care house. IEEE Trans. Robot. **23**(5), 972–980 (2007)
5. Kanoh, M., et al.: Examination of practicability of communication robot-assisted activity program for elderly people. J. Robot. Mechatron. **23**(1), 3 (2011)
6. Harper, M., Schoeman, W.J.: Influences of gender as a basic-level category in person perception on the gender belief system. Sex Roles **49**(9–10), 517–526 (2003)
7. Bussey, K., Bandura, A.: Social cognitive theory of gender development and differentiation. Psychol. Rev. **106**(4), 676 (1999)
8. Siegel, M., Breazeal, C., Norton, M.I.: Persuasive robotics: the influence of robot gender on human behavior. In: 2009 IEEE/RSJ International Conference on Intelligent Robots and Systems, Hawaii, pp. 2563–2568. IEEE (2009)
9. Eyssel, F., Kuchenbrandt, D., Hegel, F., de Ruiter, L.: Activating elicited agent knowledge: how robot and user features shape the perception of social robots. In: 2012 IEEE Roman: The 21st IEEE International Symposium on Robot and Human Interactive Communication, Hawaii, pp. 851–857. IEEE (2012)
10. Kuchenbrandt, D., Häring, M., Eichberg, J., Eyssel, F., André, E.: Keep an eye on the task! How gender typicality of tasks influence human–robot interactions. Int. J. Soc. Robot. **6**(3), 417–427 (2014)
11. Yang, J., Fang, L., Ling, W.: Major issues of performance appraisal. Chin. J. Appl. Psychol. **6**(2), 53–58 (2000)
12. St. Clair, A., Mataric, M.: How robot verbal feedback can improve team performance in human-robot task collaborations. In: Proceedings of the Tenth Annual ACM/IEEE International Conference on Human-Robot Interaction, pp. 213–220. ACM, New York City (2015)

13. Shah, J., Wiken, J., Williams, B., Breazeal, C.: Improved human-robot team performance using Chaski, a human-inspired plan execution system. In: Proceedings of the 6th International Conference on Human-Robot Interaction, pp. 29–36. ACM, New York City (2011)
14. Wang, H., Niu, X.: Multi-dimensional leader-member exchange (LMX) and its impact on task performance and contextual performance of employees. Acta Psychologica Sinica 36(2), 179–185 (2004)
15. Tsui, A.S., Pearce, J.L., Porter, L.W., Tripoli, A.M.: Alternative approaches to the employee-organization relationship: does investment in employees pay off? Acad. Manag. J. 40(5), 1089–1121 (1997)
16. Brislin, R.W.: Translation and content analysis of oral and written materials. Methodology 389–444 (1980)
17. Echterhoff, G., Bohner, G., Siebler, F.: "Social Robotics" und Mensch-Maschine-Interaktion. Zeitschrift für Sozialpsychologie 37(4), 219–231 (2006)
18. Schermerhorn, P., Scheutz, M., Crowell, C.R.: Robot social presence and gender: do females view robots differently than males? In: Proceedings of the 3rd ACM/IEEE International Conference on Human Robot Interaction, pp. 263–270. ACM, New York City (2008)
19. Alexander, E., Bank, C., Yang, J.J., Hayes, B., Scassellati, B.: Asking for help from a gendered robot. In: Proceedings of the Annual Meeting of the Cognitive Science Society, vol. 36, no. 36, pp. 2333–2338 (2014)
20. Koulouri, T., Lauria, S., Macredie, R.D., Chen, S.: Are we there yet? The role of gender on the effectiveness and efficiency of user-robot communication in navigational tasks. ACM Trans. Comput.-Hum. Interact. (TOCHI) 19(1), 1–29 (2012)
21. Dario, P., Guglielmelli, E., Laschi, C., Teti, G.: MOVAID: a personal robot in everyday life of disabled and elderly people. Technol. Disabil. 10(2), 77–93 (1999)
22. Heerink, M., Krose, B., Evers, V., Wielinga, B.: The influence of a robot's social abilities on acceptance by elderly users. In: The 15th IEEE International Symposium on Robot and Human Interactive Communication, ROMAN 2006, Hawaii, pp. 521–526. IEEE (2006)
23. Enz, S., Diruf, M., Spielhagen, C., Zoll, C., Vargas, P.A.: The social role of robots in the future—explorative measurement of hopes and fears. Int. J. Soc. Robot. 3(3), 263 (2011)
24. Powers, A., Kramer, A.D., Lim, S., Kuo, J., Lee, S.L., Kiesler, S.: Eliciting information from people with a gendered humanoid robot. In: IEEE International Workshop on Robot and Human Interactive Communication, ROMAN 2005, Hawaii, pp. 158–163. IEEE (2005)

Author Index

Ai, Chao II-423
Amalkrishna, P. S. I-475
An, Dadi III-14
Au, Tsz Wang II-230

Bao, Lin I-496
Bautista, Kevin III-319
Berque, Dave III-319
Bin Mothana, Abdulwahed II-277
Blakey, Jordyn III-319

Cao, Chenxi III-171
Cao, Huai I-288
Cao, Huajun II-423
Cao, Jiajia III-410
Cao, Jing II-53, II-289, III-80
Cao, Shuangyuan I-3
Cao, Xin I-197, II-241
Chaing, Yu-Houng I-408
Chang, Shu-Hua II-3
Chen, Chien-Chih I-234
Chen, Feng III-319
Chen, Hao Wu III-23
Chen, Hao I-100, I-486, II-18, II-119,
    II-162, II-201
Chen, Jen-Feng I-210
Chen, Jiangjie II-241
Chen, Jiaru III-267
Chen, Li-Hao I-224
Chen, Min I-288
Chen, Na III-410
Chen, Rosalina I-373
Chen, Tien-Li I-80
Chen, Yingjie (Victor) III-68
Chen, Yu II-303
Cheng, Pei-Jung I-14
Cheng, Shiying III-182
Cheng, Shuxin II-318
Cheng, Tuck-Fai III-3
Chiang, I-Ying I-244
Chiba, Hiroko III-319
Chiocchia, Irene I-27
Chiou, Wen-Ko I-100, II-18, II-119, II-162,
    II-201

Ding, Lu III-68
Ding, Shaowen I-41, I-164
Ding, Wei II-330, III-14
Ding, Zi-Hao I-262
Dittmar, Anke II-277
Dong, Fangzhou I-49
Dong, Huimin III-182
Du, Peifang III-195, III-212, III-240

Fa, Jing III-116
Fan, Hsiwen I-141
Fang, Hongyi II-265
Fu, Li-Chen III-335
Fu, Zhiyong I-185, II-303, III-396

Gamborino, Edwinn III-335
Gander, Anna Jia I-390
Gander, Pierre I-390
Gao, Qin II-183
Gao, Ting II-138
Gao, Ya Juan III-23
Gao, Yang II-150
Gu, Naixiao III-116
Guan, Shing-Sheng I-262
Guo, Haowen III-225
Guo, Shuo I-547
Guo, Weiwei I-496
Guo, Xiaolei III-68
Guo, Xiuyuan II-138

Han, Lijun II-344
Han, Xinrong I-273
Hao, Jie II-396
Herrera Ruiz, Alberto III-335
Hong, Tai-He II-119, II-162
Hosseinpanah, Adineh III-350
Hsiao, Ming-Yu I-127
Hsieh, Yu-Chen I-127
Hsu, Szu-Erh II-119, II-162
Hsu, Yen I-197, I-351, I-408, II-217, II-241
Hu, Lizhong I-288
Huang, Chiwu I-234
Huang, Ding-Hau I-100, I-311, II-162
Huang, Jianping II-150, III-103
Huang, Kuo-Liang I-127

Huang, Lan-Ling   I-141, I-262, I-340
Huang, Mei-lin   I-330
Huang, Min Ling   III-23
Huang, Tze-Fei   I-300
Huang, Wenkang   III-319
Huang, Xinyao   II-330, III-14
Hung, Chi-Sen   I-80

Imura, Shun   III-319

Ji, Xiang   III-369
Jia, Ruoyu   III-281
Jiang, Changhua   I-41, I-164

Ko, Minjeong   I-463
Komiya, Kosuke   I-420
Konomi, Shin'ichi   III-44
Krämer, Nicole C.   III-350
Kumar, Jyoti   I-475
Kuo, Tzuhsuan   I-430

Lai, Ka-Hin   I-62
Lai, Xiaojun   I-442
Lan, Feng   III-195
Laohakangvalvit, Tipporn   III-319
Lee, Yun-Chi   I-80
Lei, Xin   I-463
Li, Hao   III-212
Li, Judy Zixin   II-31
Li, Junzhang   III-225
Li, Wen   II-396
Li, Wentao   III-384
Li, Xiaohui   I-49
Li, Xiaolin   II-330, III-14
Li, Xiaoyi   I-115
Li, Yuzhen   I-49
Li, Zhengliang   II-172
Li, Zhuoran   II-362
Li, Ziyang   II-362
Liang, Jin   I-115
Liang, Yu-Chao   I-100, II-18, II-119, II-201
Liao, Chi-Meng   I-141, I-340
Liao, Yifei   I-496
Liao, Zhen   I-115
Lin, Chih-Wei   I-141, I-340
Lin, Hsuan   I-127, III-33
Lin, Po-Hsien   I-210, I-244, I-300, II-53,
       II-65, II-107, II-251, II-289, III-3
Lin, Rungtai   I-210, I-244, II-18, II-40, II-65,
       II-107, II-150, II-201, II-251, III-23,
       III-90

Lin, Ting-Yu Tony   I-486
Lin, Wei   I-127, III-33
Lindström, Nataliya Berbyuk   I-390
Liu, Bingjian   I-273, II-408
Liu, Chao   I-100, II-18, II-119, II-162, II-201
Liu, Fang   I-3
Liu, Lei   I-115
Liu, Na   III-384
Liu, Ruoxuan   III-384
Liu, Xuan   III-296
Liu, Yang   II-183
Liu, Yi-Chien   I-224
Lo, Cheng-Hung   III-116
Lo, Liang-Ming   II-119, II-162
Lu, Honglei   I-351, II-454
Lu, Jiangyan   III-68
Lu, Wen   III-56
Lu, Xuanye   I-496
Lu, Yuhong   III-44
Luan, Yimeng   III-240

Ma, Liang   II-183
Ma, Xiaolei   II-423
Murayama, Kento   III-319

Nakajima, Tatsuo   I-420
Ng, Wai Kit   II-289, III-80
Niu, Jianwei   I-41, I-164

Ohkura, Michiko   III-319
Ou, Hung-Chug   II-217
Ou, Shuang   III-90
Ouyang, Yulu   III-56

Pan, Shuyu   III-225
Peng, Xing   II-265
Pratap, Surbhi   I-475

Qian, Zhenyu (Cheryl)   III-68
Qie, Nan   I-442, I-527
Qin, Weiheng   III-195
Qiu, Boxian   III-296
Qiu, Chunman   III-195

Rao, Jiahui   II-265
Rau, Pei-Luen Patrick   I-27, I-62, I-373,
       I-442, I-463, I-486, I-527, II-374,
       II-441, III-369
Ren, Xinrui   III-240

Shi, Hanwei   I-156
Shi, Minghong   III-90

Spehlmann, Eric   III-319
Sripian, Peeraya   III-319
Sterling, Sara   I-49
Sugaya, Midori   III-319
Sun, Haoran   II-374
Sun, Lili   I-288
Sun, Xiaohua   I-496
Sun, Xu   I-273, II-408
Sun, Yikang   II-40, III-103
Sun, Yue   II-265

Tan, Hao   III-171, III-182, III-195, III-212,
    III-225, III-240, III-304
Taniguchi, Yuta   III-44
Tao, Jini   III-253
Tao, Yuheng   II-53, II-107, II-289
Teh, Pei-Lee   II-230
Teng, Hsiu-Wen   III-3
Tian, Miaoqi   II-172
Ting, Yi-Wen   II-65
Tsai, Chia-Wen   I-127
Tseng, Tsung-Yuan   III-335

Wang, Bingcheng   I-62, II-374
Wang, Bo   I-41, I-164
Wang, Chun   I-224
Wang, Chunyan   I-506
Wang, Duannaiyu   III-253
Wang, Huan   II-396
Wang, Jing-Fen   III-335
Wang, Jingzhi   III-116
Wang, Xiangyi   III-171
Wang, Xin   I-115
Wang, Yujing   III-267
Wei, Jialing   III-171
Wen, Xin   III-90
Wong, Kapo   II-230
Wright, Cade   III-319
Wu, Jiang   I-273, II-408
Wu, Jun   II-150
Wu, Lei   I-288
Wu, Shih-Kuang   I-80
Wu, Weilong   I-197, II-241
Wu, Yi-Ming   III-33

Xia, Qing   III-396
Xiang, Yunpeng   III-116
Xiao, Yu-Meng   I-311
Xie, Yufan   I-174
Xu, Fenggang   I-41, I-164

Xu, Ji   II-265
Xu, Jiayao   II-138
Xu, Rui   II-84
Xu, Wenwei   II-423
Xu, Xi   III-129
Xu, Xuhai   I-527
Xue, Shipei   II-138

Yan, Qiao   III-267
Yang, Cheng Hsiang   II-251
Yang, Hao   III-281
Yang, Tao   I-340
Yang, Yulu   III-296
Ye, Junnan   II-330
Ye, Sheng   II-95
Yeh, Su-Ling   III-335
Yu, Andrew   II-40
Yu, Sisi   III-212
Yu, Yang   I-115
Yuan, Xiaojun   I-506
Yue, Yifei   III-182

Zeng, Jingchun   II-408
Zeng, Yuqi   II-138
Zha, Yong   I-288
Zhai, Yanan   III-410
Zhang, Ang   I-547, II-362
Zhang, Chi   I-115
Zhang, Hui-Zhong   III-33
Zhang, Jingyu   II-423
Zhang, Jitao   III-281
Zhang, Kan   II-423
Zhang, Liang   I-115, II-423
Zhang, Ningxuan   II-423
Zhang, Qianyu   II-330, III-14
Zhang, Teng   I-41
Zhang, Wenlin   III-139
Zhang, Yiqiao   III-304
Zhang, Zhanshuo   I-115
Zhang, Zhen   I-164
Zhao, Jingyu   III-369
Zhao, Xue   III-212, III-240
Zhao, Yan   I-41, I-164
Zheng, Hong Qian   II-53, II-107
Zheng, Jian   I-486
Zheng, Jiatai   III-212
Zheng, Shujun   III-155
Zheng, Ya-Yi   I-311
Zhong, Runting   II-265

Zhou, Aven Le   II-31
Zhou, Tuoyang   I-115
Zhou, Xingchen   II-441

Zhu, Chunxiao   II-454
Zhu, Enyi   III-253
Zhu, Lin   I-185

Printed in the United States
by Baker & Taylor Publisher Services